Human and Machine Vision

Analogies and Divergencies

Human and Machine Vision

Analogies and Divergencies

Edited by

Virginio Cantoni

University of Pavia
Pavia, Italy

PLENUM PRESS • NEW YORK AND LONDON

Library of Congress Cataloging-in-Publication Data

On file

QP
474
.H85
1994

Proceedings of the Third International Workshop on Perception,
held September 27–30, 1993, in Pavia, Italy

ISBN 0-306-44902-1

© 1994 Plenum Press, New York
A Division of Plenum Publishing Corporation
233 Spring Street, New York, N.Y. 10013

Printed in the United States of America

In honor of the memory of
Prof. Eduardo Caianiello

PREFACE

The following are the proceedings of the Third International Workshop on Perception held in Pavia, Italy, on September 27-30, 1993, under the auspices of four institutions: the Group of Cybernetic and Biophysics (GNCB)s of the National Research Council (CNR), the Italian Association for Artificial Intelligence (AI * IA), the Italian Association of Psychology (AIP), and the Italian Chapter of the International Association for Pattern Recognition (IAPR).

The theme of this third workshop was: "Human and Machine Vision: Analogies and Divergencies."

A wide spectrum of topics was covered, ranging from neurophysiology, to computer architecture, to psychology, to image understanding, etc. For this reason the structure of this workshop was quite different from those of the first two held in Parma (1991), and Trieste (1992). This time the workshop was composed of just eight modules, each one consisting of two invited lectures (dealing with vision in nature and machines, respectively) and a common panel discussion (including the two lecturers and three invited panellists).

The lectures aimed at presenting the state-of-the-art and outlining open questions. In particular, they were intended to stress links, suggesting possible synergies between different cultural areas. The panel discussion was conceived as a forum for an open debate, briefly introduced by each panellist, and mainly aimed at a deeper investigation of "analogies and divergencies" between the different approaches to vision, including the search for possible new solutions to problems in the different fields. The panellists were asked to prepare a few statements on hot-points as a guide for discussion. These statements were delivered to the participants together with the final program, for more qualified discussion.

The number of participants to the workshop was limited to 60. Besides the 40 invited lecturers and panellists, 20 more participants were admitted. Priority for these positions was given to young researchers who made important contributions to the open discussions. The lectures and the contents of the panel discussion are included in the present proceedings.

Table 1 shows the scheduled program with the eight modules. Minor variations were made to the workshop presentations, and in the reports, to accommodate the positions that emerged from the debates. In the following paragraphs each subject is briefly introduced and commented upon.

Vision Architectures

The first module sought to compare a taxonomy of the bio-vision systems developed over eons of evolution, with the zoology of the vision machines that have been built or proposed over the last 40 years. Obviously, as was pointed out in the discussion, this comparison is promising, but mandates precaution.

During evolution, biological vision developed through a number of functional steps, each of which has supported the survival and development of the animal world. The lowest level of phylogenetic development, the simple structure of an insect compound eye, is able to detect

Table 1. The Eight Subjects

Subject	Nature	Machine	Panel
Vision Architectures	The phylogenetic evolution of the visual system	Machine vision architectures	One model for vision system?
Retinae	Neurophysiology of the retina	On chip vision	Logmap and multi-scale approaches
Visual Primitives	Neurophysiology of the retina	Visual forms and data structures	Looking for visual primitives
Camera and Eye Control System	The oculomotor system	Active vision	Allocation of attention in vision
Motion Perception and Interpretation	Time-flow and movement in human vision	Motion analysis	Dynamic perception of the environment
Visual Representations	Visual cognition and cognitive maps	Representations and models	Icons and words: metaphors and symbolism
Visual Reasoning	Mental models and spatial reasoning	Spatial reasoning	Image interpretation and ambiguities
Imagination and Learning	Imagery and imagination	Learning and neural nets	Perceptual learning and discovery

polarised light and the whole visible spectrum from IR to UV. At higher levels, a decrease in the capacity for light detection, retinal processing, adaptation in response to environmental and/or behavioural stimuli, has been accompanied by a corresponding increase in central processing.

Machine vision developments have been driven to a large extent by the evolution of processing power in the computer era. The history of computer vision architectures is the history of computer parallelism, with a last period in which VLSI design facilities have become accessible to most researchers, supporting efforts towards the integration of sensing and analog and digital processing, and guiding the development of *smart vision sensors* and *vision front ends*.

Broadly speaking, it looks as if the natural trend is towards an enrichment of high- level central processing (often to the detriment of periphery capability). In contrast, the present trend in computer vision is towards a higher, and more powerful, allocation of processing resources in periphery. Nevertheless, if we return to the panel, the hot points, despite this situation, look very attractive for both computer scientists and bio-experts. The panel highlighted several common fields of investigation, such as high-level control and hierarchies for sensor signal manipulations, flexibility and adaptability of vision systems, exploitation of illusions and drawbacks, basic hardware and biological material capabilities. In summary, the panel on architectures for vision has given the richest of contributions and position statements from a very large spectrum of competence.

Retinae

Neuroscientists have shown that the retina of vertebrates is, in origin and by functional organisation, a parallel computational unit similar to a portion of the central nervous system.

Indeed, its peripheral location, the possibility of using precisely controlled natural stimuli (without artificial electrical stimulation) and of analysing the resulting signals delivered through the optic nerve, makes the retina one of the best understood components of the nervous system.

The approximately 120 million retina photoreceptors have a decreasing distribution density going from the *fovea* (close to the visual axis, and composed of uniformly and densely distributed *cones*) towards the periphery (composed mainly of highly interagent *rods*). The retina between the receptors and the optic nerve follows a "vertical" pathway formed by the *receptor-bipolar-ganglion cells*, which is integrated with a two level 'horizontal' organisation consisting respectively of the *horizontal* and *amacrine cells*. The ratio between the number of receptors, bipolar and ganglion cells in the fovea is close to 1 : 1 : 1, whereas in the periphery it is of hierarchical nature: the ratio between rods, bipolar and ganglion cells is 100 : 17 : 1. In the two vertical layers downstream, there is a concentration of data with a 100 ÷ 50 ratio between the receptors and fibres of the optical nerve (ganglion cells). Finally, it is worthwhile mentioning that the retina receives a retrograde pathway "from cortical centres through a system of fibres that spread widely within its layers."

As was mentioned above, from an architectural viewpoint, an increasingly common solution for machine vision is based on an intelligent sensor combined with a control mechanism capable of re-directing the sensor to "just'" the right data needed to accomplish the required tasks. In particular, this mechanism must be able to switch "attention" in a pre-defined, or shortly determined, sequence between the relevant areas of the scene. The retina of vertebrates allows this with minimum effort (cost). Many ongoing research efforts in micro-electronics and computer vision try to exploit silicon technology to embed a retina into a single chip, drawing an analogy from the fovea solution. Alternative approaches to supporting sensing and processing capabilities on the same device have been investigated. One example is by integrating an array of light sensors with an associated array of simple Boolean processors by adopting wholly analog, mixed analog-digital, or wholly digital processing modalities; hence CCD versus CMOS silicon technology; ASIC or general purpose implementations; log-polar, space variant or using uniformly distributed photoreceptor arrays.

The panel confirmed that to achieve real-time performances on a large field of view, with restricted computational resources, a vision system must be capable of: rapidly selecting the regions of interest in space and time; changing, as required, the data resolution, to enable operations with local and global (or regional) features; rapidly updating the strategies to the partial results achieved step by step. The retina-foveation solution is very attractive from the cost/benefit viewpoint, but choice of its implementation and interpretation is problematic. From the computational viewpoint other proposals have been made which offer different benefits. Amongst others we debated: log-polar representations reproducing the retino-cortical mapping and supporting some invariance for scale and rotation; multi-scale, quad-tree and pyramidal approaches that support hardware implementation and planning strategies in varying degrees; wave-let transformation characterised by a compact support and a localisation in space and frequency components.

Visual Primitives

The primary visual cortex performs an "intermediate" stage of analysis: information coming from the retina is examined for various parameters of the visual stimuli (e.g. orientation, colour, velocity, etc.), and transmitted along neural pathways to higher-order visual areas. The analysis occurring in the visual cortex is performed by interacting cells that play a detector role with receptive fields of various sizes, selectively sensitive to the orientation of lines or edges or even textures. The cells also act as spatial frequency filters, and, whilst covering the range of spatial frequencies at which the animal visual system is sensitive, different cells are tuned to different spatial frequencies. The cells are architecturally arranged into columns perpendicular to the cortical surface, according to similar stimulus selectivity, ocular dominance, and spatial frequency. There is evidence, however, for horizontal long-range connections assuring communication between columns of similar orientation preference and integrating information over a large part of the visual field. In primates, colour information is processed by classes of cortical neurons with chromatically opposing receptive fields. Moreover, there is evidence of nonvisual input to the striate visual cortex, e.g. the interaction between sensory and oculomotor information to process binocular disparity that is modulated by the changes in ocular vergence and accommodation.

In machine vision much work has been done to devise a "complete" and "open" set of feature detectors that can be applied to solve a generic vision task. Completeness means that all the basic features have been taken into consideration, in the sense that it is possible to represent a generic visual form (sometimes in a hierarchical manner, following the principle of recursive decomposition, and/or approximate representations). Openness means that new, "ad hoc" detectors can be easily added if the peculiarities of the current data are not exploited in the set. These basic modules (or *primitives*) measure internal and external form characteristics, by including photometric, geometric, statistic, and morphologic, etc., operators. Knowledge of the task at hand can provide criteria for the selection of a profitable subset of primitives. Some attempts at building an expert system that automatically discriminates the subset, were also made.

If retina behaviour is easily analysed, the cortical areas are not (at least with equal effort) for physiological reasons (e.g. interfaces and interactions, higher connectivity and plasticity, neuronal information coding). Perhaps mainly for this motive, as for the machine vision case, many common questions still remain open: what are the suitable shape primitives? What role do visual primitives play subservient to the task? What is the role of knowledge and context when looking for primitives? To what extent is the use of feature descriptors necessary or effective?

Camera and Eye Control Systems

Visual sensing in man and primates is performed through a sophisticated control system for eye movement which is composed of different agents. Phylogenetic evolution has led to reflexes such as the vestibulo-ocular and the optokinetic ones, which prevent the image of the environment from slipping on the retina; to voluntary or involuntary fast position-resetting mechanisms for gaze orienting and visual scanning of the wide field of view (through *saccadic* movements); to eye rotation for focusing (by *vergence* or *version* movements in frontal-eyed animals), and velocity servomechanism for *pursuit* of the selected targets. The conflicts between these different agents, and interaction with other sensors and/or between the different activity subsystems (like head movements that may either be synergistic, as during pursuit of moving targets by means of combined eye and head movements, or antagonistic, as during fixation of stationary targets while the head is moving) are solved through a *hierarchical* control structure.

During the last decade the computer vision community introduced a paradigm called *active vision* to model the capability for dynamic and purposeful modification of the parameters of the vision sensory apparatus in order to simplify the visual task. "Smart" sensors have since been designed containing control loops which implement *alerting* mechanisms (pre-attentive, diffused attention, unpurposive scanning); windowing processes for *foveation* and attention focusing on a small region of interests; target tracking, trajectory detection, and collision predictions. These achievements have led to the visual-front-end, which, combined with a high level reasoning system that guides the observation strategy and selection between concurrent stimuli, can arrive at the interpretation of what is going on in the scene.

Both the camera and eye control systems introduced so far gain efficiency when combined with the capacity to rapidly select the regions of interest (both in space and time), to allocate resources following data resolution changes, and to update the strategies to the partial results thus achieved. The key role in the implementation of these processes is played by the "allocation of attention strategy." Note that attention processes can work at different levels: from the purposive or focused or conscious or voluntary level to the diffused or pre-attentive or covert or implicit level. Following the panel coordinator: "Solving the attention problem (understanding thoroughly how it works) seems tantamount to solving the vision problem itself."

Motion Perception and Interpretation

There is evidence that visual perception in humans is "intrinsically dynamic," in the sense that motion is one of the dominant primitives. The superiority of dynamic to static primitives is shown in the capacity to recover the spatial 3D structure (which is a well-known ill-posed problem in a static frame) and in the direct perception of the environment. Proximity is considered, for the detection of the reference frame, at the higher control levels of the hierarchical structure of perception for motion as well. Moreover, "the motion organization reflects a structure of relationships between objects in term of 'actions', not simply of displacements": dynamic events are rich in information (for instance, the kinematic pattern after a collision furnishes material on the ratio between the masses of the objects). To this end, the motor-perceptual interaction has been of great importance for human motion perception. It has been stated (at least in the case of stimuli originating from external motor activities) that knowledge of the rules governing the human motor-control-system guides the visual recognition of biological movement.

The computational approaches for machine detection and estimation of motion parameters are distinguished into two categories: those following the *correspondence* scheme in which motion is conceived as a set of displacements, and those following the *gradient* scheme that follows a velocity-based formulation. In the first case, a set of significant tokens (usually relatively sparse) is extracted from the image segments and tracked along the frame sequence; in the second case, the two-dimensional field of instantaneous pixel velocities, called *optical flow*, is estimated without considering correspondences of noticeable features. For both approaches, the monocular technique gives ambiguous solutions from motion: at least one scaling factor is unknown; instead the stereo pair approach can provide absolute values.

In order to properly interpret the motion parameters it is necessary to process the motion components (uniformly distributed or sparse) thus far detected. Besides the reliable estimation of the primitive motion parameters, events and motion perception require taking into account what is "known" about the properties that the stimuli should satisfy, such as inverse optics, transformations, invariants, salient features and functional relationships of the perceived 3D world. In dynamic perception some degree of ambiguity exists in all the quoted subjects. The panel represented an attempt to elucidate the relevant ambiguities.

Visual Representations

This module consists of a cascade of subjects: how visual mental imagery may be relevant to hypothesis formation; what representation formalisms of visual information may be effective in machine perception and artificial intelligence; the role and the efficiency of visual *metaphors* in the conceptualisation of new solutions.

We use imagery to remember, to anticipate, to answer a question, or to invent something new. A few meaningful statements on visual representations and thoughts are presented by

Penrose in *The Emperor's New Mind*. Among these, there is a quotation from a letter of Einstein to Hadamard: "The psychical entities which seem to serve as elements of thought are certain signs and more or less clear images which can be 'voluntarily' reproduced and combined ... The above mentioned elements are, in my case, of visual and some muscular type. Conventional words or other signs have to be sought for laboriously only in a second stage...." Following this quotation, Hadamard agreed with Schopenhauer's aphorism: *"Thoughts die the moment they are embodied by words."*

Nevertheless, the exploitation of visual representation is almost a novel subject for computer scientists. To model image-based hypothesis formation, *visual abduction,* and to be effective in problem solving, imagery representations must achieve the kinds of computational tools already available for propositions and non-imaginistic schemes. The computational visual representation scheme (proposed by Glasgow and Papadias), introduced in the first lecture, combines image-like representations with linguistic information processing, and makes inferences by cognitive representations similar to visual mental imagery.

Representations suitable for operating on a computer should easily describe the common regions of interest, should highlight the important information, should be stable (minor changes should produce a small variation in the representation), should be compact with no redundancy, should require simple computational procedures for managing knowledge, and finally should be applicable to a number of distinct problem domains. In the second lecture, the symbolic representations employed in artificial intelligence are grouped into three classes: *relational* representations (analogical or conceptual, based on networks, graphs or frame schemes), *propositional* representations (linguistic, based on the first order predicate logic) and *procedural* representations (dynamic and pattern-directed schemes). The first class supports the construction and organisation of internal descriptions of the world; the second provides phrase-like descriptions to enable logical reasoning and linguistic communication; the third addresses the solution of perception tasks through agents which implement special actions which modify the internal status of the system, thus creating and updating the representation.

The use of icons in the theory of representation and the relationship between language and image are debated not only among humanistic scientists: the latest computer programming paradigm is related to visual programming. Although it is not known to what extent the executor of an iconic program might support a theory centred on images as tools, one of the computer features which brought the most success to personal computing is surely the user-friendly interface based on the desk metaphor. Studies on the role of context (including, together with the environment, the user-transmitter and the observer-receiver), on the analogy with graphical approaches to the solution of scientific problems, on gesture in actor training, on what the "words induce to see" and on the integration of verbal and visual description (e.g. in brief textual poems like the Japanese Haiku) have all been considered, and discussed by the panel. This has allowed the definition of models, metaphors, and paradigms for multimedia and multimodal interfaces, and for visual language programming.

Visual Reasoning

Many scientists claim that their important discoveries were achieved by a visual approach to the generation of the proper hypothesis. Let us quote two well-known cases: Friedrich Kekule, the chemist who discovered that organic compounds, such as benzene, were closed ring structures and Richard Feynman, the physicist who discovered a visual approach to the solution of complex particle physics problems. Nevertheless, even if a great effort is put into artificial reasoning systems, research on inferences based on 3D descriptions of the spatial relations between objects, remains a relatively new and independent area.

Psychological research shows alternative views on the nature of the representations that reasoners construct to describe spatial layouts, and for relational reasoning. Following an age-old tradition, visual stimuli are distinguished into *proximal* ones and *distal* ones. In the former case a 2D image analysis is performed, mainly "looking at the retinal image" by means of a viewer-centred representation which observers achieve "naturally" without "attentional efforts or special training". The latter case concerns a 3D analysis, following an object-centred description, which is presumably invariant under observation conditions, a task that is modelled as an ill-posed problem. There is evidence that human vision engages both types of analysis in parallel, while machine vision generally focuses mainly on the distal approach. The two parallel interpretation attempts suggest another way to examine visual uncertainties deriving from intrinsic image ambiguities. The alternative interpretations can be exploited to achieve "a more reliable or deeper understanding" of the scene by managing the emergent description so that the removal of ambiguities increases its consistency, its completeness, and its reliability (see Table 1).

There is evidence that reasoners construct mental models of spatial descriptions. In particular, there is evidence of the effect of some aspects of the description, such as its continuity and determinacy, and of the effect of the integration of new information into a representation. Following the discussion started during the 1987 AAAI Workshop on Spatial Reasoning and Multisensor Data Fusion, we can conclude that spatial reasoning goes well beyond computer vision, and involves spatial task planning, navigation planning for autonomous robots, representation and indexing for large spatial databases, symbolic reasoning and the integration of reasoning with geometrical constraints, accumulation of uncertain spatial evidence, and multisensor data fusion.

The discussion concentrated on spatial reasoning for high-level computer vision. The mental model theory of spatial inference and the results of several experiments on the perception of movement and rhythmical pattern support the characteristics of stability and self-organization on visual thinking. According to the psychologists of the Gestalt and others, the perceptual organisation is characterised by a leading principle: the tendency to *prägnanz*. Visual interpretation of perceptual process is guided by a sort of "principle of minimum" that in the process of interpretation is responsible for self-organization towards stability and harmony. However, the perceptual result of phenomenally *singular* appearances is supported by the admixture of trial-by-trial variability with locality of evaluation of saliency, so as to produce a "trial-by-trial stability of outcome."

Imagination and Learning

In a very singular sense, imagination has been defined as "the use of imagery for constructing something new, like a product, an invention, a fantasy, a dream."

Some experimental studies show that spatial strategies and imagery representations are useful in the preliminary stages of thinking problems, where they facilitate the construction of mental models. Following Finke, Denis, and Andreani's formulation, mental images are *pre-inventive forms* which spring from "active exploration and manipulation of physical objects, (which) produces mental representation, passes through combination, transformation of spatial relationship, discovers unexpected things and concepts, explores possible interpretations restricting them first to a general class of objects, then to a particular object. Invention in this way is a product of combinational play that often shifts into creative thinking."

Quoted examples of this process are the apple-Newton theory of gravity, the pendulum-Galileo theory of periodic oscillation, the mould-Fleming discovery of penicillin. In these cases, observation was followed by abductive, deductive and inductive reasoning. That is, it moves from the perceptual domain to an abstract domain, which is more suitable for learning, constructing new models, and in knowledge restructuring. Nevertheless, imagery may not only be involved at the starting point, rather interaction with logical processes can be continuous. One could hardly do better than repeat Andreani's conclusion on the topic: "There is a large variety of images which lie on a continuum from perception to imagination and creative thinking: they are a form of representation which may be generated in every phase of the cognitive processes, in initial phase when they anticipate the identification of the stimulus, in working memory when they code the material by visuo-spatial code, in L.T.M. as dynamic schemata which can activate the visual or verbal code, in analogical reasoning, when the perceived similarity is the source of the relationship between objects and systems, in meta-phoric thinking which often generates models, in imagination which constructs new worlds. They can be concrete, vivid representations based on sensorial inputs, or they can be abstract as in geometrical or other symbolic form."

It has been said that children learn from images and show use of imagery in memory more than with other medium. The present situation in machine perceptual learning is quite different: up to now the vision learning paradigms do not seem to be suitable for the requirements of context adaptation and novelty detection and generalisation. As has previously been men-tioned, the promising active vision approach involves close interaction between sensor and environment, requiring a consistent model of the context in which the system is operating, to be constructed. This model must be as simple as possible but incrementally updatable and sufficiently precise. Moreover, there is the problem of the combinatorial explosion of cases in inductive learning, evolutionary processes, and novelty emergence discovery. The panel discussed all these themes, while Caianiello in his lecture presented one of the last fruits of his research, based on the integration of the fuzzy set theory with neural networks in order to accomplish pattern recognition tasks.

IMAGINA and Virtual Reality

During the workshop, two evening sessions were dedicated to image synthesis and artificial and virtual realities. In the former, the lecturers presented the international Pixel-INA awards, "IMAGINA: images beyond imagination," for videographic and cinematographic works containing one or several sequences of computer graphics or computer-generated special effects. A short report on the second evening's presentation is also included here. Virtual reality has a short history, but represents the most advanced stage of human-computer interaction and, being so centred on direct experience, it is destined to influence human perceptual aptitudes. Two different ways of creating artificial reality were presented: one in which the user, as in many video games, is identified with a character on the screen (so playing in *third* person) and another one in which, by full immersion through a helmet, the user acts in *first* person fulfilling the YAT principle of "You Are There." Research in virtual reality involves a multiplicity of disciplines, while the number of applications is growing in sectors like entertainment, architec-ture, medicine, ecology, molecular physics, and many others, with challenging perspectives.

This workshop was the last one which professor Eduardo Caianiello attended before his untimely death in December 1993. Thanks to his extraordinary dedication his contribution was instrumental, more than anyone else's, in founding some of our most important institutions: the Institute of Cybernetics of the National Research Council, the Italian Chapter of the Interna-tional Association for Pattern Recognition, and the International Institute for Advanced Scientific Studies. He was also one of the most prominent scientists in the field of perception, an inspiring colleague and teacher, and a dear friend to all the workshop participants. In recognition of his many accomplishments, I would like to dedicate this book to his memory, and I conclude this preface with a quotation taken from the conclusion of one of his most important papers, in which he discovered the theory of homogeneous hierarchical systems. One could hardly do better to describe the spirit of this book: "In conclusion, we feel that our attempt is 'exposed to destruction,' in Mendeleev's sense, in so many ways, that we find it stimulating to present it for this very reason".

Acknowledgements

The workshop, and thus indirectly this book, was made possible through the generous financial support of the university, and the research and industrial organisations that are listed separately. Their support is gratefully acknowledged.

The editor would also like to express his appreciation to the Scientific Committee of the workshop for their suggestions: G. Adorni, S. Gaglio, and V. Roberto (AI * IA), A. Buizza, and L. Moltedo (CNR), I. De Lotto, and V. Torre (GNCB), V. Di Gesù, and P. Mussio (IAPR), W. Gerbino (AIP).

Special thanks should go to Dr. Alessandra Setti, for her contribution of special quality and magnitude in organizing the workshop and in the preparation of this volume. Without her help, I could not even guess what these activities would involve! Thanks are also due to the "Centro Documentazione" of Pavia University for their help in formatting and putting the manuscript into camera ready form.

Virginio Cantoni

SPONSORING INSTITUTIONS

The following institutions are gratefully acknowledged for their contributions and support to the Workshop:

- Gruppo Nazionale di Cibernetica e Biofisica del CNR
- Comitato Nazionale Scienza e Tecnologia dell'Informazione del CNR
- Comitato Nazionale per le Ricerche Tecnologiche del CNR
- Università degli Studi di Pavia
- Dipartimento di Informatica, Sistemistica e Telematica - Università di Genova
- Cassa di Risparmio delle Provincie Lombarde
- Centro Ricerche Fiat
- Banca del Monte di Lombardia
- Dipartimento di Informatica e Sistemistica - Università di Pavia

CONTENTS

THE PHYLOGENETIC EVOLUTION OF THE VISUAL SYSTEM

Silvana Vallerga

Istituto di Cibernetica e Biofisica del CNR
Via De Marini 6, I - 16149 Genova, Italy
and International Marine Centre
Lungomare d'Arborea 22, I - 09072 Torregrande (OR), Italy

ABSTRACT

The first eye could only tell light from dark, then motion, colours and forms became visual images. Through phylogenetic evolution a large number of visual systems have been originated and shaped to increase general efficiency and to adapt to the available ecological niches. The compound eyes of invertebrates are best at short distance sight. Their field of view is almost spherical. Many invertebrates can see from ultraviolet to infrared, and detect the plane of polarisation of light. Vertebrates have developed different visual structures to achieve high acuity and accommodation. The retina of fish, the ancestors of all vertebrates, is a mini-brain performing complex visual tasks as motion detection and is an example of ongoing adaptation to different environments; during development retinal cells modify to adapt to changes in habitat and habit. In higher vertebrates retinal processing is greatly reduced and substituted by the much more efficient central processing. Extended colour vision and detection of polarised light are lost in humans, but we can tune our vision to read a book or to look at the stars.

1. INTRODUCTION

The eye to this day gives me a cold shudder to suppose that the eye with all its inimitable contrivances..could have formed by natural selection, seems, I freely confess, absurd in the highest degree, wrote Darwin in 1860. Since then, the available interpretation on the phylogeny of the visual system has been based on bits of evidence provided by paleontology, morphology and ecological considerations and on inferences made by the study of function, physiology and behaviour of living analogues[1-3].

The history of visual systems extends back in time several hundreds of millions of years. The oldest eye of which we have record belongs to the trilobite (see figure 1), an ancient marine creature which lived 500 millions years ago.

The trilobite had a compound, faceted eye that could see side but not up. The *limulus,* the

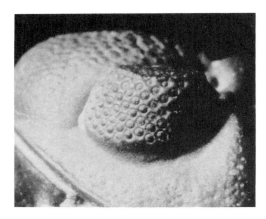

Figure 1. The eye of the trilobite, a marine creature which lived over 500 millions of years ago (reproduced from Sinclair[4], with kind permission from Croom Helm Ltd).

horseshoe crab, another ancient creature that managed to survive to the present day resembles the long-extinct trilobite allowing us a glimpse to a distant past.

Since then nature has experienced a large number of solutions to best accomplish visual tasks. Two types of evolutionary processes have produced diversity:

1. *increase in general efficiency* of the system, a slow and continuous process to raise the upper limit of efficiency from the molecular, sub-cellular, cellular and tissue grades.
2. *adaptive radiation* to populate the available ecological niches, a faster process from which a species adapt to a new environment and develop adaptive features within a relatively short time.

The first life forms that swam in ancient seas developed patches of light-sensitive cells, allowing them to tell light from dark. As life evolved these patches became eyes that could detect first motion, then form and finally colours. New habitats became available, as land and air, requiring performances that were satisfied by specific visual systems. Terrestrial animals developed different vision than aquatic animals, diurnal creatures differentiated their vision from nocturnal creatures. The different visual need of preys and their predators resulted in diverse visual strategies. As habitat and/or habits changed the visual system adapted, ocular media first, then photoreceptors and neural cells.

2. FORMS OF VISION

Three main different ways of forming images have evolved. The simplest eye scans the scene like a tv camera (see figure 2a), building a picture from successive impressions. This form of vision, the least common, is used by the smallest creature capable of forming a visual image, the copilia, a copepod a few hundred microns in size, which is thought to be the link between the primitive eye-spot and the more complex image-forming eye[4].

The compound eye is the second solution, widely adopted by invertebrates (see figure 2b). It is found in animals as different as insects, crustaceans and molluscs. This is a fixed-focus eye with individual refractive units, ommatidia, each responsible for a portion of the visual field. The compound eye allows a close view over an almost spherical field of vision free of spherical and chromatic aberration (see figure 3).

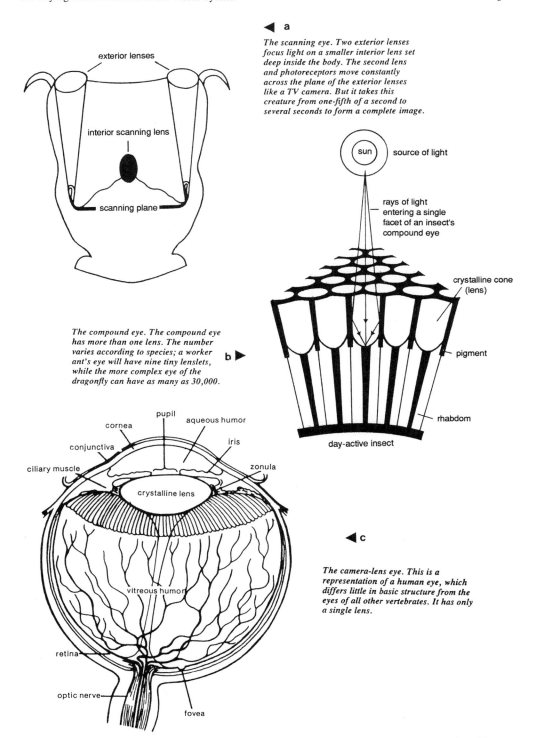

The scanning eye. Two exterior lenses focus light on a smaller interior lens set deep inside the body. The second lens and photoreceptors move constantly across the plane of the exterior lenses like a TV camera. But it takes this creature from one-fifth of a second to several seconds to form a complete image.

exterior lenses

interior scanning lens

scanning plane

sun   source of light

rays of light entering a single facet of an insect's compound eye

crystalline cone (lens)

The compound eye. The compound eye has more than one lens. The number varies according to species; a worker ant's eye will have nine tiny lenslets, while the more complex eye of the dragonfly can have as many as 30,000.

b ▶

pigment

rhabdom

day-active insect

pupil

cornea   aqueous humor

conjunctiva   iris

ciliary muscle   zonula

crystalline lens

◀ c

The camera-lens eye. This is a representation of a human eye, which differs little in basic structure from the eyes of all other vertebrates. It has only a single lens.

vitreous humor

retina

optic nerve

fovea

Figure 2. Three ways of vision. a) The scanning eye; b) the compound eye; c) the vertebrate eye (reproduced from Sinclair[4], with kind permission from Croom Helm Ltd).

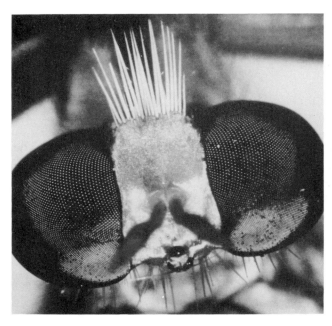

Figure 3. The eye of the robber fly provides this insect with an almost spherical field of view.

A third form of vision is found in vertebrates (see figure 2c). The vertebrate eye is a camera-like device that brings an image into focus on the retina, a continuous receptor surface, apposed onto neural layers. The photoreceptors, rods and cones, lie in the innermost part of the eye, thus the light crosses the whole retina before being absorbed by the visual pigment molecules. For this reason the vertebrate retina is called "reverse", in contrast with the "direct" retina of invertebrates (see figure 4).

3. COLOUR VISION

Colour vision may first have evolved as early as 440 millions of years ago[5]. Some of the ostracoderm fish gave up their benthic habits at that time and become free-swimming. Their paired dorsally located eyes moved away from the median eye, down onto the sides of the head. No doubt vision became more important for the feeding of these animals. Away from the bottom of the sea the whole visible spectrum became available. Multiple receptor mechanisms were developed for daytime use, when light intensity permits the separate sensory pathways required for high visual acuity and for colour vision, which requires two or more classes of photoreceptor cells each with a different spectral sensitivity.

One can easily image that colour vision has evolved in terrestrial animals in response to the same factors. The advantage of colour vision in providing better photo contrast is so evident that it is difficult to image its gradual development in an evolutionary sequence[6]. Strongest contrast would be the most useful to solve the need to recognise objects with the same size and form, but differing in important properties as ripe or unripe, flowers with nectar or not, sexes of the same or of close species.

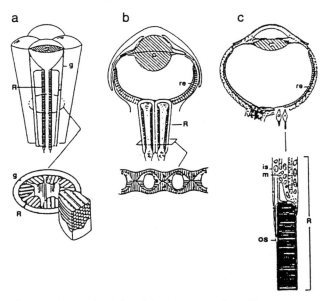

Figure 4. Comparison between the eye of cephalopods b), the compound eye of insects a), and the vertebrate eye c). R, receptors; re, retina; g, pigment granules; is, inner segment; os, outer segment; m, mitochondria (reproduced from Wolken[8]).

4. VISION IN INVERTEBRATES

There is great diversity in the visual systems of invertebrates, from the pigment depression that serve as eye for the earthworm to the largest eye of any animal, that of *architeuthis*, the giant squid, which eyes are 370 cm in diameter. Photoreception is widespread but acute vision is restricted mostly to cephalopod and various arthropods.

4.1. Spiders

Nearly 300 millions of years ago, during the Devonian period spiders appeared. They have simple eyes, to distinguish them by the compound eyes of other invertebrates, even if they are far from being simple.

The eyes of spiders occur in clusters of six to eight, they have a form of colour vision and detect UV and polarised light. Spiders, as do the majority of insects can see clearly up to 40 cm, the best image is at about 8-10 cm. The frontal pair of eyes form the visual image and detect the plane of polarisation of light, used as an aid for orientation. The three lateral pairs control the periphery of the visual field, estimate distance and motion. Accommodation is performed shifting the retina back and forth to bring the image in focus, the retina can be also moved sideways to increase the field of vision. Many spiders are diurnal, including the jumping spider which is considered as having the best vision of all arachnids.

4.2. Insects

Insects have been very successful in evolution. The number of known insects species is over 700,000. The other groups, except arachnids (30,000 species) do not exceed 5,000. Insects have compound eyes (see figure 3) with a number of units that can reach 28,000 as in the dragonfly giving the animal an almost spherical field of view.

Colour vision is typical of most insects. Their photoreceptors are tuned to biologically important signals such as feeding and mating and to the secondary emission of the "blue" sky which shows pronounced peaks in the blue and UV because of the scattering effect of the atmosphere. The distribution of UV photoreceptors allows for the orientation of insects with the so called sky compass, though the detection of polarised light is also an helpful aid to navigation[7].

4.3. Cephalopods

The eye of cephalopods has features common to both invertebrate and vertebrate eyes (see figure 4). The photoreceptors are rhabdom facing the incoming light as in invertebrates, but they are organised in a retina layered on an eye-cup where the image is focused through a single lens as in vertebrates.

5. VERTEBRATES

The eyes of most vertebrates conform to a similar pattern. They have transparent media, the cornea, the aqueous, lens and vitreous, through which the light passes before it reaches the retina. The retina of all vertebrate has a common structure (see figure 5).

Photoreceptors input signals to the neural network in which information is processed through a vertical pathway, photoreceptors - bipolar - ganglion cells, and two lateral pathways, photoreceptors - horizontal - bipolar cells and bipolar - amacrine - ganglion cells. From the ganglion cells visual information is sent to the brain via the optic fibres. The human eye as well as all other vertebrates' eyes work on this principle.

Horizontal cells control the general sensitivity of the retina, mediate laterally the signal, are colour coded and provide surround information to bipolar cells. Bipolar cells begin the coding of forms, which adds to the coding of amplitude and colour and feed such signals directly to ganglion cells and through feedback circuits, to amacrine cells. Amacrine cells provide the surround information to ganglion cells, are colour coded and initiate the detection of movement. Ganglion cells code intensity, movement, colours and shapes and feed the visual information to the brain through long axons. Local feedback are present at all levels. In high vertebrates the more complex visual tasks, as motion detection are performed by the brain.

5.1. Fishes

Fish appeared 600 millions years ago and from them originated all the other vertebrate classes. Some ancestral fish are still dwelling present oceans to prove the success of this group of vertebrates.

For fish, vision is undoubtedly the most important sense. They have good perception of form and colour. The eyes are well developed, and the optic lobe, which also controls the learned behaviour, is the largest part of fish's brain. The eye of fish grows throughout life. Photoreceptors and retinal neurones are continuously added as peripheral rings. The adult fish retina maintains the capacity to adapt to changing environment that other groups have only in juveniles.

All fish have a flattened cornea, in contrast with the curved cornea of land creatures. An important reason for that is that the refractive index of the cornea is almost exactly the same as water's. The fish lens is the most important light-gather. Large and spherical, the fish lens has

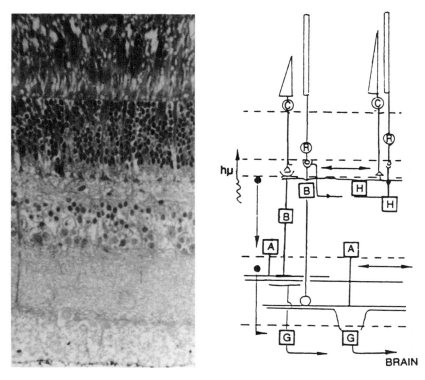

Figure 5. The vertebrate retina. Left, photomicrograph of a cross section of a fish retina. The total width of the retina is about 250 µm. Right. Schematic diagram of the retina. C, cones; R, rods; H, horizontal cells; B, bipolar cells; G, ganglion cells.

the highest refractive index of any vertebrate lens. Due to the round shape of the lens the fish are nearsighted, but the eye is partly able to accommodate for far vision by pushing the lens away from the retina.

The evolutionary adaptation of visual system of fish is of particular interest since fishes adapted to the widest variety of photic environments[9], from the sparkling waters of coral reefs, to most turbid waters, to oceanic depths. The type of vision a fish has is influenced primarily by the type of water and depth at which it lives. The visual pigments, the sensitive elements in photoreceptors, are strongly correlated with the photic environment, deep-sea species have rod pigments maximally sensitive to shorter wavelength than surface-living species, and freshwater fish have their vision shifted toward the longer wavelengths[10]. The neural retina adapts to photic environments. Bottom dwelling fish have the inner retina more developed, indicating a higher degree of neural processing, while clear water animals rely more on colour vision for contrast discrimination than on inner retina processing, therefore their outer retina is more developed[11,12].

Sharks, rays and mantas have the same ancestor as other fish, but their eyes have distinct variations. They are farsighted, their lenses are so perfectly graded that they suffer from very little astigmatism. Their vision is probably inferior for they have fewer photoreceptors and hardly any cones, and most see only in black and white. Their visual world is a grainy and indistinct picture, though they see forms at a distance more clearly than do the other fish.

5.2. Amphibians

The amphibians developed from ancestral fish that reached shores to spend on land part of their life. These animals are aquatic as juveniles and terrestrial as adults, amphibian means with a double life. Their visual system adjusts during metamorphosis to adapt to the change of habitat. Amphibian visual neurophysiology has provided a great deal of research on the visual system, starting with the work of Lettvin, Maturana, McCulloch and Pitts in 1959[13], that showed that the retina of anurans can process a great deal of visual analysis: a) sharp edges and contrast, b) movement of edges, c) local dimming produced by movements, d) the curvature edge of a dark object. Vision, by means of the lateral eye is intimately involved in prey catching behaviour, in prey-avoidance and with the extra optic receptors in orientation[14, 15]. Further processing of the visual information is accomplished by neurones in the optic tectum and elsewhere.

In general the visual system of many amphibians is characterised by a high degree of peripheral coding. It is supposed that as we ascend the phylogenetic scale the central nervous system becomes better developed, behaviour more flexible, and the need for peripheral coding decreases[14]. While there may be some truth in this view, it must be remembered that even in the primates' retina there is some degree of peripheral coding.

5.3. Reptiles

Reptiles were the first vertebrates to live on land, in many ways their visual problem became much simpler. To them air offered a clear medium for vision with great advantages secured by the rectilinear passage of light through it. The difference in the refractive indexes of air and water resulted in having the cornea becoming the major focusing element and the lens a fine focusing device. Furthermore since water absorbs heavily in both the ultraviolet and the infrared end of the spectrum, terrestrial vertebrates had a much broader band of electromagnetic radiation available for the transmission of visual information[16].

The higher standards of vision required by life on land, allowed for new anatomy and function to develop. The first fovea developed in lizards. The environmental adaptation of the reptile visual system, especially of the retina, is an example of the evolutionary process. The studies are associated with the work of Walls[3], he believed that the primitive reptiles were all diurnal and he supposed that they had inherited the cones of their ancestors, and lacked rods.

As competition on land increased, many reptiles were forced to become partially or completely nocturnal. This process was accompanied by an increase in the size of the outer segments of some cones, so they gradually became rod-like both in function and appearance. Walls summarises this hyphotesis as *"the first rods in the world were produced by the transmutation of cones, and the process has been occasionally repeated, whenever needed, ever since the vertebrate came on land"*.

Many lizards have on top of their head the ancestral third eye or parietal eye, which contains photoreceptors that resemble the cone and rods of the lateral eye and is connected through a nerve to the diencephalon. The development of the parietal eye differs from the lateral eye: the retina is not cupped by invagination and lens is formed from the same outgrowth of the brain as the retina, as a result the photoreceptors point towards the entering light, unlike the situation in the lateral eyes. The function of the parietal eye is to detect solar radiation, and is probably specialised to detect changes in ambient light at sunrise and sunset, which could be important for the control of the diurnal rhythms.

Lizards, geckoes and snakes show intermediate stages of the evolutionary process. The snakes probably evolved from an ancestral lizard, that became fossorial. As a result the lateral eyes developed a protective spectacle, beneath which the eye became slowly atrophied. When later the

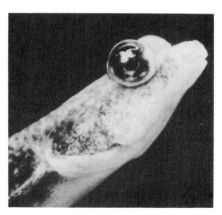

Figure 6. The *Anableps* a Caribbean fish is sometimes called the four-eyed fish. It has a double retina and a double iris. The top retina watches out for predatory sea birds while the bottom retina searches the water for food (reproduced from Sinclair[4], with kind permission from Croom Helm Ltd).

snake came back to the surface the eye was reconstructed and the eye of the present day diurnal snakes, is very different from those of other reptiles showing an example of convergent evolution.

5.4. Birds

Vision is a major sensory modality in birds. The avian eye, in general, combines a remarkably large size with a superb accommodation mechanism and a thick, neatly stratified retina devoid of blood vessels and capable of high visual acuity, colour discrimination, and movement detection. Characteristic features of such retina include a large retinal image, a fascinating array of visual cells (rods, single and double cones), an efferent centrifugal system, elaborate and diverse areae of acute vision (with one or two foveae) and an additional nutritional source, the pecten.

Both diurnal and nocturnal birds have a duplex retina, but the proportion of cones is low in nocturnal forms. Many birds have a deep foveal pit, the convexiclivate fovea (see figure 7), whose function is thought to magnify the image due to the fact that the refractive index of the retina is slightly greater than that of the vitreous (ratio 1.006). This refraction would magnify the image, Walls[3] suggested that this magnification effect could increase the foveal acuity of some hawks and eagles to at least eight times that of man.

The photoreceptors of birds contain coloured oil droplets which role is probably associated with colour vision. Bowmaker and Knowles[17] studying the chicken retina pointed out that hue discrimination within the broad spectral band dominated by double cones, which are nearly half of the cones in the chicken retina, would be very poor were it not for the presence of two types of single cones, one combining a red oil droplet with yellow visual pigment and the other combining a yellowish droplet with a green pigment. This results in having a narrow band spectral sensitivity with only a little overlap.

Birds, as a number of invertebrates and some fishes, are able to detect UV light. This is in contrast to man and many other vertebrates which have lenses that absorbs UV light. UV vision might be important in bird's navigation. UV light penetrates haze better than longer wavelengths and sky light away from the sun peaks in the UV. The polarisation pattern of the sky is thus more pronounced at these wavelengths, and if a bird can perceive the plane of

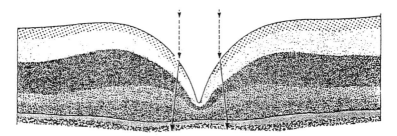

Figure 7. The convexiclivate fovea of birds.

polarisation of light at very short wavelengths it will have the best opportunity to use this information as an aid in direction finding[18].

5.5. Mammals

Two completely new developments occurred in the vision of mammals: consensual pupillary movements and co-ordinated eye movements. Both are the result of the increased visuomotor organisation commanded by the brain.

Mammals exhibit a variety of sensory adaptation for specialised ecological niches. The specialisation for nocturnal and diurnal vision implies that whereas diurnal species can sacrifice visual sensitivity to improve acuity, nocturnal animals are limited in their visual acuity by the necessity of achieving maximum sensitivity.

Nocturnal mammal were the first to appear. Most creatures active at night are either the descendants of creatures who fled to the relative safety of the dark or pursued them. The dark-adapted eye (see figure 8) has wide apertures, the cornea and lens are huge in comparison to those of diurnal animals. The retina consists of all, or nearly all, rods, which have higher sensitivity than cones and are frequently pooled to connect to the second order neurones. The sensitivity is shifted towards shorter wavelength. For diurnal mammals the peak of sensitivity is in the yellow region whilst for nocturnal animals green objects appears brighter.

The evidence of the long mammalian night is present in the structure of the mammalian eye. When mammals returned to daylight they had lost the coloured oil droplets found in more primitive vertebrates to protect photoreceptors from UV damage and reduce chromatic aberration. Their lens adapted becoming opaque to UV and chromatic aberration was reduced filtering out shorter wavelength.

The diurnal mammalian's eye retains the advantage of the bird's eye. The lens is pushed toward the cornea increasing the size of the image falling on the retina. The cornea provides the first focusing on the retina, and some mammals rely entirely on the cornea for focusing.

In all sub-mammalian species, including marsupials and monotremes, the visual information from one eye crosses over to the opposite hemisphere. In primitive mammals starts a new development and optic fibres of each eye split at the optic chiasm to reach both hemispheres. The result of the split of information is responsible for a peculiarity of primate, three dimensional stereoscopic vision. The brain compares the two retinal images of each eye and judges distances, comparing the angle between the sighted object and the eye. The movement of the oculomotor muscles to bring the object into focus help in judging the exact distance. As the distance increases the pupil widens to let in more light. The process requires extensive co-ordination between the sensory-motor system and the visual system.

Figure 8. The dourocouli, a monkey of South America, is the only known nocturnal primate, and its eyes are so sensitive that if they are exposed to daylight for any length of time the animal can go blind (reproduced from Sinclair[4], with kind permission from Croom Helm Ltd).

6. ADAPTABILITY

The extraordinary adaptability of the eye and its ability to respond in the most varied conditions are without question. This ability is observed in many vertebrate classes, including mammals. Ungulates that spend their life scanning the horizon for carnivorous predator become horizontally astigmatic by the third year. Domesticated birds become nearsighted even if birds are naturally farsighted.

The most interesting adaptation occurs in fish. Visual systems of fish are highly dynamic in their functional organisation, whereby fish vision may undergo greater levels of adjustment and fine tuning than any other vertebrate. They may change their visual pigments within a few weeks, and change cones for rods, and vice versa, whenever needed by the photic environment[10, 19]. The synaptic organisation of the neural retina also undergoes significant change during light adaptation. In particular, external horizontal cells, which are intimately involved in colour processing, go through major structural changes, extending and retracting finger-like processes called "spinules"[20]. Recent evidence suggests that similarly plastic mechanisms are active in the inner retina, which must remembered is part of the central nervous system[11].

7. CONCLUSION

Natural visual systems are the results of a dynamic process of adaptation to the visual tasks to be performed. This process resulted in a large number of diverse systems which respond optimally to different visual tasks. Adaptation is an in built mechanism in vision and can be observed, not only during phylogeny and ontogeny, but also during the life of individuals as they are exposed to changing environment. The wide choice given by the heterogeneity of natural visual systems should be kept in mind when planning to mimic natural vision.

REFERENCES

1. M.A. Ali, *Sensory Physiology: Review and Perspectives,* NATO ASI, Plenum Press, New York, NY (1977).
2. F. Crescitelli, The visual pigments of geckos and other vertebrates, in *The Handbook of Sensory Physiology*, Springer-Verlag, Berlin, Heidelberg, New York (1977).
3. G.L. Walls, *The Vertebrate Eye and Its Adaptive Radiation,* Cranbrook Inst. of Sci. Bull., Vol.19, Bloomfield Hills, Mich (1942).
4. S. Sinclair, *How Animals See. Other Visions of Our World*, Croom Helm Ltd, London and Sidney (1985).
5. W.N. McFarland and W.R.A. Munz, Part III: The evolution of photopic visual pigments in fishes, *Vision Res.,* Vol.15, pp.1071-1080 (1975).
6. W.R.A. Munz and W.N. McFarland, Evolutionary adaption of fishes to photic environment, in *The Visual System of Vertebrates*, F. Crescitelli ed., Springer-Verlag, Berlin, Heidelberg, New York, pp.194-274 (1977).
7. M. Gogala, Ecosensory functions in insects, in *Sensory Physiology: Review and Perspectives*, NATO ASI, M.A. Ali ed., Plenum Press, New York, NY, pp.123-154 (1977).
8. J.J. Wolken, *Photoprocesses, Photoreceptors and Evolution*, Academic Press, New York, NY (1975), in Photoreception, M.A. Ali, M. Anctil, and L. Cervetto, *Sensory Ecology: Review and Perspectives*, NATO ASI and M.A. Ali, Plenum Press, New York, NY, p.480 (1977).
9. E.R. Loew and H.J.A. Dartnall, Vitamin A1/A2 based visual pigment mixtures in the cones of the rudd, *Vision Res.,* Vol.16, pp.891-896 (1976).
10. J.N. Lythgoe, The adaptation of visual pigments to the photic environment, in *Handbook of Sensory Physiology. Vol. VII/1*. Springer-Verlag, Berlin, Heidelberg, New York (1972).
11. S. Vallerga, A. Arru, and D. Lupi, Retinal plasticity during eye migration in flatfish, *Invest. Ophthal. Visual Sci.*, Vol.31, p.159 (1990).
12. S. Vallerga and S. Deplano, Differentiation, extent and layering of amacrine cell dendrites in the retina of a sparid fish, *Proc. R. Soc. Lond. B,* Vol.221, pp.465-477 (1984).
13. J.Y. Lettvin, W.S. Maturana, W.S. McCulloch, and W.H. Pitts, What the frog's eye tells the frog's brain, *Brain. Proc. Inst. Radio Eng.,* Vol.47, pp.1940-1951 (1959).
14. R.G. Jaeger, Sensory functions in amphibians, in *Sensory Physiology: Review and Perspectives*, NATO ASI, M.A. Ali ed., Plenum Press, New York, NY, pp.169-196 (1977).
15. E. Kicliter, Flux, wavelength and movement discrimination in frogs: forebrain and midbrain contributions, *Brain Behav. Evol.,* Vol.8, pp.340-365 (1973).
16. W.R.A. Munz, The reptile sensory system, in *Sensory Physiology: Review and Perspectives*, NATO ASI, M.A. Ali ed., Plenum Press, New York, NY, pp.197-216 (1977).
17. J.K. Bowmaker and A. Knowles, The visual pigments and oil droplets of the chicken retina, *Vision Res.,* Vol.17, pp.755-764 (1977).
18. R.A. Suthers, Sensory ecology of birds, in *Sensory Physiology: Review and Perspectives*, NATO ASI, M.A. Ali ed., Plenum Press, New York, NY, pp.217-251 (1977).
19. J.N. Lythgoe, *The Ecology of Vision*, Academic Press, London, UK (1977).
20. M.B.A. Djamgoz and M. Yamada, *The Visual System of Fish*, Chapman and Hall, London, UK, pp.159-210 (1990).

A CONTROL VIEW TO VISION ARCHITECTURES

Bertrand Zavidovique[1] and Pierre Fiorini[2]

[1] Institut d'Electronique Fondamentale, Université Paris XI
Centre d'Orsay, Bat 220, F - 91405 Paris Orsay, France
[2] Etablissement Technique Central de l'Armement
16 bis Avenue Prieur de la Côte d'Or, Arcueil cedex, F - 94114 Paris, France

ABSTRACT

Trying to compare robot vision architectures to a mythic reference like the human system might appear somewhat ambitious and asks for precautions. First two major difficulties are outlined. Drawing analogies between systems from their outputs is risky: such a limit is illustrated through one formalizing example. So, better compare major features. But a zoology of vision machines is questioned when, aiming to a well informed architectural feature choice, a rapid presentation of trends in the field is proposed.

Then an approach closer to physics prompts to a classification from a control point of view: it reveals some duality between operations and communications.

A few visual operations are distinguished provided technology is not trailing behind. But a fair stress must be put on communication networks if properties likely suitable for comparison are to be found.

1. INTRODUCTION

It might appear somewhat ambitious to compare a mythic reference like the human or animal system to machines that we build and call robots. Restricting to vision does not make it simpler in any way as perception is a better posed problem than vision in introducing other sensors to compare data from, or in providing actions to close loops eg. in regularising. Therefore if a machine reflects some optimal ability to process data and extract information, and then its design is meant to be somehow optimal too, the architecture of a perception machine including fusion and control parts, or interfaces to actuators, is very likely more a representative candidate for comparison with the optimum reached by nature over million years than a vision system in isolation. In that respect, one forecasts a poor benefit from attempting to draw inspiration from devices we do not actually know even whether they are machines or not, devices we do not know well how they are made and how they work.

In fact, an opposite point of view comes from physics. In physics, we have been used to build models, to conjecture properties of systems when they were not strictly delimited in their intrinsic nature, when they were not completely circumscribed in their effects (internal or external). It is so with a thermodynamical view of phenomena where mechanics is involved too. It is true again with a chemical approach where nuclear physics provides decisive supplementary explanations. Many other examples can be proposed coupling classical versus relativistic mechanics or micro versus macroscopic explanations. A partial view to a problem does not hurt as long as one remains aware that it is necessary to build a theory, to check on its coherency, to face it to "reality" and notice scrupulously every unexpected event, occurring paradox... to explain it in enlarging the theory (or trashing it) if necessary, and so on.

The present analysis of "machine vision architectures" will draw on physics in two consequent manners, to start:
• first we guess some traps to avoid when attempting to compare brains and computers or nervous systems and controllers;
• second we make choices, restricting architectures to one or two features which properties (hopefully suitable) can be derived from.

The organisation of the paper stems from that hope for scientific steps, in our case inspired from physics. The first sections will be on two major difficulties: limits of system comparisons are illustrated by one formalising example. Then we try to explain briefly why a zoology of vision machines is not that helpful. Indeed, aiming to a better-informed architectural-feature choice, a rapid presentation of trends in the field is proposed.

Therefore the third part focuses on a classification from the control point of view since control appears a major feature. It makes evident some duality between operations and communications.

The fourth part is then on some visual operations that may be distinguished if technology is not trailing behind. Finally a fair stress will be put on communication networks and their likely important properties as for comparison.

2. PRIMARY PARADOXES OF ANTHROPOMORPHY

From an intuitive point of view, yet scientific enough, when looking for a model it sounds better try to get interested in foundations and evolution of computer architectures rather than describe them one by one, or just compare them. Group theory, linear algebra, optimal control theory, Maxwell's equations ..., and $E=mc^2$, among other federating results are good to remember in that respect. So are Lamark and Darwin on other grounds! But how to prove it? Would the model be the human being, some major differences pop out as potential principal axes to be investigated more in depth.

As for raw material, silicon is different enough from carbon and water so that it should influence at some point.

As for the fabrication process, (re)production is not quite comparable. It is seemingly more enjoyable (a male is writing of course) and far more automated in the human case. According to a theory put forward by Y. Burnod[1], the behaviour of a biological system can be explained at some level through a unity, maybe unknown but mandatory, since the whole system is grown from a unique cell. To the opposite, the current state of vision machines and more generally computers, shows a plurality of functional units at many levels. Unless one wants to derive analogy between a basic biological cell and a transistor! Whether it is memory vs. ALU, I/O devices vs. inner-communication means, operating system vs. programming language and so on, technology is different, assembly is different, the way each of them has been designed is different too.

As for the operating process, we do not know much about the exact human vision process although a lot of progress has been made for the last decade[2]. Surprisingly enough we do not know much too about the exact way a computer processes data. It is well known that the great majority of programs are not proved correct and then so are not the computers when they run these programs. But along with the advent of parallelism, respective impacts of hardware and software, namely on control and programming, become so hard to evaluate that one can say we have to experiment on these machines to know them better. Even if man built them, man does not master them anymore.

In a word from three main a priori features - technology, organisation, functioning - the only reason why animal vision systems and robot vision systems are comparable might be: we do not know precisely the way they are designed and generated, but they act comparably. Now this idea, or its contrary, holds only because in the analysis so far, systems can not be envisioned else than from their outside. Actually one tends to parallel their outputs, i.e. a potential ability to process visual information and trigger an appropriate action from a scene. Hooking to these facts - information processing and output comparison - we get to leave the somewhat philosophical debate above by attempting a simple model of perception systems (information) to derive a warning against superficial analysis or comparison (from outputs).

The theory is based on Bayesian inference and information systems are modelled as Markov processes, suitably known to model well enough perception means like sensors, processors and scenes. Suppose we are interested only in the number of jump-over of a process X_t with continuous time and numerable state-space, and only particular jumps are observed (e.g.. taking place in a given subspace J of the transition space). This means for instance a threshold has been set, and each time the output gets greater than this threshold a counter is incremented*. The evolution model is known and the Mean Transition Flow (MTF) has to be estimated. It is a macroscopic indicator, far from the system state in most cases, as are actions triggered by such perception.

P_t^i is the conditional probability that our system is in i state at time t, over the observed jump history, M_i^t is the conditional-mean transition ratio from i to rewarding states, then:

$$MTF = \Sigma_i \, P_t^i \, M_i \tag{1}$$

Question: " Can MTF be estimated without full knowledge of P_t? " It is interesting since P_t, the information state, is updated under the Bayesian formalism through a non linear filter that requires to solve a numerable quantity of equations:

$$P_t^i = P(x_t = i \, / \text{ all observed jumps}) \tag{2}$$

Olivier[3] has shown in his thesis the existence of peculiar queuing processes with (non-finite) numerable state space that cannot be modelled by any finite dimension queuing system in the following sense: no finite state-number Markov chain can have the same MTF. These processes have a poissonian traffic modulated by a Markov chain for inputs, which is a good enough model of occurring edges under tracking for instance.

Now to make an infinite story short, let us assume the following analogy: men are infinite state space systems, if ever they are systems, would it be only because of learning abilities; this allows to avoid any contradiction - although not providing a proof - in the fact they come to an appropriate action after a finite sample of the observed state space: their infinite built-in proper state space stands supplying missing information to the price of a certain risk, to be estimated in turn, from ascertaining probable-only facts. Would a manufactured system, finite state-number device in the sense above, mimic human beings in given circumstances, there cannot

---

* We can relate this for instance with the equivalent surface of a bright coming object during a visual tracking phase: noise, sensor limits, occlusions, points of view and attention focusing make the partial observation.

be any evidence, after Olivier's counter-example as for any common internal structure.

Of course we already knew that planes do not wave wings, yet they fly. But an additional clue from this model of systems prompts to look more carefully for system states and then compare them more globally. Aiming to go deeper into architecture analysis, let us first try a synthesis which follows the actual way systems are built, and check whether currently known components reveal a suitable state representation.

3. LIMITS OF MACHINE ZOOLOGY

Since already-built machines are dealt with, let us start with a technology-driven view to them. Architectures may be dedicated on two different levels:
• the basic components, usually VLSI circuits;
• the complete systems, gathering either specific or general purpose components.

3.1. Basic Components

Two different types may be distinguished, so called operators and processors.

Operators are meant to be more specialized than others, achieving very specific image processing and then being hardly programmable. Machines that are built out of operators are fully optimized and dedicated to one application. Implementing each an optimized function, most often in VLSI (ASIC)*, several operators need to be assembled and to communicate for this application. So many such commercially available operators process a digital pixel flow on the fly, and they are in great need to be more easily organized according to some data flow graph. Among classical functions already on the shelf: NxN convolvers, mathematical morphology circuits, Hough and Radon transforms, histogram makers, digital filters and gradient operators in particular.

Processors are more programmable and corresponding machines are intended for a wider class of applications. Yet this class strongly depends on the processor architecture and granularity, and on the way it is embedded into a machine with respect to communications, parallelism type and memory. Most of these processors were designed in view of a specialized (massively) parallel machine. They can be again data-flow, but more likely they are to be operated in the so called SIMD (Single Instruction Multiple Data) mode. This means that every processor runs the same instruction on different data. Although the burst of ASIC technologies enabled quite a lot of research teams to propose their own, not so many such circuits are commercially available. Typical granularities are 1, 8 or 16 bits. Specific instructions support (low level) image processing operations like, again, filtering, convolution, histogramming, projections, Fourier and some other orthogonal transforms, mathematical morphology. And the increasing generality causes communication modes to become explicit between processors or towards memory and I/O devices. Actually, the trade off performance/generality depends strictly on these specific instructions (or built in macros) and communication modes. In that matter universality kills performance!

3.2. Complete Systems

Going on with the same logic, one can distinguish between three types of image processing machines:

---

* VLSI: Very Large Scale Integration, ASIC: Application Specific Integrated Circuit.

- heterogeneous machines: assembly of operators;
- homogeneous machines: parallel arrangement of specific processors;
- parallel machines: based on standard processors.

3.2.1. Heterogeneous Machines. Here several printed circuit boards are assembled. Home- made or bought, they usually obey a standard like the VME bus, for sake of easier reusability. Again available functionalities of the boards are in the range of arithmetical operations on various data structures (lines, arrays, multiwindows ...), convolution, linear and non linear filtering, morphology, look-up tables ... up to edge detection and tracking. According to the progress in circuit integration more and more primitives are gathered on a board. "Building" procedures results from interconnecting several boards in hardware. Changing procedures: the limited offer in terms of operator types leads to reconsider the whole interconnecting hardware or to multiply glue circuits as multiplexors.

Fortunately technology improvements help in that too: VLSI favours complex routing chips like cross-bars to inter-configure boards in a more software manner. However this new flexibility aggravates an already known limitation of such machines: programming efficiency. No formalism exists for representing all these various boards under a unique software model. This lack involves unstructured programming then any system integration cannot be else than complex and very expertise demanding, in contradiction with automatization. Nevertheless this type of machine is well adapted and economically optimal in the case of a very specific and tailored application.

3.2.2. Homogeneous Machines. These machines are bound to the incoming of processors (see above) and they depend a lot on the growth of ASIC technology. Until recently they could not be made, both technically and economically. Nowadays, integrating the whole machine seems reachable thanks to hybrid or mosaic technologies or wafer scale integration. They feature a regular architecture, centered around a communication network like mesh array or multidimensional grid, pyramid, hypercube ... Each network shows a priori advantages in terms of privileged communications or robustness against contention. This key question will be investigated in more details in the last section of the paper. The regularity implies a rather easy design compared with the pick computing power to be expected - up to several dozen billion operations per second for a few thousand processors.

They can involve different types of processors (2 or 3 at most) dedicated each to different image processing steps; typically so called low and high level operations are considered separately, as low means topologically close to the image array while high is rather evocating complex decision graphs. Due to the already mentioned SIMD or data flow programming modes, such machines are easier to run than heterogeneous ones.

Now the present technology stands already the design of processors, that are homogeneous machines themselves. Due to the massive nature of inputs - the images - there was no other solution in such attempts than to make them sensitive to light. They appear as true silicon retinas and the rapprochement between the sensitive coat and the processing layer is so important from a vision point of view that a full section will be devoted to such machines.

3.2.3. Parallel Machines. Here processors are general purpose and commercially available. Whence a large granularity, up to 128 bits, and a big computing power (around tens of MFlops). Among these processors some are signal processing, i.e. they are built around a couple multiplier/ALU which implements the inner-product and they feature other minor facilities to make easier access to data in sequence like built in pointers. Other processors may have been designed to incorporate communication means and prepare concurrent work.

However parallelism is lower than in the previous section, typically 4 to 64 processors and the machine originality relies at least as much on the programming model as on the network itself. Although such machines are most natural to think of when trying to increase the computing power, not so many have proved successful. The fairly powerful processors are incentive to run concurrently rather long local sequences of instructions, with relatively rare resynchronizations. These modes are called MIMD or SPMD for Multiple Instruction Multiple Data and Single Program Multiple Data. They turn out to be extremely difficult to put efficiently in practice, mainly because programmers cannot be given a global enough and abstract enough view of the machine. Different instructions on different data make a system heterogeneous again. However these machines are carefully considered recently for two main reasons:

- the commercial availability of amazingly powerful processors (in the order of two hundred million operations per second - i.e. 10 times an average personal computer);
- the evolution of programming. It leads to emulate thousands of (virtual) simple tailored processors on these general purpose processors, to reach a global behaviour of homogeneous machine rather than to build one.

Whether they are homogeneous or parallel, vision machines can incorporate both dedicated and standard processors. In such cases specialized processors are used for low level processing in a massively parallel fashion and general processors solve high level vision problems with moderate parallelism.

Finally we can derive three leading ideas from this technological explanation of vision machines:

- the main word is "performance". Performance had to be increased, so engineers thought to put together several computing units, more or less "specialized". And then everybody fell into the programming problem, whether it is for associating operators or to preset their operation to run;
- as for this "specialization" of processors a few operations can be distinguished, but they are not easily bound to vision at that stage. It looks like they were showing up quite often in algorithms written for conventional computers and then appeared worth to be implemented. This approach is not specific to vision either and known, after Patterson[4], more universally in computer science, as the RISC (Reduced Instruction Set Computing) approach[5];
- a vision specificity comes back under the form of a difference between low and high-level vision. Respective topologies of data are not comparable and would algorithm implementations follow them, adequate machine structures are different too.

Such conclusions remain far from allowing any guess on plausible internal states of vision machines to be suitably compared to any reference. Again the main reason is because, technologically speaking, the vision specificity is extremely poor. Conventional architectures used in information processing did not meet vision algorithm requirements: a colour image sequence with average resolution (512 x 512 pixels) and usual time-sampling (25 pictures/s) implies a data rate in the order of 20 Mbytes per second. The least feature extraction involves a convolution or a filter of the kind accounting for a dozen multiplication/addition and as much addressing, then some result cleaning and shaping up, accounting for a few dozen tests and local computation again, and finally an output dressing like rewriting and code changing costs a few additional memory read/write. Altogether one gets around a hundred operations per pixel, which makes a total in the order of a Giga operations per second, beyond possibilities of best current available processors like the Dec Alpha or the last Texas or HP. To reach a realistic application we still need decision techniques, either structural or combinatorial, and several changes of representation e.g. hundreds of times more operations per piece of data. Even if the number of data pieces is decreasing with semantics, the total amount in such calculation comes to the order of hundreds giga operations per second.

As vision is a dynamic process most often and it triggers an action which in turn defines a new recognition problem based on a new image analysis, there is little hope that the loop be closed. Some super computers are beginning to reach this range of a computing power like the CM-5 or a CRAY-T3D but they make it in economical conditions that eliminate them. For instance a 1024 processors CM-5 provides around 60 Gflops: it is worth 30M$, asks for a 400 square feet room with a power consumption in accordance! This economical flavour is efficient to conclude with, it shows well enough how weak was the image processing prompting. Dedicated architectures appear necessary in practice for reasons of cost or cumbersomeness/ performance ratio. In terms of evolution this could be a reasonable engine but in the present case a large discontinuity becomes evident as for the software: parallel machines do not benefit the large program bases developed over years and tested by million users. In such conditions we have to put the vision specificity back into prospective for our architecture analysis. As parallelism is involved a good way to do so should be to focus on vision specific communications or data structure and movements, if any.

4. A SOCIAL VIEW OVER MACHINES

Let us come back to pictures. What should be kept from them to specialize architectures?
- They are multidimensional arrays from the very acquisition. At least two dimensions are involved, but motion (time-sequence) or colour (wavelength-vector) imply a third one. Not all directions are as important.
- An image is very structured and often redundant. Extracting information from data leads to reveal the underlying organization while eliminating noise. As every piece of data is a priori equivalent, first operators are to be local (processing a neighbourhood (n)) and identical for all pixels (p).
- Many actions may be triggered from a single image and there are many features anyway to be extracted from a single image, like points of interest, edges, regions, motion, up to shape, stability or agressivity. This means a lot of transformations (o) to be applied to raw data progressively or simultaneously.

If one adds the somewhat technological factor that is the signal amplitude ie. for digital machines the word width or number of bits per pixel (b), an expression of the processing complexity in the case of pictures comes as a function $f(b,p,n,o)$. Actually the number of items in a neighbourhood is around few tens (beyond that, noise considerations make physical (not semantic) interaction very unlikely), current word lengths are in the same range. The image size and the number of operators to be involved are rather in the order of thousands. Then, if parallelism were again a major feature of specialized architectures it should occur in two possible ways:
- to process many pixels in parallel. By continuity from the image topology, architectures will first mimic the data organization. Let be "data-structure machines" the corresponding architectures. We already saw that a technological view tends to assimilate the latter machines to low level processing even though some of them handle already some privileged data movements with a flavour of image analysis[6];
- to run in parallel various operators applied onto images. As some of these operators need to be fed with results from others in a systematic manner, what matters here is data circulation for information extracting. Let be "information-propagation machines" the corresponding architectures. At some point the dynamic organization of processors may become essential to the understanding and their representation as a graph leads naturally to a new type of control focused on data flows.

In the next sections, we describe both classes in more details based on examples to make key features rise up. Then additional examples will be given that mix different types in aiming to "vision" rather than to mere image processing. This highlights the importance of programming before to summarize with a more abstract (social) view to machines.

4.1. Data-Structure Machines

They are designed to run simultaneously a same processing on many structure elements and to make easier exchanges adapted to a given data organization. Most of them are thus operated in the SIMD mode or a multi-SIMD mode. Two types can be distinguished: arrays and pyramids, although other structures like hypercubes or rings have been so far proposed that are not as closely related to the image structure[7, 8].

4.1.1. Arrays. These machines consist of a bidimensional processor network which structure is meant to be similar to images. The mesh-connection is then square- or hexagon-based, and each processor gets straight connections to 4 or 8 (resp 6) nearest neighbours. Some machines allow two types of connectivity (CLIP 3[9] below). Each processor has its own local memory to store pieces of input pictures and (intermediate) results. As the image loading time could artificially limit performances, most often built-in communication abilities are taken advantage of by shifting pixel rows or columns for massive input/output. A last common feature of many arrays is to gather monobit processors, for sake of density. Thus, data are transmitted and processed bit after bit in a sequential (pipelined most often) manner. This model suits well a spatial parallelism based on regular image splitting. Communication locality and SIMD control tend to restrict array applications to low level processing.

The first attempts of that kind come back to the fifties with Unger[10] in 1958. The SOLOMON (Simultaneous Operation Linked Ordinal Modular Network)[11] was proposed at NASA since '62. It was a 32x32 array of 1-bit processors with 4K bit RAM each and could never be achieved due to the technology at that time. In '66, the study is reconsidered to build Illiac IV[12], delivered to NASA in '72. Concurrently an Illiac III[12] was tried for image analysis in physics and never achieved. From '70 to '75 ICL developed the DAP (Distributed Array Processor)[13]. It was a 64 x 64 1bit processor that was commercialized by INTEL in a 32 x 32 version called AMT500. More recently other interesting examples are MPP[14-16], GAPP[17], CLIP[9, 18-20] or GRID[21,22]. We just highlight in the following some key features in some of them.

CLIP 1 to 7 (Cellular Logic Image Processor)

CLIP 4, the best known, belongs to a family of SIMD processors designed at University College of London by Duff and his colleagues[23]. The initial aim was to process binary images in adapting the machine to the nature of processed data. The architecture is centered around a 96 x 96 array. Each elementary processor is bit serial with two inputs, two outputs. Six long shift registers, one per bit plane, delay and carry out the input-output process. Storing relies on three registers (one for carry) and a 32 bit-memory. Processing relies on a double Boolean ALU able to compute any two variable logical function. Variables into the ALU are any register or a weighted sum of the neighbours, to accelerate some peculiar operations like erosion or dilation. Processors may be configured as adders to process grey level images. Two other features stem from a more-global-behaviour interest to be underlined:
- asynchronous information transfers are enabled by straight emission of a processor result to its neighbours. This allows to accelerate iterative operations[24];
- to get global results, like activity gauges to end processes or associative-like queries, a bit counter that is in fact a tree adder is connected to one edge of the array, profiting by bit-plane shift abilities.

CLIP 5 did focus on more elaborate local functions as histogramming, sorting, complex thresholding... CLIP 6 and 7 (up to an integrated version LSI CLIP7) were significant of a trial to stick even better to data in processing 8 bit words.

MPP (Massively Parallel Processor)[14]

It was developed by Good Year Aerospace (see figure 1a) for NASA around '83. The intended application was remote sensing whence a more "arithmetical" fashion. The central array is 132 x 128 (4 columns make redundancy in case of failure). Processors (see figure 1b) are bit-serial with 1K bit memory and have direct access to their four nearest neighbours. An auxiliary 32 Megabyte memory is connected to the network via an 80 Mbyte/s rapid bus. A meshed control unit spreads command-signals to all processors and a host VAX 780 supports program developments and displays. Each elementary processor gets 6 1-bit registers (A, B, C, P, G, S). P register with related functions performs any two-variable logical operation and manages the communication with the four neighbours (N, E, W, S). The processing unit is made of A, B, C, a programmable length shifter and an ALU (adder + P/logic). G is a control register to run masked (conditioned) operations. S is a shift register to serialize data towards the auxiliary memory without disturbing other processes. Altogether in 100 ns cycle time a processor runs simultaneously:
- add between A and P, with C. The shifter allows to multiply;
- logic operation with neighbours. P is loaded either with the result of a logical function of P and the data bus, or by the value of P in a neighbour;
- equivalence between P and G (test and masking);
- shift on S from west to east for rapid I/O at 320 Mbyte/s.

An or-sum tree shows the global "or" of all data buses, supporting global tests. Results in the range of 10^6 to 10^9 pixels per second for histogramming, correlation, convolution or region growing prove a good fit to local and regular transforms.

DAP (Distributed Array Processor)[25]

It was not designed for image processing, although a great deal of such applications have been completed[26]. Therefore the processing element gets no image specificity and appears simpler than both previous ones with three registers (an accumulator Q, an activity index A for inhibition, and C for carry). But the interest is real on communication features.

Highways (Rows and Columns) are buses spread over the whole memory, giving fast access to any line in the network. "Ripple carry" are transfers to use the array for operating as

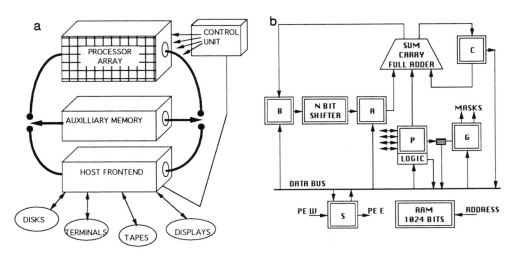

Figure 1. a) MPP overview; b) MPP: a P.E.

well on N x N arrays as on N N-bit words. The latter case requires N-bit adders, made out of N 1-bit adders thanks to these "ripple carry", and one word is shared by N processors.

A16 K-bit local memory allows considerable storage, limiting input/output flows.

AAP (Adaptative Array Processor)[27, 28]

It has been made by NTT as a general purpose super computer but one of the very first applications was Kanji recognition. The processor offers two independent data paths. One way is devoted to vertical transmissions while the other supports the regular 8-connectivity. Whence data shifts (input-output) are done concurrently with computations, or peculiar data movements like array transpose become basic. An additional ability along the data path consists in hierarchical deviations (by pass), since one processor or a cluster of processors can be skipped. It provides these long distance communications like in DAP and some additional facility: one out of four elements can send or receive independently and concurrently.

The one bit ALU is given 2 memories, 32 and 64 bits. The second one stores data and information for adaptive control. Thanks to that device, multiplexors are locally managed: some processors are transmitters, others are receivers and block transmissions. This may result in isolating clusters (regions in images) where to operate asynchronous and independent transforms. Hence different processing types cohabit in the same machine, as more as it is logically an N^2 1-bit processor or an N N-bit processor machine, with this asynchronism ability in both cases.

As an intermediate conclusion about this type of machines, let us underline that computing on lots of simple processors rather than on few complex ones was experimented positively in the case of image processing. Not only performance is increased but adapted interconnections avoid most overhead like address calculation. Moreover, bit serial processors reduce silicon surfaces for communication and processing, increasing the density which favours the emergence of a collective behaviour to be studied. But this does not go up to any semantic of data movements as more as privileged communications are fixed and thus, not always well suited to the algorithm. The array size is also smaller than the image size, involving artificial manipulations due to "paging" or "rendez-vous on the edge" (according to the image splitting). Generally speaking random exchanges between processors are costly and this is why some added tricks have been used to global gauges for instance. At some point, in accordance with advances in technology, both hardware and software, it appeared not so clear anymore whether it was not more efficient to emulate thousands of elementary processors on tens of powerful processors, at least as long as machines have to be programmed. This is again part of the CM-5[29] story.

4.1.2. Pyramids. To mitigate 2D machine limits, some teams have proposed a communication structure making easier global information transfers while securing a comparable processing quality. To that purpose, on top of the 2D structure, a pyramid of identical processors centralises data thanks to a tree like additional structure. That way a flavour of information propagation is naturally introduced and makes a transition to the next section.

Such machines can be considered a stack of processing layers with decreasing size. They mimic a pyramidal data structure and then allow multi-resolution image processing[30-33]. In the same way they may evocate a process in which information is progressively refined and compressed up to a decision, or conversely a parameter to be spread over an entire scene for improved local processing or for divide and conquer strategies.

Vertically a processor in the nth layer (a father) is linked to a given number of neighbouring and connected processors in the $(n-1)^{th}$ layer, generally two or four. Some models have been proposed where the number of sons may be random or even adapted, but they rather belong to software. An approximation could be: if there are enough sons, some overlapping could be organized so that a son gets several potential fathers, being assigned during a next process according to dynamic conditions. So a regular pyramid of which the lower layer (basis) has n^2 processors gets Log_2 n+1 layers. The ith layer has $n^2 / 4^i$ or $n^2 / 2^i$ processors. At each layer, a processor has horizontal connections to neighbours and vertical connections to sons and father(s). The vertical paths make a tree, which is known to be the fastest way to send and receive top down or bottom up messages. Bottom up operations result from a dychotomical decomposition of global processes. They cover global property computing like counting, extremum finding or more general measures based on some integral (momentums, histograms, clustering) to be expressed in a systematic manner as, for instance: sum (I) = sum (I/2) + sum (I - I/2). Or they help input/output conflict-solving in logarithmic time. Top down operations support specific communications with processors predetermined according to their address (marking) or their content (masking). This very dissymetry between vertical and horizontal communication speeds amounts to an associative-memory behaviour. Therefore it is not surprizing that this type of an architecture was envisioned in the eighties for database machines and information management[34, 35]. And such applications have proved successful on other massively parallel machines like the CM-2.

While each layer is operated in the SIMD mode an additional kind of parallelism is made possible since different layers can work independently. Their respective controllers have then to synchronize in some way. In that respect pipeline is a peculiarly simple (mechanical) synchronization process and it suits particularly well in the present case the bit serial processing fashion. The successive execution of several routines in pipeline reduces the time between two results to the maximum run time for a layer.

Nevertheless input-output have proved difficult to be made efficient and the network, after its density and irregularity, is electromechanically complex to be implemented in practice, compared to other types which have been tested. This explains, together with the already mentioned more general tendency to emulate such massively parallel devices on coarser grain machines thanks to software advances*, that very few projects have been actually completed. Let us mention PAPIA[36], MPP Pyramid[37], PCLIP and HCL Pyramid[38, 39] and SPHINX[40], that we now evocate briefly to conclude this part.

SPHINX (Système Pyramidal Hiérarchisé pour le Traitement d'Images Numeriques)[41, 42]

It is a joint development from ETCA and IEF/PARIS XI. Each processor is horizontally linked to four neighbours and vertically to one father and two sons (see figure 2a). The polarity of connections to fathers, x and y, switch between consecutive layers to occupy at best the 3D volume while keeping a binary tree. The three types of parallelism as mentioned above are available: SIMD array per layer, pipeline between consecutive layers, MIMD when a process is distributed on several sub pyramids. Processors (see figure 2b) are bit serial with a tagged ALU, 256 bit memory, built in pointers and access to a paged 8K bit external memory. An or-sum and an I/O scan path provide global facilities. They require about 1800 transistors which would allow about one hundred processors per chip in a CMOS two-metal 1 μ technology considering available yield. The actual chip has 16 processors in a 2 μ technology.

PAPIA from Pavia and PCLIP from Seattle are quadtrees with respectively 8 and 4 connectivities in planes. The three-machine processors include control registers (inhibition) that carry out processor masking, re-introducing some flexibility and desynchronisation with respect to strict SIMD operating.

---

* For instance SPHINX has an efficient simulator on CM-2 and 5, HCL Pyramid was first a model of cellular operator and Cantoni has found a clever way to back project a pyramid provided an adapted "hierarchical programming" model.

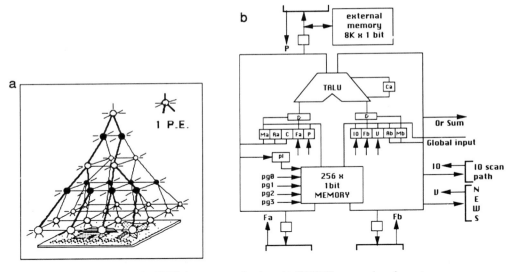

Figure 2. a) SPHINX: intercommunications; b) SPHINX: processing element.

4.2. Information Propagation Machines

In this class of machines the leading idea is to observe information flows. They cross physically a sequence of operators to finally make a result. Two different prompting may be considered here:
- a rather technological one, again, is the adaptation to usual cameras that scan images;
- a likely more fundamental one is considering that successive transformations provide data movements with semantics; this point was already made for hierarchical networks.

Two main types can be distinguished - pipeline and systolic - the main difference being in synchronization which obviously matters here. A third one, sort of an hybrid, called wave front array will be mentioned in conclusion of this section.

4.2.1. Pipeline Machines. The pipeline techniques emerged very soon in computer design (IBM 360 series for instance). They are widely used in super computers and vector processors both for instruction fetch and processing. The principle is to break a task into a sequence of several shorter subtasks, possibly to the price of additional intermediate storage. On a long enough data flow, where the set up time becomes negligible, results outcome with a period equal to the duration of the longest subtask.

In image processing one can split a segmentation process for instance according to some progressively changing representation like local processing, thresholding, binary processing, blob feature extraction, etc... A likely drawback is then to freeze an operator chain which may be not universal. As a matter of fact most pipeline machines are bound to a restricted class of image processing, or based on some unity of data representation or leading idea which renders the architecture well adapted.

CYTOCOMPUTER[43]

It is a chain of 88 operators. Each of them performs a local function on a 3 x 3 neighbourhood, either 2 dimensional for Boolean processing or 3 dimensional for grey level transforms. Choice is given for each cell between four, six or eight connectivity. The local shaping up that actually makes the neighbourhood is completed by shifting data along a fix-size register. Then image and neighbourhood sizes are imposed. But today's technology would make it possible to dimension them dynamically thanks to inner pointers in memories. Logical transforms are stored in look-up tables. Except for set up time, the cytocomputer was applying mathematical morphology processing for cell automatic classification in real time.

PIFEX[44, 45]

It has been developed at Jet Propulsion Laboratory, under the form of a set of identical programmable operators interconnected via a toroidal bus. This enables a repartition between length and parallelism of paths. Each operator (see figure 3) offers two 3 x 3 convolvers which outputs are fed into a bidimensional lookup and then a linear interpolator. An additional comparator on a 3 x 3 neighbourhood performs a non linear transform to detect zero crossing for instance. In the current version PIFEX allows up to 80 operators (in 5 x 16) for low level feature extraction (like an Hessian matrix computation and Laplacian zero crossing for instance).

The difficulty in this approach, and the jackpot in accordance, is to know whether the basic cell to be duplicated allows a sufficient coverage of the whole intended image processing. If it is so, concentrating on interconnection-network programming independant of node programming should turn out to be a real success.

PIPE (Pipeline Image Processing Engine)[46]

It was designed by the Sensory Interactive Robotics Group at NBS to process video on the fly. It mixes point and neighbourhood operations in a chain of identical processing stages. For each stage a set of transformations is run on the data stored at this stage. A control unit sets the operation sequence, that can be modified between frames by the host processor. In addition to the standard pipeline mode, an image can be injected back into the current stage or the former one (see figure 4a). Such backwards modes enable processing iterations on any number of stages. Point operations are implemented by look-up tables and ALUs. At each stage two arithmetic or logic local operations are performed (see figure 4b). A so called region of interest operator commutes two sets of operations, pixel to pixel, based on a stored operation map. This feature helps real time for stereo or motion matching.

A purely linear transmission of data, merely implementing a map product of operators chosen from a limited set, limits the processing variety even if the operator set is general enough (see concluding remark on PIFEX). Pipe offers more flexible interconnections as each operator gets two locally sequencible image buffers, before transmission to next stages. That way a first attempt of data flow control is introduced. Although global variable computing is tough on this machine, it appeared very well adapted to most image segmentations on the fly.

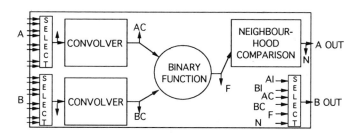

Figure 3. A PIFEX module.

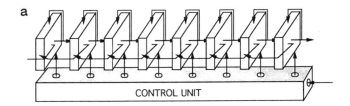

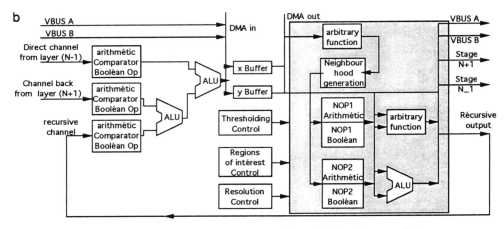

Figure 4. a) Main connections in the PIPE machine; b) PIPE: architecture of a layer.

4.2.2. Systolic Arrays[47-51]. Unlike the previous class of machines that was addressing the implementation of a complete image processing job, the intended improvement here targets elementary operators. The difference may be stated as following:
- the pipeline consists in chaining rather meaningful and possibly complex operations, hence a quasi monodirectional propagation, an adhoc synchronization and some attempt to introduce a data flow control for by-passing these latter constraints;
- the systolic approach concerns more elementary operators, to define a cellular organization well adapted to one (or very few) operation, taking advantage from a structural parallelism of this operation. The data flow can be then multi dimensional and organized as a steady and synchronous pipeline.

The leading idea in systolic is to mix intimately very simple operations with communications in a way that guaranties no contention. Many operators have been implemented[52] ranging from multipliers to some elementary syntactic pattern recognition, passing by linear filters (transverse, recursive ...), various matrix-vector multiply (see figure 5), Lu-decomposition and 1 or 2D convolution in 4 or 6 connectivity.

Data are pumped into the network through its edges at each cycle time. Accordingly each PE inputs its data subset from its ports, performs the calculations and outputs the results on its ports. Since the network was tailored for data transfers just required by the algorithm, allowing an intensive use of pipeline in several directions, performances are amazing especially in the case of large amounts of data. Moreover simple operators, replicated many times, make easier VLSI implementation. Now the price to pay is in programming. The local functions are either frozen in hardware or implemented as programmable functional units. But the programming

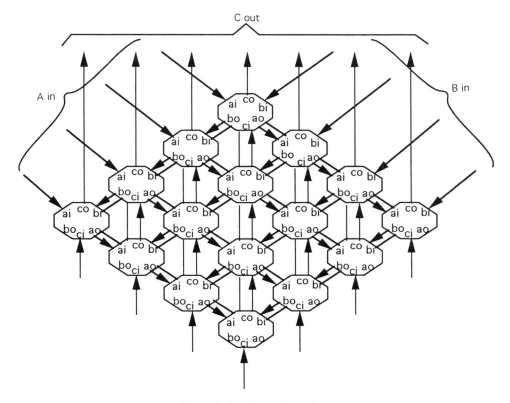

Figure 5. Systolic matrix product.

part engraved in the network pattern is likely not to be easily modified. Despite a great deal of work on automatic programming and synthesis of systolic architectures[53-56], a complex data sequencing, a lack of reconfigurability and some I/O mechanism specificity leave users with a lot of difficulties when they implement such architectures in practice. Main examples of this type that could tackle image processing are WARP from CMU[57] with a more recent VLSI extension iWARP[58] by Intel, and GAPP.

WARP[59-63]

It is a systolic linear architecture (see figure 6) of 10 to 20 cells. Each cell performs 10 MFlops and has its private program memory and local data memory. Inter-cell communication relies on two wide bandwidth channels (40 Mo/s). The software has been designed to literally hide the systolic model to users. Cells are programmed in W2, a kind of Pascal that stands asynchronous communications between procedures. The global model is an implicit data partitioning, implemented in the Apply language. A library (so called WEB) of 140 image processing routines is available (edge extraction, data conversion, orthogonal transforms...) and was even ported on the Image Understanding System (see section after).

GAPP (Geometric Arithmetic Parallel Processor)[17]

It was designed and made by Martin Marietta Aerospace in '82. The chip is distributed by NCR since '84. It is an array of 6 x 12 1-bit-processors that execute the same instruction on different data. Each cell is rather classical with a bus towards the NEWS neighbours, a bus for

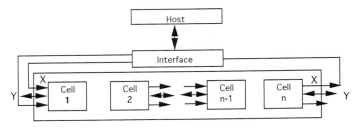

Figure 6. The WARP processors array.

external communication N/S, a 1-bit ALU and a 128-bit RAM. The originality comes from programming which deals mainly with communications. The ALU performs always the same operations at each cycle, and different behaviours and results come from data moves inside the structure. This particularity enables up to five simultaneous instructions per elementary cell, in exploiting the 13 bit instruction word with 5 independant fields*. One gets here, in a restricted case, this balance between communication and operation programming which was aimed at in arrays. Although a 32 x 64 version was announced, the lack of software probably makes such devices doomed, except for very specific vision automata, may be.

4.2.3. Wave Front Array Processors and the Data Flow Concept. Trying to draw inspiration from images by their structure or by a likely (hopefully generic) organization of image transforms progressively emerged on a different way to control and program subsequent specialized machines. Not only operations have to be determined and well organized in time and space, but communications too, as more as both architectural efficiency and operational semantics are incentive to a better balance between communication and actual processing. Most described tentative architectures so far try in fact to skip this problem by a predetermined and fix network supposedly near-optimal to convey major communications (in fact most frequent ones) and by a predetermined synchronization: for instance SIMD means every body operates the same except masked processors that do nothing, or pipeline or systolic imply no synchronization at all because processors are just passive data crunchers cadenced by a global clock along a flow. But every application where choices come out ill-adapted forces program writhing and, anyway, sends the most elementary automatization back to dreams.

We have already encountered a potential solution: the "data flow" approach**. For instance in '87, S.Y. Kung[65] proposed an extension to the systolic concept in which the global synchronization is replaced by a local one based on data availability. Calculations are then fully data driven and directed by a mechanism known in software as the "producer-consumer's handshake". He called that "Wavefront Array", but such machines are not quite easy to design with respect to the current technology. Standard Telecom in UK has built a multi DSP[66] called DDM1. For ten years at ETCA we have been looking for a common representation between algorithms and architectures in view of the automatic generation of vision automata. This is nothing

---

* The same idea is supporting a gender of general purpose machines called VLIW (for Very Large Instruction Words) were several pipeline units, memories and register banks communicate through a crossbar. This very fine parallelism appears extremely difficult to install. Here again only the extreme regularity allows some hope for compilers.

** Known since '74 after a semantical model proposed by Dennis[64].

but another formulation of the same problem, if one accepts, at least to start with, that existing algorithms are independant enough of any machine and therefore they have a flavour of universality.

Based on a natural decomposition of image processing into segmentation (cf.5), representation changing, and decision, that induces trivially a multi pipelined architecture, the Hecate[67] project turned out to be hard to program and harder to maintain. Nevertheless many lessons as for computer architectures were drawn from this effort, among which a likely common representation through functional[68] expression of algorithms[69, 70]: data flow graphs. Although it is primarily a true programming problem a massively parallel machine had to be designed for experimental proofs of the concept[71, 72]. It is described now to conclude with the information propagation machines.

FUNCTIONAL COMPUTER[71-75]

A functional parallelism where all the operations corresponding to one iteration are executed in parallel for only one element at every time step (MISD) is used, instead of a data parallelism where only one operation is performed for all the elements at every time step (SIMD as this is the case for the Image Understanding Architecture[76, 77], the CM-2[7, 78] and others already mentioned). There is a one-to-one correspondence between physical processors and operations in the algorithm instead of a one-to-one correspondence between physical processors and data elements.

The core of the architecture is a mesh-connected three-dimensional network of processing elements. The network is organized in 16 interconnected 8 x 8 processor planes (see figure 7a).

It is physically supported by a set of 8 stacked circuit boards, each including 8 x 8 biprocessor chips (total amount of 1024 processors). A single processing element can perform up to 50 millions 8- or 16-bit operations per second. Thus the theoretical peak power is 50 billions 8- or 16-bit operations per second. Due to the mesh-connected structure of the processor network, only local interconnections are involved. Communication links are 10-bit wide: 9 bits are reserved for data-flow passing, the last bit is a flag indicating whether the receiver is ready to accept data or not.

From Hecate's experience[68], the set of available operations has been constituted in order to satisfy most of the low-level image processing implementation requirements (presence of a hardwired histogramer for example). The internal structure of one processor (80,000 transistors) is made of three main blocks (see figure 7b):
• the input/output ports, which are an interface between the processor heart and the outside world. Each port can be used either as an input port or as an output one;
• the input/output stacks which are respectively entry and exit points in the datapath;
• the datapath and memories where all the computations are performed.

The processor datapath is organized as a three-stage pipeline. The first stage decodes input data and generates control signals for the following stages. The second stage is associated with a 8-bit datapath (8 x 8 multiplier and input shifters); 16-bit operations (ALU operations, absolute value, minimum or maximum and output shifter) are performed at the third stage. Moreover a 256-byte memory is provided and can be used either as a dual-port memory , as a FIFO or as histogram memory.

An additional bidimensional network of 12 Transputers (T800) with 1 Mbyte memory each runs intermediate-level processing (see figure 7a). The bandwidth difference (25MByte/s vs. 1MByte/s) between both networks is the only trace of heterogeneity in this model. Thus a data flow mechanism was necessary to enable communications. The computer also incorporates digital B/W and color video input/output subsystems, a global controller and low bandwidth interfaces coupled with a SPARC[tm] workstation.

For each processing element, the model of execution is a data-flow model. A data-flow operator is viewed as a sequence of elementary actions. The execution of a given operator involves a 64 32-bit programmable state machine (a state of which corresponds to an elementary action), a static programming register and a configuration register. The current state of the programmable state machine specifies the data needed on each input stack and the computation results to be placed on output stacks, for each clock cycle. According to a "dynamic firing rule", the processor controller validates operator execution if all needed data are available on input stacks and if output stacks are not full.

From the functional programming expression, operator execution is controlled by parentheses inserted in the incoming data-flow and is completely independent of what may occur in other processors. The operator is therefore fully "data-driven". Data-flows between stacks and input/output ports are routed through a full crossbar according to a configuration register. Input and output crossbar configurations are static and independent of the operator function. Full crossbars enable any connection scheme (obviously, as long as input ports and output ports are physically distinct). Programming is based on the "data-flow graph" (DFG) formalism (see figure 7a). An image processing algorithm is described using a DFG where nodes represent elementary operations involved in the algorithm. Graph arcs represent communication links between physical processors . The DFG formalism is naturally well-suited to data-flow structures and to low-level image processing where the two-dimensional image topology still prevails and where local computations are performed. It has been shown too[65] that the DFG formalism is an adapted formalism to program wavefront array architectures.

Thanks to the data-flow programming model, there are no side effects, allowing a modular programming style. This is a very important feature when describing complex algorithms involving several hundred operations per pixel. It is thus possible to design "macro-functions" reusable in a number of different algorithms. Examples are convolution, windowing, large histogram computation, gradient direction computation, dynamic look-up tables... Data-flow principle meaning data-driven processors, the same algorithm runs with different image size or format without modifying anything. An elementary operator library has been designed so that image processing applications are built using predefined data-flow processor configurations (static and dynamic). Each configuration corresponds to an identified low-level image processing operator. The current elementary-operator library consists of about 150 operators. . Another library including "macro-functions" is available.

To facilitate complex DFGs mapping, some elementary processors are used for routing purpose so that a physical node in the network can be considered as a processor with routing capabilities or as a routing element with processing capabilities. Token balancing problems are solved using processing elements as FIFOs.

4.3. Towards Vision Machines

Previous sections confirm that, whatever the architecture model, data movements have to be handled differently in accordance with the processing level. It might be depending only on the way image processing is understood today or it might be technological. Possible ways to come back to a better homogeneity have been proposed for every model. L. Uhr[79] for instance had considered increasing the computing power and word-length of processors in a pyramid along with their altitude in the pyramid. In the previous paragraph it was evocated an heterogeneity in the functional computer based on the two processors DFP and Transputers. Thanks to the data flow concept the difference never appears to programmers, although programming tools, if not programming, had to be complexified. This greater importance and difficulty of the software with image processing verticality and closeness to vision seems to be general. Two last machines are now described to put a stress on this open problem in vision architectures at the edge between hardware and software.

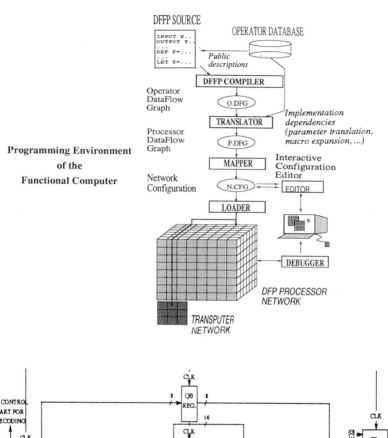

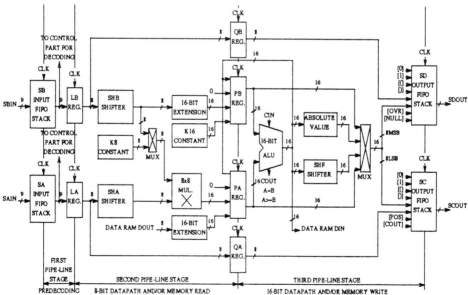

Figure 7. Programming Environment of the Functional Computer and Functional Processor Outline.

SYMPATI[80]

It was designed at the CERFIA lab to embed a SIMD part and a MIMD part in a same structure (see figure 8a) while guarantying the best processing continuity.

The SIMD kernel includes two modules or image memories. Each module is a ring of 16 blocks. Each block gets a memory part, a processor part and a shift register to communicate with neighbours. A micro programmed controller carries out the sequencing. Operators are connected via a 10 MHz rapid bus. Processors (see figure 8b) offer an 8-bit ALU, an 8 register scratch-pad to lower memory accesses (intermediate results for instance), 4 global indicators from the ALU or the shifting loop (mask). Internal communications associated with different classes of operations are supported by a set of logical gates. Data movements on a bus and processing are executed fully in parallel.

The MIMD part is a group of conventional processors with private memory on a bus. This bus is connected to the rapid bus through an exchange module: a transfer loop spares the global

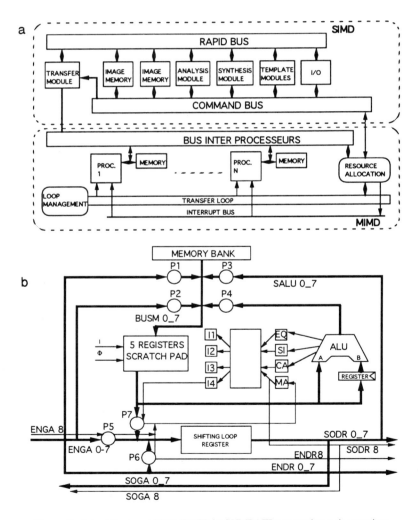

Figure 8. a) Architecture of SYMPATI; b) SYMPATI: processing unit overview.

bus short communications between processors. A resource allocation distributes tasks on processors thanks to an interrupt bus. It manages also the control bus to activate specialized modules and the inter-memory transfers.

Image processings are oriented towards one or the other part according to their underlying topology. A global understanding of the machine at work remains hard and led to provide users with a graphical "language" based on the KHOROS[81] environment. Such programming facilities had been already found necessary for Hecate, including a graph compiler[82] that shows up again in the Functional Computer[83].

IMAGE UNDERSTANDING MACHINE[76, 77]

Here three representation and processing levels are intended to be implemented in hardware (see figure 9). There are three layers, each with respectively different granularity processors and operating independently. Layers communicate through double-port memories.

The lower level is implemented by an array named CAAPP for Content Addressable Array Parallel Processor. 512 x 512 1-bit processors are operated in a SIMD manner for associative or multi associative processing to extract features from raw data.

The intermediate level is a 64 x 64 array of 16-bit processors, ICAP (Intermediate Communication and Associative Processor). The control is both SPMD and MIMD for feature handling. For instance matching, geometrical relations between items, and result fusion into abstract entities are performed at this level.

The high level layer is made of 64 symbolic processors (SPA for symbolic processing array) programmed in LISP and run in a MIMD fashion. All inferences, hypothesis checking, model matching belong to this part. Moreover the two other levels are controlled by SPA.

Although this machine is still in a prototyping phase (1/64th of the machine is being built) the whole programming environment has been developed including an architecture simulator, a hardware resource monitoring and result display tools. Several languages are involved like extensions of Forth, C or Apply (general image processing language). The Web library is available under the Apply compiler as well as a low level image processing library of C++ classes.

4.4. Conclusion: a More Social View onto Machines

Several ideas are to be drawn from the above trial to find a few principal features in relation

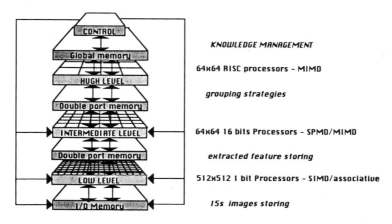

Figure 9. Image Understanding Architecture.

with the vision process, out of existing image processing machines:

- whether it is in one, eight or more bits the operative parts get to resemble one another, never reaching real vision specificity. The technology plays a prominent part and really efficient devices rely on VLSI one way or another, but it does not look like much helpful in that respect. The main originality is often in the way registers and links are organized;
- communications between pixels or operators, depending on the model, are shown a key feature of image processing architectures. They involve a different way of programming, and a good balance between operations and transfers seems to benefit performances not only in terms of processing speed but also in terms of processing generality;
- programming remains artificial and external to the device. Even in the data flow concept, any self programming from data or from goals remains illusory, would it be only because of intermediate representations like graphs or of prefixed communication patterns. In that respect, the associative concept ghosts here or there under the form of a privileged communication in cellular machines or of efficient mask and tag processing, never really emerging in a straight forward implementation. As for interconnections, their automatic changes according to results are at best simulated and most often synchronizations are skipped or made mechanical.

Whether it is for operations or communication definitely dual after the analysis above, the problem is the following:

- either adequate basic operations and communication patterns, specific of vision tasks, are found and implemented in hardware or basic software. Then manipulated data structures and classes of algorithms fit the architecture or it may adapt to them. This looks like much work to be done before;
- or algorithms need to be transposed or even rewritten to fit architectural features, whether the latter do mean something or not for the application. Then even elementary functions with regards to vision might require to be simulated. This looks like a lot of work to be done after.

This latter case is the situation robots are in now, and the reason why programming never gets natural. One can measure the distance we are from suitable architectures in the character of the steps to "parallelize" algorithms.

On massively parallel machines major factors are:

- from the algorithm side: the operator locality and the neighbourhood patterns, the context dependency, the iterative nature and the data types;
- from the communication side: the properties of the network to support data movements, not to forget memories and processor topology which the data organization should be projected on, the local resources available for communications and the characteristics of exchange facilities towards secondary storage;
- from the processing side: the importance of the instruction set, the functionality of instructions in terms of vision to be opposed to instruction types like vectorial, scalar and conditional... important too. For instance, conditional sometimes refers to associativity through masking.

The factors above are representative, in their whole, of an idea that massively parallel machines are run in a rather synchronous fashion and communications do not require intensive dynamic adaptation. On the contrary in shared memory or distributed - say coarser grain - machines, the asynchronous execution is meant to reflect a given non determinism of steps, with evolving data structures and synchronizations on events. Execution reminds of a general relaxation scheme and the main concern here is to avoid "superfluous" data exchanges due to control repartition. Thus major "parallelization" factors are:

- decomposability: algorithm ability to split into (balanced) cooperative processes;
- assignability: programmability of these processes on the processing units;
- mimetism: processor topology and network reconfigurability with regards to the function

connectivity (for instance in terms of shared data like images, graphs...);
- concurrency: simultaneous accessibility of shared data, like images or more abstract objects;
- versatility: dynamic creation and deletion of processes and related interconnections.

The underlying issue, in this analysis, appears at last to make several processors, or many according to their power, cooperate to transform strongly correlated and redundant data. While data links may create processor relations in the beginning, communication and processing natures change along with processing advances, very much alike in any school cooperating for task achieving. This phenomenon is as marked as images are not merely multidimensional signals and their features resort at least as much to semantics as to structure. It is worth to be taken advantage of, both for communication and processing as both communication and processing could be indicative of visual or understanding powers. So, these principal vision architecture axes that we are in great need of for a valid state representation to support model building or comparison, can be:

Sequencing type: might be either from instructions or data.
- If instructions are in question, what matters is the dependency between instructions and data. This is summarized by the famous Flynn's classification[84] where S stands for single, M for multiple, I for instructions and D for data.

  SISD covers Von Neuman's machines where data, instructions and results are stored in the memory and everything passes through the ALU.

  MISD was encountered in our machine survey under the form of "pipeline" which was signalled already not to reach a real image processing meaning.

  SIMD was, somewhat improperly, assimilated to low level processing and topology similar to images. The most prominent feature appeared to be the conditions to inhibit processors and thus desynchronize processes in an intelligent manner. This is announcing local operations adapted to results and data types that could be profitably synchronous.

  MIMD was devoted similarly to high level processing. And the atomic computing power intuitively (and technologically!) related to this sequencing leads to carefully design specialized buffer managements and communication controllers. The burden is then on processor designing (see individual type).

  Obviously this sequencing type matters but it does not relate so strongly to the granularity and the more or less massive character of parallelism. A good counter example was the ability to organize adaptively data clusters* in massively parallel machines, behaving independently. It was the case for some arrays provided special control memories and path gating inside processors, or for pyramids where it is part of the model[85]. A conclusion in that area is to require both sequencing types and the facilities to shift dynamically from one to the other. And this is coming in general purpose massively parallel machines like the CM-5. As for the MISD semantics, it was already discussed in front of the data flow concept listed now.
- If data are in question we call "data flow" this sequencing. It was proved more dynamic and, although there could be many facilities for self-programming in this model, no such tendency is evident currently. On the contrary, this type of programming asks for intermediate representations - data flow graphs - somehow orthogonal to image data properties, although it benefits well from topological properties and supports masking. In that respect, it is most interesting to parallel the saga of Artificial Intelligence (AI) specialized architectures and the current efforts on general purpose data flow machines.

  AI machines emerged on memory management never dissociating themselves from languages. This vein dried up with the burst of the RISC fashion already mentioned in 3.2.3. RISC means implementing most frequent instructions under the form of hardware facilities,

---

* regions for instance, but it could be any back projection of a concept on to the image as well.

so it was well suited to a search for performance in data management, apparently necessitated by special languages. A bit of LISP normalization completed the burial. As efficient it could be, this evolution did not make machines more intelligent, better fitted with expertise handling etc… Research in architecture had enriched all machines with garbage collecting mechanisms or special memory caches for instance but it failed as for specialization, starting again with expert systems in software up to multi-agents… and neural networks in hardware. In the same way data flow machines might turn out to be real efficient implementations of vision mechanisms. This is all the more true as a great progress is to notice mainly in software management which is, again, a major problem. Languages do evolve from VAL[86] or Id which tried to translate the data flow concept; SISAL[87], Idnouveau[88, 89] or Lucid (with streams)[90] and Haskell could fit better image processing needs. Hardware is in progress too, featuring data driven processor arrays with programmable interconnections (Snyder's CHIP[91] or Koren's machine[92, 93]) or more static and less integrated like Cornish's DDP (by Texas)[94] or a signal processing machine at Hughes[95], not to forget the NEC µPD7281[96] quite well adapted to image processing (filtering in cascade). A probably major step should be cleared soon with the implementation of coloured tokens to more dynamic devices. A completely different question is to know whether such machines are representative of vision, which they are especially intended for, but again: "submarines" do not wave fins or flippers, yet they swim.

For important the sequencing type could be in our processor gathering, it is not alone.

Control type: is a major issue at least for programming. It might be centralized, and then the controller power and its communication means have a straight influence on the number of considered features in a given context, then on the extent to which a global reconstruction of an object is possible. If the control is cooperative, are there hierarchies? In that case they should relate for instance a progression like points of interest, edges, regions or facets, objects, scenes; or they may be indicative of order relations like coming slower or faster or in given directions, for data to be processed accordingly. An other major point here is the way communications are precised. Is it message passing or circuit reservation? Usually the choice is bound to the data packet length but, since that, it gets here a straight image semantics. Are messages just left on a systematically read blackboard or announced to destinations? Actually such control features prove more important, from the current advances in parallel programming, than SIMD or MIMD differences which tend to be hidden to users anyway.

Society type: in question here, the private or public character of resources. When all are private, it sends back to communications and control. And the way neighbouring pixels, variables or concepts can be accessed when they belong to an other processor might change your vision! If some resources are public (sensory parts for instance) are they central and shared, making a common context or culture, or distributed like view angles or any information back from actuators which belongs primarily to interfaces? In the central case, access modes influence procedures as well as the nature of links (different from control links): multiple buses, crossbar, hierarchical routing… In the distributed case priorities need to be installed and do matter.

Individual type: it was already quite well mentioned. Either elementary processors are specialized. Then implemented functions and the performances are fully constrained even in the case of massive parallelism with collective behaviour emergence. Programming is also judged in advance of course, would it be only because of the level of expression. Or elementary processors are general purpose. The question is then how to install some differences between them to specialize momentarily, and in what conditions from mere inhibition to associative programming.

Of course these four types do overlap. They are, and are only, principal axes. In that sense they are based on subsets of machine features that we encountered here or there and tried to underline to the detriment of others in the described machine examples of previous sections.

Such features may be difficult to isolate and describe in a precise way, independant of technology. But such a characterisation suggests an analog typology of algorithms, and should help understanding better special architectural visual functions if any. This true pattern recognition process - matching algorithm specificities and architectural characteristics - relies on combinatorial optimization among which methods are relaxation (social model), simulated annealing extensively used in high level synthesis of VLSI, and genetic algorithms sending back to evolution. Another avenue is a common expression of architectures and algorithms which we have proposed functional programming for.

Finally we are left with the confirmed equation:

$$\text{ARCHITECTURE} = \text{OPERATIONS} + \text{COMMUNICATIONS}$$

This means, by the way, that memory is implemented somewhere between both entities. Would memorizing cells require intrinsically any specialization to vision, this could be a lack in the "theory". But there is no technological or algorithmical evidence for that so far.

If operations need to be particularized, technology is likely to be a crux of the matter. This fact was obvious from section 3 where boards and VLSI were described inflating in the absence of specialization, for sake of programmability and communication overhead. It was mentioned again about arrays and pyramids. It will be dealt with briefly in the next section devoted to silicon "retinas".

Now tailoring communications to vision favourite data movements, if any, requires to better understand interconnections first. This is the reason why the last section is fully devoted to interconnection networks in general, not strictly related to vision anymore, since their properties need to be known in an abstract enough way before one tries to apply it to image processing.

5. SILICON RETINAS: BASIC ARTIFICIAL-VISION OPERATIONS

Some operations usually encountered in image processing algorithms and, upon that score, implemented in hardware have been already mentioned in section 3 and to a least extent in section 4. The general idea got there-from was that either the operative level was wrong or the algorithm representation was wrong. Either operators have so little a meaning in terms of vision that the least true visual operation, like shape description to recognition or motion detection for tracking, would force many such hardwares to be assembled causing a complexity beyond any technological feasibility. Or these functions which were appearing in algorithms come from an ill-view to them, a bad description that does not capture the right operative clusters. Filters and histogram makers do not look like atomic cells one would consider straight in a library of basic information-processing tools to be replicated thousand times and assembled in a systematic manner.

In the HECATE project[67] a somewhat different - more hardware concerned - description of vision algorithms was attempted. The aim was to reach a better trade off between real-time vision algorithmic complexity and VLSI implementability. The machine was a set of several dedicated hardware modules, in two layers: one for feature extraction, the other for decision techniques and recognition (see figure 10a). An interface between these layers was a huge image buffer[97, 98] able to secure data-type conversions and representation changes, like a sequence of raster-scanned edge points becomes a "word" to be compared with a similarly coded reference through dynamic programming, and so on. Each module was having its own controller (Motorola 68000) and all controllers were gathered into a parallel machine, multi-bus based, run under Chorus and Unix (see figure 10b). It turned out that a single feature like edges or regions was asking for a complete rack of hardware (see figure 11) and used to get to the complexity of a small parallel machine in itself (few thousands of LSI chips, not to forget integrated key parts under the form of proprietary VLSI chips). This was again the case with

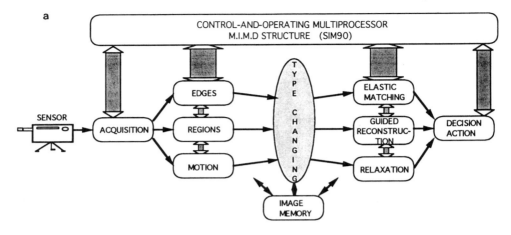

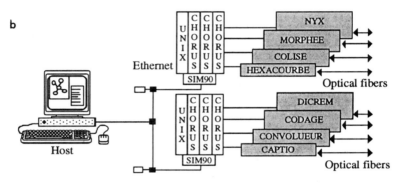

Figure 10. a) General Outline of HECATE; b) HECATE's machine organization.

decision techniques for them to be general enough. For instance a dedicated hardware to dynamic programming[99] required 800 circuits, among which several TMS 320 DSPs, in 5 large printed circuit boards. At the same time, less general VLSI versions were being designed[100]. As an other example, the "region extraction" module (see figure 11) was including identified functions like a parallel sorter, a labeller, a statistics maker to be assembled following a given graph thanks to two full crossbars[101, 102]. Again, the difficulty of programming such an heterogeneous structure despite the better image processing functionality of atomic operators and the complete programming environment based on graphics and advanced user's interface management ... led to reconsider the whole problem under the form of the Functional Computer; and this Functional Computer was already found unsatisfactory as for the meaning of basic operations implemented in the VLSI hardware.

An interesting enough comparison is between the proprietary ASICs designed in view of Hecate and other circuits or chip sets concurrently proposed. At ETCA[103], standard cell circuits were made for data-rearrangement (neighbourhood shaping up), generalized convolution (inner-product of any two basic arithmetic or logic operations), curve building and handling (histograms, Radon transform) and an associative crossbar to route data from some features of them[103-107]. At the same time TRW was making the TDC 1028-6 multiplier/accumulators on a chip - for digital filtering. Nathan at JPL was studying a family of convolvers. A French

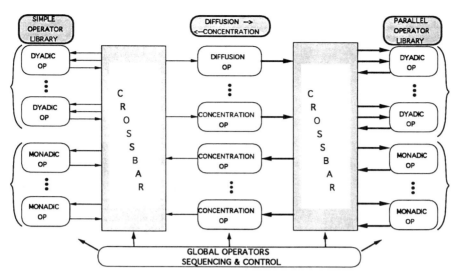

Figure 11. Region module: COLISE.

company, TRT among many others in the world, proposed a 2D-correlator (4x8), a memory with several parallel lines of bytes and image processing special addressing modes, a linear filter (5x10) including some history of results, a registration-oriented statistical filter[108]. At Berkeley, Ruetz and Brodersen[109] tried a chip set including pre-stored linear filtering, histogramming, and edge tracing.

But the most illustrative example remains the evolution of the Digital Signal Processors (DSP) mentioned in section 3.2.3. The inner product shows up in most filters but in the correlation as well and in many orthogonal transforms. This was enough to design "dedicated" processors centered around its straight hardware implementation: a pipeline multiplier-ALU. The technological advances together with the RISC fashion involved progressively different kinds of parallelism inside the chip. Addresses and operations are determined in parallel, different instructions are executed in parallel. From the TMS 320 (10, 20, C25, C40) by Texas, to 64 bits floating point DSPs like the ADSP 8110-8120 (in ECL) or 3212-3222 (in CMOS) by Analog Device, programming, languages and direct applications have been systematically privileged to the detriment of the signal or image natures, and neglecting inter processor communications. For instance Analog Device offers a multiport register memory for communication (ADSP 3128) that is in the tradition of I/O devices to external memories in computers, and remains very difficult to handle in the case of limited pipeline.

To the opposite the "Retina" concept emerges on a device which intimately associates an optoelectronic layer with some processing facility. The closeness definitely suggests a VLSI implementation approach, possibly monolithic. Moreover, current robotics is not only moving towards involving complicated senses such as vision or aerial acoustics but it aims at associating several of them within sensor fusion schemes. Theoretical results like the so called "multiarmed bandit" theorem[110] tend to prove that it is worth implementing some local computing power closer to sensors in that case. But in such conditions, only elementary feature extraction, up to limited object identification, could be proved technologically feasible. For instance line orientation recognitions, or limited detections and tracking have been implemented. In these examples, the outside world is simplified (either exhaustively described or in translation). Then, a bit of analog processing followed by a uniform result gathering performs the intended task, and only one or two global outputs are produced. However, the experiences suggest potential benefits from "analog thinking" when an algorithmic concept comes to cohabit with analog implementations of earlier vision processes. Descriptions of analog phenomena inside the system provide a language which helps to drastically compact any design, and enforces some interesting improvements at the algorithmic level. This fact is illustrated by comparisons between various proposed implementations of the convolution or other basic operations like differentiation. Indeed, in less toy-like cases current robot vision does not allow routine actions in such a direct manner and anyway, such actions would be triggered on a larger set of parameters. This shows integrating is not enough, even associated with analog thinking. There still remains an additional price to pay: either to deal with very specific applications or to particularize vision in some other manner like restricting it to a rough type. We have introduced the concept of "rough vision"[111-113], based on separating the structure of the image from the semantics it refers to. It applies first to object recognition thanks to neighbourhood combinatorial logic which is easy enough to implement on retinas. Logical implies binary, but in this process the adapted binarization is made a true processing operation, possibly a feature extraction and not only an A/D conversion.

More precisely, let us define "smart retinas" as tentative "human-size" vision machines, intimately associating:
- optoelectronic devices for image formation;
- analog-to-digital converters, then cooperative and shared between processors;
- digital processors, most often monobit for sake of density, to be integrated on monolithic (CMOS) circuits.
    On top of clear advantages like:
- maintaining the image topology;
- improving the input output balance (only rare meaningful results are forwarded in accordance with the claim of smartness);
- exploiting in an easy and natural way the spatial and temporal correlations between pixels;
- increasing operative density and rapidity;
there is anyway a need for a fair share of analog contribution to meet the constraints of rapidity and compacity as imposed by real time robot vision. This fourth analog processing layer right after image acquisition provides specialized operators as mentioned in the previous sections and above, but with a singularly increased dimension.

Currently the list of examples in silicon vision is continuously extending. Among many other operativities:
- grey scale modifications (log, exp, square) based on elementary behaviours of transistor for vision adaptation;
- classical operators popular in neural networks (sum, product, transfer, thresholding (sigmoid for instance)...) based on simple transistor associations for collective decision;
- convolution by Bessel or Gaussian like kernels based on resistive networks or basic properties of charge-coupled devices, for local communication and noise removal;
- feature extractions:
    - velocity fields as the optical flow;
    - edges as the Laplacian of the kernels above;
    - regions in collectively saturating amplifiers;
- tracking based for instance on median/mean combined filtering, thresholding and vote inside a single function;
- local Boolean operations for recognition, based on a shift register network and programmable.

One illustrative functionality is now explained to give a flavour of what is going on in this field. More complete informations are in Zavidovique and Bernard[114], Bernard and al.[115, 116] and in chapter 5.

5. DIFFUSION-BASED SPATIAL CONVOLUTION

Static image processing is fundamentally based on spatial interactions between pixels or sub-structures that are more or less far apart in the processed image. This corresponds to the structural approach of vision, which can actually take place at every level of vision. When performed at the lowest level anyway, these spatial interactions are extremely computationally intensive and would definitely benefit from "natural" physical interaction phenomena.

When statistically considered, images have to be processed in a shift-invariant manner, without privileging any particular direction. Moreover, it makes sense to weaken their interaction as pixels get further apart from each other. We were thus looking for a shift-invariant phenomenon allowing the isotropic but radially decreasing diffusion of a physical quantity towards its neighbourhood. This can be implemented thanks to current diffusion in resistive materials (see figure 12), which is a linear process: if a current is injected at some point of a resistive sheet of conductive material featuring a uniform superficial leakage resistance towards some source of potential (e.g. ground), the induced voltage profile or impulse response is indeed a rotation-invariant kernel whose radial shape is given by the first modified Bessel function: $V(r) \propto K_0(r)$, where r is an absolute normalized radius. Before discussing the relevance of the "diffusion kernel" shape for vision purposes, let us characterize it more precisely. To get some physical intuition about $K_0(r)$, we can consider the current diffusion in the adjacent dimensions: 1-D and 3-D. For a resistive line $V(r) \propto \exp(-r)$, and for a resistive volume $V(r) \propto \exp(-r)/r$. As expected $K_0(r)$ shows an intermediate behaviour that we can precise thanks to equivalent forms for small and large arguments*: $K_0(r) \sim_0 -\log(r)$ and $K_0(r) \sim_\infty \exp(-r)/\sqrt{r}$.

In a VLSI circuit however, we are bound to spatially discretize this current diffusion process onto a resistive ladder network of the type shown on below. This network is shift-invariant. Horizontal resistors are called diffusion resistors with value R_d. Vertical resistors are connected to ground, and called leakage resistors with value R_l. Input injected currents X_i diffuse all over the network contributing to the output node voltages V_j. This process is linear such that we get $V=K*X$, where K is a characteristic convolution kernel depending on the sole ratio R_l/R_d. If this ratio is variable, this is truly a multiresolution facility which is available to the analog vision algorithm designer! Recent developments about the use of wavelets[117, 118] in image processing still enhance the importance of such a feature.

In the 1-D case, the kernel voltage profile is simply exponential (as in the continuous model), that is $K(r) \propto \exp(-r)$ or $K(x) \propto \exp(-|x|)$ because Kirschoff laws can be written in a recurrent manner. In the 2-D case however, there is no closed form giving K(x,y). There are actually at least 2 network topologies that can be used: either rectangular or hexagonal. The continuous model proves useful to understand the asymptotic behaviour (towards ∞). Unlikewise, close to 0, that is for the central pixel on which the unity current is injected and for its neighbours, infinite voltages forecasted by the continuous model vanish; the node voltages are finite and have to be estimated thanks to iterative algorithms.

It is fairly easy however to derive analytically K^{*-1}, the inverse of K for convolution (regardless of the dimension or the network topology) which in turn yields FT(K), the Fourier

---

* $\sim_x$ means equivalent in x.

Transform of K (with K considered as a distribution). This is a door to understanding the effect of the discrete current diffusion in terms of frequential analysis.

By expressing Kirschoff laws for each node of a rectangular 2-D extension of the network shown on figure 12, we get

$$(i \in Z) \; (j \in Z) \; X_{i,j} = (1/R_1 + 4/R_d) \, V_{i,j} - 1/R_d \, (V_{i-1,j} + V_{i+1,j} + V_{i,j-1} + V_{i,j+1}). \tag{3}$$

By using the dirac distribution d and the rectangular laplacian $D_r = 4 \, d_{0,0} - d_{-1,0} - d_{1,0} - d_{0,-1} - d_{0,1}$, Kirschoff laws yield: $X = (d/R_1 + D/R_d)*V$, where $*$ stands for convolution. But $V = K*X$, so:

$$K^{*-1} = (\, d/R_1 + D/R_d \,) \tag{4}$$

We can now switch to the frequency domain to get the periodic Fourier transform of K^{*-1} and finally K, with frequency coordinates w_x and w_y:

$$FT(K^{*-1}) = 1/R_1 + 4/R_d \, [\sin^2(w_x/2) + \sin^2(w_y/2)] \tag{5}$$

$$\Rightarrow FT(K) = (1/R_1 + 4/R_d \, [\sin^2(w_x/2) + \sin^2(w_y/2)])^{-1} \tag{6}$$

We have just been characterizing 2-D "diffusion kernels" in many aspects. We have now gathered enough information about them to show their relevance for vision purposes.

Within recent years, much work has been devoted to the optimization of smoothing diffusion kernels allowing the removal of noise before edge detection. Beside the "Gaussian hegemony", exponential filters have also been proved in Shen and Castan[119], to be optimal for a multiedge model. Now, when a straight edge is convolved by a 2-D diffusion kernel K, K is actually projected according to the direction perpendicular to the edge into ... an exponential filter! The edge detection capabilities of the "silicon retina" described in Mead and Mahowald[120, 121] are the straightforward application of this property. We have also shown that diffusion kernels are particularly suited to the halftoning problem, that is the analog-to-binary conversion of images, as mentioned in Bernard and al.[122-125].

Though the fully 2-D parallel implementation of diffusion kernels seems much more "natural" than that of gaussian kernels, there remains a few difficulties to map it into silicon. Using transistors in the weak inversion region is a potential solution to lower current values, as explained and applied in Mead and Ismail[126, 127]. In that case, resistors are controlled thanks to the tunable transconductance of a CMOS differential amplifier used as a unity-gain follower. Yet, the linearity range is not larger than 200mV. When the device gets saturated, it turns out to perform a simple but automatic segmentation of the input image by preventing two neighbour pixels from exchanging more than a fixed current upper bound.

However, considering the uncertainty on each transistor characteristics in the weak inversion region (up to an equivalent gate voltage uncertainty of a few tens of millivolts), the linear range narrowness seems more undergone than desired: it requires dynamic selfcorrecting circuitry or static a posteriori analog compensation by EPROM-like techniques*, all of which may be area-consuming. Further more, such analog voltage precision seems to prevent the cohabitation with digital layers which requires external clocks, inducing significant amounts of noise.

An alternative solution was studied at ETCA to the implementation of diffusion filters based on an unconventional use of switched capacitors[122, 124]. This approach leads to reasonable power consumption: To give an order of magnitude, if a (fairly large) 1pF capacitor was to be charged and discharged from 0V to 5V at a 1MHz frequency at every pixel site, a 100x100 pixels retina would demand a power of about 0.1W. However, either it requires an analog

---

* Such techniques provide longer term analog storage of charges[128].

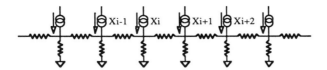

Figure 12. A Resistive Diffusion Network (1-D version).

CMOS process providing a double polysilicon layer, or "slightly" non-linear p-n junction capacitances have to be used[129]. In the latter case, it is amazing to notice how many roles the same simple device can play: a strip of n-diffusion over the p-substrate will be used a) to connect two pixels, b) to act as a switched capacitor and c) to convert light into current.

Finally, a globally better precision can be achieved with comparable silicon area, partially because capacitors are really easy-to-use bidirectional media to perform "type conversion" between charges and voltages.

The convolution is only asymptotically obtained after a sufficient number of elementary switching cycles. About 10 are necessary to reach a 0.1% precision if all capacitors are equal. The ouput voltages are somewhat immaterial since only half of them are available in one clock cycle. "Neurons" (pixels) are indeed separated according to their parity. This iterative aspect allows to share a single leakage capacitor between a pair of odd and even neurones. This neatly generalizes to 2-D, where neurones are separated in a checkerboard fashion. Though capacitances have static values, a discrete multiresolution facility is recovered thanks to the use of more complex cycles in order to obtain narrower diffusion kernels or even different types (e.g. gaussian-like) of kernels at no further implementation cost!

We have just been comparing different implementations of regular diffusion networks. However, when resistors can be separately and dynamically controlled, resistive networks can have much broader early vision applications[130-134]. The price to pay is area, but also algorithm complexity: for example, negative resistors, which are area-consumming, can also pose convergence problems.

We have illustrated here, how a peculiar architecture naturally (physically) constrains the solution of a pattern recognition problem, in privileging communications by its topology and operations by the "instruction set" hidden inside the components organization. The analog behaviour of devices favours fancy solutions literally imposed by silicon. The example above can be enlarged in minimizing a quadratic energy à la Hopfield thanks to a diffusion network. The non linearity to be installed at the network nodes sets the global behaviour to edge, region or motion detection. Thousands of programs, huge efforts in numerical analysis, years of controversy about respective advantages of the Gaussian kernel compared to other convolution kernels... are ruled by mastering some physics of the silicon that favours typical convergences up to make them trivial.

However it is very unlikely that all vision problems be solved that way. On the contrary such approaches as silicon vision rely on accumulating imprecise results of various and independant kinds, namely the results that come naturally out of neat silicon implementations. To overcome the somewhat arbitrary choices and the limitations of elementary operations with respect to more universal vision, augmented communications between devices are thus necessary.

6. COMMUNICATION NETWORKS

6.1. Interconnection Networks for Parallel Architectures

Moving information around appeared a critical part of parallel processing. Therefore, a parallel machine efficiency heavily depends on the network design. In this part, foundations of such design will be considered from a theoretical point of view before explaining the current evolution of architectural implementations. Only "symmetrical" architectures where exchanging data is a pure algorithmical activity are studied here, excluding dataflow machines or dedicated architectures where data transfers are a specific part of control.

If interconnection networks and nervous systems are to be compared, a first question arises naturally from their different speeds. Transmissions appear to be several orders of magnitude faster in a machine network than in the human brain. Actually, interconnection networks for parallel machines are tightly linked with the traditional concepts of memory in computers and of data mapping. It will be explained in the following that networks provide access to data between processes or from a bulky storage, but only data-processing conditions significant output at last.

So, a further insight for comparison will expose first formal parameters which an absolute measure of a network capacity can be founded on. Then, the evolution of physical networks will be described to determine significant, and perhaps new, features of modern and future machines. Such results should help reconsider functional differences between nervous systems and machine communication systems.

6.2. Memory Models of Parallel Computers

Most often, from an algorithmic point of view, it is enough considering memory accesses to obey Parallel Random Access Memory rules, that is: all accesses can be concurrent and take a constant time. But this model remains a computational characterization and covers different types of implementation with more complex behaviours. As for practical implementation, two models of memory can be distinguished:
- distributed memory machines consist of a set of processors linked by an interconnection network. Every processor gets a local memory and exchanges explicitly informations with other processors by means of messages. Data accesses to or from a local memory are fast, while accessing data in a distant processor memory requires passing through the network according to a more or less complex communication protocol;
- shared memory machines try to emulate a global memory to support concurrent accesses. Traditionally, this type of machines shows two distinct parts : a set of processors and a set of memory banks (although, current trends do modify this form of an architecture up to its model). The role of the interconnection network is to give all processors independent access to all memory banks. The structure of one bank do not permit a concurrent access but for a very small number of processors. Even in the case of valid memory addressing patterns, full concurrency is not always reached, as it is up to the network to provide enough data paths to serve requests.

6.3. Data Transfer Organization

Data transfers in distributed parallel machines consist of exchanging messages between processors, while, in shared memory machines, processors exchange data through the medium of memory banks. Anyway, in either case, different problem levels are to be distinguished as for managing communications:
- routing. Given origins and destinations, it is necessary to compute the path of data pieces

along the interconnection network, if processors are not direct neighbours in the network topology. Computation can take place before message sending or during the message progression through the network. Achieving a pattern of transfer between processors and memory banks in a shared memory machine is most commonly referred to as the network control;
- communication protocol. Establishing communications and formatting data necessitate several operations. For instance, messages may have to be separated in several smaller flits. Such flits have to contain enough information to reach destination and, then, be reorganised into a full message;
- physical transmission. Finally, the hardware capacity has to be considered : how many communication links could be used at the same time by a processor, how conflicting communications can be handled, and so on.

Both protocol and transmission are tightly linked in parallel architectures, which is not the case in Large Area Networks, for example. They depend on the hardware implementation. Routing relates more to the network topology.

6.3.1. Routing. Classifying different types of routing leads to consider various aspects of the problem that is computing a path from one point to another in a graph topology. First, this computation can either be centralised or distributed. In the first case, every transfer of data in the network is under control of a central agent. While this allows global optimization of network resources, it appears difficult to extend for large machines as the central controller turns into a communication bottleneck. A solution is to rely on a compiler for computing communication patterns from the code before run-time. A second approach, better suited to large scale architectures, is to rely on local optimization and distributed computation of routing paths.

Distributed routing can be further subdivided in two classes. Paths may be completely precomputed by sender processors. But it is more common to make the path incrementally computed by intermediate processors. For this last class, routing algorithms are generally oblivious, that is they only take the destination address into account to compute a routing path from an intermediate position.

Another routing-strategy classification is more interested in the type of path computed by routing algorithms. Paths may be static, that is for a given sender-receiver pair, the routing algorithm produces always the same path. This path may be a shortest path between the two processors or not, routing may be symmetrical (going path is equal to back path) or not... On the contrary, routing may be adaptive, i.e. it is bound to the network global state. Such routing is intended to optimize the bandwidth with respect to the load.

Either adaptive routing is deterministic: for a given network state, routing paths are always the same. It fits centralized routing where the controller has a global view over the network. Or routing is stochastic: the idea is then to convert a special communication pattern into a mean one, which the network is efficient for, by use of intermediate random paths. The main drawback in stochastic routing is that sometimes any notion of communication locality is lost. Messages between direct neighbours might have to follow lengthy paths.

6.3.2. Switching techniques for interconnection networks. This paragraph covers hardware specificities of the network (physical transmission) as well as data organizing for message exchanges (communication protocol). Different strategies may be used to enable the control of data flows by processors, and more specifically by their hardware part dedicated to routing - the router.

One way is to introduce this control from an historical analogy between parallel-computer interconnection networks and telephone networks. As for a telephone exchange, the control strategy must establish a physical connection of reserved links between two processors which, then, are able to communicate as long as line is open. The so-called circuit switching technique, requires some time to connect on, but, then data exchanges are very fast and do not need intermediate storage resources. Routing and conflict resolution to link-reservation are completed while establishing connections. Such a protocol happens to be very inefficient in the case of small messages. Indeed, links are reserved as long as the message does not reach destination. Short messages imply that effective transmission time is small compared with total reservation time. Start-up time does not depend on the message size, making balance even worse.

Another control strategy consists of dealing with packets of information : packet switching. Messages are cut in equally sized "flits" (flow control digits). These flits follow the same path in the network, which restricts the path calculation to the leading packet. In the case of a distributed routing, there is no start-up cost, routing is processed while the flits cross routers. Conflicts may occur when several leading flits from different messages compete for taking the same physical link. In this case, all messages but one have to stop until the path has been freed, meaning extra storage resources to buffer waiting messages. The limited buffer size makes deadlocks likely to occur. The way flits behave one towards the other during their progression and in case of conflicts defines different types of message commutation (store-and-forward, virtual-cut-through and wormhole).

The aim of this evolution is to make communication resources as free as possible. Considering that there is no reason to keep a resource locked if it is not busy, it should be worth dealing with routing packets of data rather than establishing a communication. Equally for sake of economy, it is necessary to reduce the memory used for message bufferization to its minimum.

6.4. Characteristics for Excellency

Beyond technological communication issues, topological features of a network have an impact on its efficiency. We list in this section few parameters coming from graph theory and describe their role in data transmission.

6.4.1. Diameter and mean distance.
The diameter of an interconnection network is the longest of the shortest paths between any two vertices in the network. In distributed-memory computers, such a parameter describes a worst case for a shortest paths routing. If a cost is associated to passing through an intermediate vertex, the more a message must cross intermediate vertices, the more costly communication is. Whence, the diameter measures an upper bound of this cost, again, for a shortest paths routing.

Of course, a mean value would be more interesting. Consequently, one can introduce the mean distance in routing a network as the mean distance travelled by a set of messages. Choosing this set amounts to some hypothesis on the structure of communication patterns. For instance the set of messages can be between randomly chosen vertices, not assuming any structure in the communications; or the set of messages can be between near neighbours, implying a property of locality in information interchange; mixed choices are possible too. Another method is to weight the different possible communication patterns. This makes again an hypothesis on communication frequencies. Now, this parameter integrates the choice of a routing strategy and changes accordingly.

Anyway, in the case of large start-up time as imposed by the initialization, the communication model turns into a model where any communication takes a constant time. The current technological evolution seems to favour such a situation. Furthermore, in shared memory machines (BBN butterfly machines[135]) or, more generally, for machines (e.g. CM-5[136]) with multistage interconnection networks based on switches (like butterfly, omega and fat-tree[137] networks), the distance between any input and output appears to be constant for the simplest routing strategies.

6.4.2. Bandwidth. Both preceding parameters actually measure the latency of a message inside a network when nothing conflicts with its routing. But when information traffic grows, conflicts arise. If topological bottlenecks exist in the network structure (consider, for instance, the case of a tree), the network throughput decreases and the whole machine slows down.

One way to get an insight into such topological drawbacks is to consider the bisection width of the network. The bisection width is equal to the minimal number of links which must be cut in order to split the network in two parts with equal numbers of vertices. In the hypothesis of unstructured communication patterns, every message has a chance over two to cross this bisection. Hence, such links are peculiarly good candidates for bottleneck. A high ratio between the bisection width and the number of vertices in the network, indicates a high probability of conflicts.

A more precise view can be obtained by considering the number of messages crossing a given edge or a given vertex for a particular routing strategy. These functions and their maxima over the set of edges -edge transmission index- or over the set of vertices -the vertex transmission index- provide a precise localization of potential bottlenecks and their importance. More specific versions of these gauges do exist. Instead of considering the entire set of possible communications, evaluations can be restricted to the frequency of visit for a particular set of communications or equivalently for any weighting on the set of communications. Again, it introduces in these values some hypothesis on the occurring communication patterns.

In the case of multistage networks for shared memory machines, bandwidth means establishing communications from one to the other end of the network. Consequently, bandwidth is equivalent to the capability of routing permutations through without conflict. This parameter refers to the network universality[138], i.e. its capability to route any input permutation to the outputs in a finite number of cycles around the network. When a conflict arises in an inner switch, communication demands must be rejected until a feasible pattern is reached, causing a situation called an hot-spot situation. Fortunately, hardware tricks have been found and can be added to a multistage interconnection network, resolving hot-spot situations not to reject a demand anymore[139].

6.4.3. Algorithmical parameters. The previous parameters are relevant while the network is not especially adapted to particular computing schemes. For example, the quadtree structure in some image processing algorithms fits very well the pyramid topology. In this case, while pyramids present serious bandwidth drawbacks, they still make very good candidates as dedicated computers.

More generally, the capability to map algorithms on the topological structure of a network may sometimes constitute one of the best reasons why to choose a particular network. The success of hypercubes (see figure 13) stems for a part from their ability to emulate efficiently

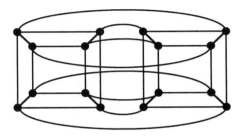

Figure 13. Hypercubes of dimension 4 H(4).

multi-dimensional grids and binary trees[140]. Conversely, this can explain why de Bruijn networks[140, 141] have been rarely implemented (see figure 14). While information broadcasting relies often on some special hardware, it may constitute an important property, too.

Gather and scatter operations are, indeed, basic features of several global operators. The efficient data-sorting ability turns out to be a basic feature which gives up to a flavour of association, and makes a network suitable for global property extraction from images. Reordering and packing can provide long range information. Histogram structure computations are good examples, but, with no doubt, the best one is the Fourier transform. Multistage networks appear to fit perfectly interconnection needs for such data exchanges. On the other hand, local operations behave poorly on these tree-like structures where horizontal neighbours may have to communicate through a far away least common ancestor. On the contrary, grids are ideally suited to perform filtering, edge connecting and so on, while region labelling appears to constitute an intermediate local-global operation[142]. In this case, inhomogeneity should be a solution : pyramids, for instance, make a tree-like structures with embedded grids; but the topological richness of the hypercube is a good enough response too.

6.4.4. Technological limits and degree. If we consider previous parameters, it is easy to see that the complete graph, where every vertex is neighbour of all others, is an ideal network. But such a network topology cannot be implemented for large machines, as its degree, that is the number of direct neighbours of a vertex, grows at the same rate as the number of vertices and the number of actual links bursts in accordance. A VLSI circuit has only a limited number of pins for physical transfers to another circuit. For a large degree, only a small number of pins can be used per logical link. This implies bit-serial data exchanges and a low bandwidth. The same technological reasons make a full cross-bar unfeasible for massively parallel machines.

On the other hand, fixed-degree networks keep a constant bandwidth while the size of the machine grows. Such a scalability is not possible to networks with variable degree. For instance, in the range of large orders, two and three-dimensional meshes turn out to get better bandwidths than equivalent-order hypercubes at given technology[143].

We see here how much their implementation can practically limit theoretically excellent networks. This description would thus be incomplete if, now, technological trends were not considered.

6.5. Impact of the Technological Evolution

Preceding criteria were relevant not so long ago, but recent machines feature improved networks. One important point is the fact that the inner network latency was only dealt with so far. Such a latency keeps decreasing sharply, and it appears that the limiting factor is the latency which occurs in the transfer of data between a processor local memory and the network.

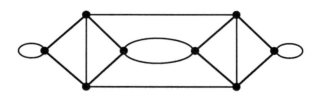

Figure 14. Binary de Bruijn graph of dimension 3 DB(2,3).

6.5.1. The interface between processor and network. Nowadays, leading machine designers, like Cray or Thinking Machines Corporation, do specialize in network implementation. Off-the-shelf processors are integrated in a proprietary architecture and the know-how consists mainly in the interface between these independent computing parts and the hardware which ties them up together into a parallel computer. But, as most often computing elements were not especially crafted for this particular architecture, such an integration becomes a difficult task.

The interface is in charge of transforming data inside the processor memory into a packet or a series of packets which the network could route to destination. On the one hand, this interface has to be well fitted to the processor to optimize the transfer, but, on the other hand, the mechanism and the hardware should be general enough to stand easy upgrade along with new processor generations. These contradicting requirements are difficult enough to fulfil so that, in recent networks, the main latency occurs at the interface between the memory and the network and not inside the network anymore. In other words, every communication is of constant cost : the start-up time spent in the interface. The more powerful the processors and the quicker the network, the more important this bottleneck. This causes real performances to be far away from peak performances.

One key future evolution should be the overlap between communication and computing, here discovered again. Recent switching techniques like wormhole[144, 145] allow processing in a pipe-line manner to mask memory accesses and data packing in taking advantage from regular structures. Such a trend, which stems from vector processing, could lead to virtually no-cost communications (the cost is limited to the initialization time for the pipe-line of operations) provided adapted architectures and algorithms.

6.5.2. Network and memory models. Yet, two different concepts of memory models have been envisioned : shared memory and distributed memory. The evolution of the respective implementations under the pressure of technological improvement is worth studying.

Shared memory machines have passed the introduction of the memory hierarchy concept. For usual machines, this concept consists of putting faster and faster, but smaller and smaller memories in a hierarchy, from disk to cache and registers. A cache consists of a small and fast memory where a copy of recently accessed data is maintained in order to allow faster further accesses. Requested data go up through the hierarchy to the cache and to the registers and replace not recently used data. This mechanism works well under the assumption of spatial and temporal locality of the data accesses. In a shared memory machine, caches are local memories to processors which are accessed by other processors and got copies of data blocks from the main shared memory. As the main memory is seen through copies in a local memory, a logical evolution of memory hierarchy consists in suppressing the shared memory and emulating it through communications between local caches. This describes the basic principles under the all-cache architecture of Kendall Square Research KSR-1[146, 147].

The overlap of communication and computation in distributed memory machines represents an explicit way of taking advantage from regularities of data and code, the same way caches are implicitly taking such an advantage in shared memory machines. Another possible mechanism to run-time optimizing is to "prefetch" the sending of messages thanks to a specific hardware as some caches are able to prefetch data accesses, i.e.. forecast future operations. At last, static patterns make it possible to introduce the concept of compiled communications. All these optimizations are not heavily used yet, but no doubt these mechanisms will be developed in years to come.

The main guiding line of these two evolutions is regularity, looking for maximizing the bandwidth consumption On the one hand, shared memory architectures are emulated on physically distributed memory in an adaptation of memory hierarchy principles. This emulation provides a solution to the inherent limitations of access bandwidth to shared memory banks. On the other hand, distributed memory architectures begin to implement hardware mechanisms in order to balance the time load of the network. In massively parallel computing, clever algorithm embedding makes a good deal of this regularity. Furthermore, a lack of locality can be considered a waste of bandwidth as the number of links travelled by packets increases and the probabilities of conflicts grows. In this case, the distance would consequently appear again as a significant parameter.

6.5.3. Router functionnalities. The router is this part of a network which has to establish a connection between the network and a processor. Optimization of bandwidth occupation and interface latency are directly dependent on the router "cleverness". What are thus requirements for an effective router?

- Technological Independence. Modularity is a key factor of evolution. An architecture should stand updates along with new generation processors. As an interface, the router must provide general mechanisms, independent of a given processor technology.
- Control Independence. The router must be able to behave independently of the processor in order to process communications during computations. Their respective control flows must be concurrent.
- Resource Independence. If the router and the processor compete for a shared critical resource, the result is a serialisation of their activities. Here, again, concurrent behaviour has to be secured.

The functionnalities of the iWarp processor of Intel[148] provide a good example of a powerful router specification. In this processor, a Communication Pathway (the router) carries out the routing capacity with several available switching techniques. Either, routing is transparent or a copy is taken for local use before forwarding. The exchange of data between the Communication Pathway and the Computation Engine (the processor) is performed by the Memory Interface Spooler and Streamer (an extended Direct Memory Access and Memory Management Unit) and makes use of a local memory. One gets an excellent example here of a modular design of functional parts as well as of an effort to make processing and data routing independent.

6.6. Interconnection Networks and Nervous System

Trends are to coarser parallel architectures where powerful processors can emulate massively parallel operations. This emulation puts a stress on the efficiency of the interface between the network and the processors. Compared with this bottleneck, the network itself appears to deliver messages anywhere to a marginal cost only. Bandwidth remains a central parameter and the architectural evolution tends to optimize its use through different concepts, namely in expliciting regularities.

To conclude this part, let us underline some communication specificities. These characteristics could finally be compared to those of the nervous system behaviour, aiming to provide some further understanding. A striking point is the complete lack of semantics in the communication progress. Routing and switching are not part of computing, but internal mechanisms of the interconnection networks. The role of the network is to exchange data between processors or between processors and memory banks, there is no special meaning added to data by the network. The lack of real understanding of the specificities in image processing communications structures, especially from low-level to high-level processing, favours the implementation of networks which efficiently emulate the complete interconnection topology. This is, probably, a potential limit to the scalability of current architectures.

Finally, we want to stress the fact that the only meaningful part of a parallel program remains computations on processors. The interconnection network has so far only an impact on the overall efficiency. More generally, every parallel machine can be simulated on a mono-processor machine, the only drawback is then the execution length.

7. CONCLUSION

In this chapter, we have tried to illustrate a current evolution of image-processing dedicated architectures. Of course, in this case, evolution gets a more technological meaning than really temporal; but it still allows a likely better comparison with biological visual systems than architectures themselves. In that respect, two warnings have been proposed: one takes advantage from control theory and from a related model of system – queues and Markov chains. It shows how questionable may be a presumptive similarity of structure based only on a similarity of outputs. And output similarity is what one gets here: only actions triggered by a vision process can be compared, even intermediate features like edges or motion are not quite proved identical and satisfactory to the other system. This situation makes it even harder to really compare "architectures" since only a part of it was in question: the vision part in isolation, to the exclusion of a control part and of a loop support part like, for instance, in attention focusing. The second warning is more experimental and comes from a purely technological description of vision machines: it brings out a concern about accelerating computations, about the computing power as the main prompting. Several computing units are put together to work in parallel and their communication topology obeys rules coming from algorithms as developed on sequential universal machines, or from languages when it is not from the sequential manner data are delivered by cameras. Basic operations are the most frequently encountered in formulas and algorithmic expressions, like the multiply-ALU, never reaching any true vision semantics. Communications and memory accesses do stem again from this accelerating need, like intermediate registers for pipeline or pointers for memory management.

Nevertheless, such an abstracted description of the whole architectural corpus suggests by itself that putting the stress on control, and more precisely trying to separate between communication and operation features, should help understand better main trends in vision architectural specificities, if any real. This guess is proved correct through a sample description of machines where such particularities are searched and underlined in a systematic manner. Then, special registers get a meaning of neighbours for local processing according to a primary image topology, given links get different ranges and speeds allowing globally adapted operations, a third dimension gives access to an associative-memory behaviour thanks to specific instructions made easier to the extreme. Up to the same pipelining process to be viewed a semantical progression in vision levels . Eventually, for lack of being comparable to human vision systems, artificial vision machines end evocating human societies: main factors for a comparative study are thus the links between control and data flows, the privacy of resources – memory mainly – as for accesses, the existence of hierarchies and whether individuals do specialize or not. And the whole study appears in great need of a deeper technological understanding of evolutions at least on to major points: operative specialization to be tightly bound to silicon and to technology, communication specialization related to network properties. If some recent results in analog VLSI for vision let hope that some real visual specifity might be obtained and proved suitable for robots, it is not the same in the area of communications: one important conclusion still is that communications in current parallel machines do not get to any semantics. On the contrary, a current tendency to emulate massively parallel machines on coarse grain machines makes all peculiarities come from processors, physical information transfers remaining not really wanted and as rare as possible, to be immediate and transparent. By the same token, association abilities get further and out of reach yet, and this whole evolution raises as many questions to living vision-system specialists. Does Nature appear to have cared for an adapted balance between contradicting factors like hardware vs. software, low-level vs. high-level processing or more generally any semantics translated into the architecture, bit-serial vs. numerical-intensive processing, direct-neighbour vs. all equal-cost communications? Computers could be now in a phase where designers would agree that such an equilibrium on that many facets is worth finding, where the technological swing at the very present time is more in favour of software labours, and where computer makers have at best slight hints about the eventual result anyway.

Thus, probably a more efficient lesson to be drawn from this opposition between the current states of operative and communication technologies are in the importance of the software which actually makes balance more even between these two other fundamental aspects of dedicated computer architectures. Although this is yet another story (not exactly independent), one can think of it in the following terms: what would be proved if ever two dead brains were found identical? Certainly not that the human beings they used to belong to were the same, or all the psychiatrists in the world should be looking for employment. Now, this again might not hold as human beings may not be machines themselves: but, perhaps, this last question is to be asked to computers some day.

ACKNOWLEDGEMENTS

We are indebted to G. Quenot whom we borrowed some technological views from, to J. Serot and M. Eccher for providing us with some more bibliography and with machine overviews, to A. Merigot and T. Bernard who are usual Zavidovique's complices respectively for hyperparallelism and retina surveys and lectures and to V. Cantoni for a final touch. Generally speaking several ideas in this paper (not all of them!) have been on the air at the "Laboratoire Systeme de Perception of the ETCA" for many years.

REFERENCES

1. Y. Burnod, *An Adaptative Neural Network: the Cerebral Cortex*, Masson ed. (1988).
2. J. Dowling, *The Vertebrate Retina: an Approchable Part of the Brain*, Harvard University (1985).
3. C. Olivier, *Stratégies d'acquisition, de traitement et de prise en compte d'informations en vue du contrôle d'un robot en environnement non structuré*, Thèse Université Paris-sud, centre d'Orsay, F (1993).
4. D.A. Patterson and C.H. Sequin, RISC I: a Reduced Instruction Set VLSI Computer, *8th Int. Symposium on Computer Architecture IEEE* (1983).
5. J.C. Heudin and C. Panetto, *Les Architectures RISC*, Dunod, Paris, F (1990).
6. A. Merigot and B. Zavidovique, Image Analysis on Massively Parallel Computers: an Architectural Point of View, *International Journal of Pattern Recognition and Artificial Intelligence*, Vol.6, No.2-3 (1992).
7. W.D. Hillis, *The Connection Machine*, MIT Press, Cambridge, MA (1985).
8. J.F. Palmer and G. Fox, The NCUBE Family of High-Performance Parallel Computer Systems, *Third Conference on Hypercube Concurrent Computers and Applications*, Vol.1, pp.847-851 (1988).
9. M.J.B. Duff, Review of the CLIP Image Processing System, *National Computer Conference*, Anaheim, CA (1978).
10. S.H. Unger, A Computer Oriented Toward Spatial Problems, *Proc. IRE*, Vol.46, pp.1744-1750 (1958).
11. D.L. Slotnick, W.C. Borck, and R.C. McReynolds, The SOLOMON Computer, *Proc. Western Joint Comp. Conf.*, pp.87-107 (1962).
12. G.M. Barnes et al., The Illiac Computer, *IEEE Trans. Comp.*, Vol.C.17, pp.746-757 (1968).
13. P.M. Flanders, A Unified Approach to a Class of Data Movements on an Array Processor, *IEEE Trans. Comp.*, Vol.C31, No.9, pp.809-819 (1982).
14. K.E. Batcher, Design of a Massively Parallel Processor, *IEEE Trans. Comp.*, Vol.29, pp.836-840 (1980).
15. J.L. Potter, Image Processing on the Massively Parallel Processor, *IEEE Computer*, pp.62-67 (1983).
16. K.E. Batcher, Bit Serial Processing Systems, *IEEE Trans. Comp.*, Vol.C51, No.5, pp.377-384 (1982).
17. R. Davis and D. Thomas, Geometric Arithmetic Parallel Processor; Systolic Array Chip Meets the Demand of Heavy Duty Processing, *Electronic Design*, pp.207-218 (1984).
18. T.J. Fountain and M. Phil, Towards CLIP 6 an Extra Dimension, *CAPAIDM* (1981).

19. M.J.B. Duff, The Elements of Digital Picture Processing, in *Real Time Parallel Computing*, Onoe, Preston, and Rosenfeld eds., Plenum Press (1981).

20. M.J.B. Duff, *Array Automata, Modern Cellular Automata*, K. Preston and M.B.J. Duff eds., Plenum Press, pp.259-274 (1984).

21. K.P. Arvind, D.K. Robinson, and I.N. Parker, A VLSI Chip for real time image processing, *IEEE Int. Symp. on Circuits and Systems*, pp.405-408 (1983).

22. K.P. Arvind and R.A. Ianucci, A Critique Of MultiProcessing Von Neumann Style Programming, *Proc. of 10th annual Int. Conf. Comp. Arch.* (1983).

23. M.J.B. Duff, Clip 4, a Large Scale Integrated Circuit Array Parallel Processor, *Proc. 3rd Int. Conf. on Pattern Recognition*, pp.728-733 (1976).

24. C.D. Stamopoulos, Parallel Image Processing, *IEEE Trans. Comp.*, Vol.C24, No.4, pp.424-433 (1975).

25. P.M. Flanders, Efficient High Speed Computing with the Distributed Array Processor, *High Speed Computer and Algorithm Organization,* Kuck et al. eds., Academic Press (1978).

26. P. Marks, Low Level Vision Using an Array Processor, *Computer Graphics and Image Processing*, Vol.14, pp.281-292 (1980).

27. T. Kondo et al., An LSI Adaptative Array Processor, *IEEE Jour. of Solid State Circuits*, Vol.SC18, No.2, pp.147-156 (1983).

28. M. Kidode, Image Processing Machines in Japan, *IEEE Computer*, Vol.16, No.1, pp.68-80 (1983).

29. *The Connection Machine CM-5 Technical Summary*, Thinking Machines Corp., Cambridge, MA (1991).

30. A.R. Hanson and E.M. Riseman, Segmentation of Natural Scenes, in *Computer Vision System,* Hanson-Riseman ed., Academic Press (1978).

31. S.L. Tanimoto, Regular Hierarchical Image and Processing Structures in Machine Vision, *Computer Vision Systems,* Hanson-Risemann ed., Academic Press (1978).

32. C.R. Dyer, A VLSI, Pyramid Machine for Hierarchical Image Processing, *Proceeding PRIP*, pp.381-386 (1981).

33. J.M. Jolion, Analyse d'Image: Le Modèle Pyramidal, *Traitement du Signal*, Vol.7, No.1 (1990).

34. J.L. Bentley and H.T. Kung, A Tree Machine for Searching Problems, *Conference on Parallel Processing*, pp.257-266 (1979).

35. G. Son, A Highly Concurrent Tree Machine for Data Base Applications, *Proc. Conf. on Parallel Processing,* pp.259-268 (1980).

36. V. Cantoni, S. Levialdi, M. Ferreti, and F. Maloberti, A Pyramid Project Using Integrated Technology, in *Integrated Technology for Parallel Image Processing*, S. Levialdi ed., Academic Press, London, UK, pp.121-132 (1985).

37. D.H. Schaefner, D.H. Wilcox, and G.C. Harris, A Pyramid of MPP Processing Elements; Experiences and Plans, *Hawaii Int. Conf. on System Sciences*, pp.178-184 (1985).

38. S.L. Tanimoto, A Hierarchical Cellular Logic for Pyramid Computers, *Journal of Parallel and Distributed Processing*, Vol.1, pp.105-132 (1984).

39. S.L. Tanimoto, T.J. Logocki, and R. Ling, A Prototype Pyramid Machine for Hierarchical Cellular Logic, *Parallel Computer Vision,* L. Uhr ed., Academic Press, New York, NY, pp.43-83 (1987).

40. A. Merigot, B. Zavidovique, and F. Devos, SPHINX, A Pyramidal Approach to Parallel Image Processing, *IEEE Workshop on Computer Architecture for Pattern Analysis and Image Database Management*, pp.107-111 (1985).

41. A. Merigot, Une Architecture Pyramidale d'un Multi-processeur Cellulaire pour le Traitement d'Images, *Thèse à l'Univ. de Paris sud Orsay,* F (1983).

42. A. Merigot, S. Bouaziz, F. Devos, J. Mehat, Ni Y, P. Clermont, and M. Eccher, Sphinx, un Processeur Pyramidal Massivement Parallèle pour la Vision Artificielle, *7ème Congrès Afcet Inria Paris* (1989).

43. S.R. Sternberg, Biomedical Image Processing, *IEEE Computer*, Vol.16, No.1, pp.22-34 (1983).

44. D.B. Gennery and B. Wilcox, A Pipelined Processor for Low Level Vision, *IEEE* (1985).

45. D.B. Gennery, T. Littwin, B. Wilcox, and B. Bon, Sensing and Perception Research for Space Telerobotics at JPL, *CH2413-3/87/0000/0311$01.00 © IEEE* (1987).

46. E.W. Kent, R. Lumia, and M.O. Shneier, PIPE, *Tech. Report of the Sensory Interactive Robotics Group at NBS*, Washington (1986).

47. H.T. Kung, Let's Design Algorithms for VLSI, *Proc. Caltech Conf. on VLSI 79*, pp.65-90 (1979).

48. H.T. Kung, The Structure of Parallel Algorithms, *Advances in Computers*, Vol.19, pp.65-112 (1980).

49. H.T. Kung, Special Purpose Devices for Signal and Image Processing, *Int. Report CMU-CS-80-132* (1980).

50. H.T. Kung, Special Purpose Devices for Signal and Image Processing: an Opportunity in VLSI, *Proc. SPIE, Real Time Signal Processing*, Vol.241, pp.76-84 (1981).

51. H.T. Kung, Why Systolic Architectures?, *IEEE Computer*, Vol.15, No.1, pp.37-46 (1982).

52. P.M. Dew, A tutorial on Systolic Architectures for High Performance Processors, *2nd Internationnal Electronic Image Week*, CESTA, Nice, F (1986).

53. U. Weiser and A. Davis, A Wavefront Notation Tool for VLSI Array Design, in *VLSI Design and Computations,* H.T. Kung ed., CS Press, Carnegie Mellon Univ, pp.226-234 (1981).

54. C.E. Leiserson and J.B. Saxe, Optimizing Synchronous Systems, *Proc. 22nd Annual Symp. Foundations of Computer Science*, IEEE Computer Society, pp.23-36 (1981).

55. G.J. Li and B. Wah, Optimal Design of Systolic Arrays for Image Processing, *Proc. Workshop on Computer Arch. for Pattern Analysis and Image Database* (1983).

56. P. Quinton and Y. Robert, *Algorithmes et Architectures Systoliques*, Masson ed. (1989).

57. H.T. Kung, Systolic Algorithms for the CMU Warp Processor, *Tech. Report CMU-CS-84-158,* Dept. of Comp. Sc., CMU Pittsbugh, PA (1984).

58. S. Borkar et al., iWARP: an Integrated Solution to High Speed Computing, *Proc. of Supercomputing Cong.*, Orlando, FL, pp.330-339 (1988).

59. M. Annaratone, E. Amoult, T.H. Gross, H.T. Kung et al., *Warp Architecture and Implementation*, 0884-7495/86/0000/0346$01.00 © *IEEE* (1986).

60. M. Annaratone, E. Amoult, T.H. Gross, H.T. Kung et al., *Architecture of Warp*, CH2409-1/87/0000/0264$01.00 © *IEEE* (1987).

61. M. Annaratone, E. Amoult, T.H. Gross, H.T. Kung et al., Applications Experience on Warp, *National Computer Conference*, pp.149-159 (1987).

62. M. Annaratone et al., The WARP Computer: Architecture, Implementation and Performances, *IEEE Trans. Computers* 32, Vol.12, pp.523-538 (1987).

63. C. Peterson, J. Sutton, and P. Wiley, iWARP: A 100 Mops LIW Microprocessor for Multicomputers, *IEEE Micro*, pp.26-29 (1991).

64. J.B. Dennis, Dataflow Supercomputers, *IEEE Computers*, No.1 (1980).

65. S.Y. Kung, S.C. Lo, S.N. Jean, and J.N. Hwang, Wavefront Array Processors: Concept to Implementation, *IEEE Computer 20(11)*, pp.18-33 (1987).

66. E.B. Davie, D.G. Higgins, and C.D. Cawthorn, An Advanced Adaptative Antenna Test Bed Based on a Wavefront Array Processor System, *Proc. Int. Workshop on Systolic arrays* (1986).

67. B. Zavidovique, G. Quenot, A. Safir, C. Fortunel, F. Verdier, and J. Serot, Automatic Synthesis of Vision Automata, M. Bayoumi ed., *VLSI Design Methodologies for DSP Architectures*, Kluwer Academic Publishers (1993).

68. J. Backus, Can Programming be Liberated from the Von Neumann Style? A Functional Style and its Algebra of Programs, *Comm. of ACM*, Vol.21, No.8 (1978).

69. B. Zavidovique, V. Serfaty, and C. Fortunel, Mechanism to Capture and Communicate Image-Processing Expertise, *Journal of IEEE Software,* Vol.8, No.6, pp.37-50 (1991).

70. B. Zavidovique and P.L. Wendel, Computer Architecture for Machine Perception, *Proc. CAMP '91,* B. Zavidovique ed., Paris, F (1991).

71. G. Quenot and B. Zavidovique, A Data-Flow Processor for Real-Time Low-Level Image Processing, *IEEE Custom Integrated Circuits Conference*, San-Diego, CA (1991).

72. E. Allart and B. Zavidovique, Functional Image Processing through Implementation of Regular Data Flow Graphs, *21st Annual Asilomar Conf. on Signals*, Systems on Computers Pacific Grove CA (1987).

73. E. Allart and B. Zavidovique, Image Processing VLSI Design through Functional Match between Algorithms and Architecture, *IEEE Int. Symposium on Circuits and System*, Espoo, Finland (1988).

74. E. Allart and B. Zavidovique, Functional Mapping for Low-level Image Processing Algorithms, *IEEE Workshop on VLSI Signal Processing Systems*, Monterey, USA (1988).

75. G. Quenot and B. Zavidovique, The ETCA Massively Parallel Data-Flow Computer for Real Time Image Processing, *IEEE Int. Conf. on Computer Design,* Cambridge, MA (1992).

76. C.C. Weems and J.R. Burrill, The Image Understanding Architecture and its Programming Environment, in *Parallel Architectures and Algorithms for Image Understanding,* V.K. Prusanar-Kumar ed., Academic Press, Orlando, FL, pp.525-562 (1991).

77. C.C. Weems et al., Image Understanding Architecture: Exploiting Potential Parallelism in Machine Vision, *IEEE Computer*, pp.65-68 (1992).

78. *The Connection Machine CM-2 Technical Summary*, Thinking Machines Corporation (1990).

79. L. Uhr, Converging Pyramids of Arrays, *Proc. CAPAIDM*, pp.31-34 (1981).

80. J.L. Basille, *Structures parallèles et traitement d'image*, Thèse d'état Univ. Paul Sabatier Toulouse (1985).

81. C. Williams and J. Rasure, (Khoros) A Visual Language for Image Processing, *Proc. IEEE Computer Society Workshop on Visual Languages*, Skokie, Illinois (1990).

82. S. Dacic, *Conception et Exploitation Intuitive de Systemes Informatiques Complexes*, These de Doctorat, Univ de Paris XI, Orsay, F (1990).

83. J. Serot, G. Quenot, and B. Zavidovique, Functional Programming on a Data-Flow Architecture: Applications in Real Time Image Processing, *Int. Journal of Machine Vision and Applications* (1993).

84. M.J. Flynn, Some Computer Organizations and their Effectiveness, *IEEE Trans. Computers C21(9)*, pp.948-960 (1972).

85. P. Clermont, *Méthodes de Programmation de Machine Parallèle Pyramidale. Applications en Segmentation d'Image*, Thèse Paris 7 (1990).

86. J.R. McGraw, The VAL Language: Description and Analysis, *ACM Trans. Prog. Languages and Systems*, No.4(1), pp.44-82 (1982).

87. J.R. McGraw et al., SISAL: Streams and Iterations in a Single Assignment Language, *Language Ref. Manual*, Vers 1.0, Lawrence Livermore Nat Lab, Livermore, CA (1983).

88. R.S. Nikhil, K. Pingali, and K.P. Arvind, Id Nouveau, *CSG Tech. Memo 265*, LCS, MIT (1986).

89. R.S. Nikhil, Id Reference Manual, *Tech. Report CSG Memo 284*, MIT lab CS, Cambridge, MA (1988).

90. W.W. Wadge and E.A. Ashcroft, *Lucid, the Dataflow Programming Language*, Academic Press, London, UK (1985).

91. L. Snyder, Introduction to the Configurable Highly Parallel Computer, *IEEE Computer*, Vol.15, No.1, pp.47-64 (1982).

92. I. Koren and G.B. Silberman, A Direct Mapping of Algorithms onto VLSI Processor Arrays Based on the Data Flow Approach, *Proc. Int. Conf. Parallel Processing*, pp.335-337 (1983).

93. I. Koren, B. Mendelson, I. Peled, and G.B. Silberman, A Data-Driven VLSI Array for Arbitrary Algorithms, *IEEE Computer* (1988).

94. M. Cornish et al., The TI Dataflow Architecture: the Power of Concurrency for Avionics, *Proc. 3rd Int. Conf. Digital Avionics Systems*, IEEE, Fort-Worth, TX, pp.19-25 (1979).

95. R. Vedder and D. Finn, The Hughes Data Flow Multiprocessor: Architecture and Efficient Signal and Data Processing, *Proc. 12th Symp. Computer Architecture*, Boston, MA, pp.324-332 (1985).

96. T. Jeffery, The µPD7281 Processor, *Byte*, pp.237-246 (1985).

97. P. Kajfasz and B. Zavidovique, Hardware Implementation of Geometric Transforms of Images for Real Time Processing, *Proc. 5th Int. Symp. Electronics and Information Sciences*, Kobe, Japan (1986).

98. P. Kajfasz and B. Zavidovique, MORPHEE: A Multi-access Memory Unit for on the Fly Image Processing Applications, *Proc. 2nd Int. Conf. Computers and Applications*, Beijing, China (1987).

99. P. Missakian, M. Milgram, and B. Zavidovique, A Special Architecture for Dynamic Programming, *ICCASSP/IEEE/ASI*, Tokyo, Japan (1986).

100. G. Quenot, J. Mariani, J.J. Gangolf, and J.L. Gauvin, Dynamic Programming Processor for Speach Recognition, *IEEE Journal of Solid State Circuits* (1989).

101. M. Eccher and B. Zavidovique, C.O.L.I.S.E. Real-Time Région Détector Based on Algorithmic Decomposition, *Twentieth Asilomar Conference on Signals, Systems and Computers*, Pacific Grove, CA (1986).

102. M. Eccher, *Architecture Parallèle Dédiée à l'Étude d'Automates de Vision en Temps Réel*, Thèse de Doctorat, Univ de Franche-Comté (1992).

103. L. Palmier, *Conception Fonctionnelle de Circuits Intégrés de Traitement d'Image*, Thèse Université de Paris-Sud, Orsay, F (1985).

104. L. Palmier, C. Legrand, F. Devos, and B. Zavidovique, Le Circuit de Mise Sous Forme Locale, *Colloque National de Conception de Circuits à la demande*, Grenoble, F (1985).

105. L. Palmier, F. Devos, and B. Zavidovique, A Functional Operation Architecture For Image Processing, *International Symposium On Circuits and Systems*, Kyoto (1985).

106. L. Palmier, F. Devos, and B. Zavidovique, Les Circuits de Traitements du Signal: Une Approche Fonctionnelle, *IASTED International Symposium*, Paris, F (1985).

107. L. Palmier, M.P. Gayrard, and B. Zavidovique, VLSI Architecture of the "CURVE" Function for Image Processing, *Computer Architecture for Pattern Analysis and Image Data Base Management*, Miami Beach, FL (1985).

108. H. Waldburger, J.Y. Dufour, and B. Kruse, A Real Time System for Image Processing, *Proc. of 1991 Workshop on Computer Arch. and Machine Perception*, Paris, F (1991).

109. P.A. Ruetz and R.W. Brodersen, An Image-Recognition System Using Algorithmically Dedicated Integrated Circuits, *Int. Journal of Machine Vision and Applications*, No.1, pp.3-22 (1988).

110. J.C. Gittins, *Multi-armed Bandit Allocation Indices*, John Wiley and Sons, Chichester (1989).

111. B. Zavidovique, Contribution à la vision des robots, *Thèse d'état Université de Besancon*, F (1981).

112. B. Zavidovique and G. Stammon, Bilevel Processing of Multilevel Pictures, *Pattern Recognition and Image Processing Conference*, Dallas (1981).

113. T. Bernard, P. Nguyen, F. Devos, and B. Zavidovique, A 65x76 VLSI Retina for Rough Vision, *Workshop on Comp. Arch. for Mach. Perception 91*, Paris, F (1991).

114. B. Zavidovique and T. Bernard, Generic functions for on-chip vision, *Invited Conference at IEEE Int. Conf. on Pattern Recognition*, La Hague, NL (1992).

115. T. Bernard, *Des Rétines Artificielles Intelligentes*, Thèse Université Paris-Sud (1992).

116. T. Bernard, B. Zavidovique, and F. Devos, A Programmable Artificial Retina, *IEEE Journal of Solid-state Circuits* (1993).

117. S. Mallat, Review of Multifrequency Channel Decomposition of Images and Wavelets Models, *IEEE Trans. on Acoustic Speech and Signal Processing* (1989).

118. S. Mallat and S. Zhong, Signal Characterization from Multiscale Edges, *Proc. of the 1990 IEEE ICPR* (1990).

119. J. Shen and S. Castan, Edge Detection Based on Multiedge Models, *Proc. of the SPIE'87,* Cannes, F (1987).

120. C.A. Mead and M.A. Mahowald, A Silicon Model of Early Visual Processing, *Neural Networks*, Vol.1, pp.91-97 (1988).

121. C.A. Mead, *Analog VLSI and Neural Systems*, Addison Wesley (1988).

122. T. Bernard, P. Garda, A. Reichart, B. Zavidovique, and F. Devos, Design of a Half-toning Integrated Circuit Based on Analog Quadratic Minimization by Non-linear Multistage Switched Capacitor Network, *IEEE Int. Symp. on Circuits and Systems*, Helsinki, Finland (1988).

123. T. Bernard, P. Garda, A. Reichart, B. Zavidovique, and F. Devos, A Family of Analog Neural Half-toning Techniques, *4th European Signal Processing Conference*, Grenoble, F (1988).

124. T. Bernard, P. Garda, B. Zavidovique, and F. Devos, Ultra-small Implementation of a Neural Half-toning Technique, *EURASIP Workshop on Neural Networks 90*, Sesimbra, P (1990).

125. T. Bernard, From Sigma-Delta Modulation to Digital Half-toning of Images, *IEEE ICASSP 1991*, pp.2805-2808, Toronto, Ontario, CDN (1991).

126. C.A. Mead, A Sensitive Electronic Photoreceptor, *Chapell Hill Conf. on VLSI*, Comp. Sc. Press, Rockville, Maryland, pp.463-471 (1985).

127. C.A. Mead and M. Ismail ed., *Analog VLSI Implementations of Neural Systems*, Kluwer Academic Publishers, Norwell, MA (1989).

128. V. Hu and A. Kramer, P.K. Ko, EEPROMS as Analog Storage Devices for Neural Nets, *Neural Networks*, 1 Supll.I 385 (1988).

129. T. Bernard, Convolueur imageur électronique, *French licence n°89*, 16552, 14 December 1989.

130. C. Koch, J. Marroquin, and A. Yuille, Analog "Neuronal" Networks in Early Vision, *Proc. Natl. Acad. Sci. USA83*, pp.4263-4267 (1986).

131. J. Hutchinson, C. Koch, J. Luo, and C. Mead, Computing Motion Using Analog and Binary Resistive Networks, *IEEE Computers*, Vol.21, No.3, pp.52-63 (1988).

132. C. Koch, Seeing Chips: Analog VLSI Circuits for Computer Vision, *Neural Compt.*, Vol.1, pp.184-200 (1989).

133. C. Koch, J. Harris, T. Horiuchi, A. Hsu, and J. Luo, Real-time Computer Vision and Robotics Using Analog VLSI Circuits, *Advances in Neural Information and Processing Systems*, D. Touretsky ed., Morgan Kaufmann (1989).

134. C. Koch, Resistive Networks for Computer Vision: An Overview, *An introduction to neural and electronic networks,* S.F. Zornetzer, J.L. Davis, and C. Lau eds., Academic Press (1990).

135. C.M. Brown, Parallel Vision with the Butterfly Computer, *Supercomputing*, pp.54-68 (1988).

136. *The Connection Machine CM-5 Technical Summary*, Thinking Machines Corporation, Cambridge, MA (1992).

137. C.E. Leiserson, Fat Trees: Universal Networks for Hardware-efficient Supercomp, *IEEE Comp.*, pp.892-901 (1985).

138. T.H. Szymanski and V.C. Hamacher, On the Universality of Multipath Multistage Interconnection Networks, *Journal of Parallel and Distributed Computing*, Vol.7, pp.541-569 (1989).

139. G.F. Pfister and V.A. Norton, "Hot spot" contention and combining in multistage interconnection networks, *IEEE Transactions on Computers*, Vol.C34, No.10 (1985).

140. F.T. Leighton, *Introduction to Parallel Algorithms and Architectures: Arrays, Trees, Hypercubes*, Morgan Kaufmann, San Mateo, CA (1992).

141. J.C. Bermond and C. Peyrat, de Bruijn and Kautz networks: a competitor for the hypercube?, in *Hypercube and Distributed Computers*, J.P. Verjus and F. André eds., North Holland, pp.279-294 (1989).

142. S. Cartier, Parallélisme de Contrôle et Fusions de Régions, *5èmes Rencontres sur le Parallélisme*, Brest, pp.243-246 (1993).

143. W.J. Dally, Performance Analysis of k-ary n-cube Interconnection Networks, *IEEE Trans Computers*, Vol.39, pp.775-785 (1990).

144. W.J. Dally and C.L. Seitz, Deadlock-free Message Routing in Multiprocessor Interconnection Networks, *IEEE Trans Computers*, Vol.36, pp.547-553 (1987).

145. L.M. Ni and P.K. McKinley, A Survey of Wormhole Routing Techniques in Direct Networks, IEEE Computer, Vol.26, pp.62-76 (1993).

146. J. Rothnie, Overview of the KSR1 Computer System, *Kendall Square Research Report TR 9202001* (1992).

147. KSR1 Principles of Operations, *Kendall Square Research* (1992).

148. INTEL, Introduction to iWarp (1991).

PANEL SUMMARY
ONE MODEL FOR VISION SYSTEMS?

Vito Di Gesù[1] *(chairman),* Serge Castan[2], Silvana Vallerga[3], Hezy Yeshurun[4], Mario Zanforlin[5], and Bertrand Zavidovique[6]

[1] Dipartimento di Matematica ed Applicazioni, Università di Palermo
   Via Archirafi 34, I - 90123 Palermo, Italy
[2] Institut de Recherche en Informatique, Université Paul Sabatier
   118, route de Narbonne, F - 31062 Toulouse, France
[3] Istituto di Cibernetica e Biofisica del CNR
   Via Dodecanneso 33, I - 16100 Genova, Italy
   and International Marine Centre
   Lungomare d'Arborea 22, I - 09072 Torregrande (OR), Italy
[4] Department of Computer Science
   Tel Aviv University, IL - 69978 Tel Aviv, Israel
[5] Dipartimento di Psicologia Generale, Università di Padova
   P.zza Capitaniato 3, I - 35100 Padova, Italy
[6] Institut d'Electronique Fondamentale, Université Paris XI
   Centre d'Orsay, Bat 220, F - 91405 Paris Orsay, France

ABSTRACT

This panel reports some considerations about the definition of *vision-models.* The panellists are scientists working on vision problems from different perspectives. The concept of model in vision seems to remain still open. In fact, it is dynamic, and context dependent. There exists the need for a better exchange of information, among biologists, engineers, physicists, and psychologists in order to improve our knowledge.

ABOUT THE DEFINITION OF A VISUAL MODEL (V. Di Gesù)

Today we will discuss the meaning of model in *computer vision* and the implications deriving in the design of artificial visual systems. Webster's dictionary provides three different definitions for the word model:
- *miniature representation of something;*
- *a pattern of something to be made;*
- *an example for imitation or emulation.*

This indicates that its semantic has some vagueness, that depends on the context of the discourse. It is interesting to note that each of them is related to a particular design paradigm, adopted by computer scientists.

For example if we replace the word *miniature* with *schematic and simplified* and the word *something* with *natural vision* the first definition reflects the *anthropomorphic*, (described or thought of as having a human form or with human attributes (Webster's dictionary)), approach to computer vision. The third definition still reflect the anthropomorphic system design approach. The second definition provides a different meaning of model. In fact, the term *pattern* can be referred to *behaviour* of a system, does matter how it is realized, the goal is to produce correct outputs for given inputs. Let us call the last approach *pragmatic*: a practical approach to problems and affairs (Webster's dictionary).

In Vallerga's presentation we have seen that nature is pragmatics, solutions are goal oriented, and well optimized. For example, the complexity of the visual acquisition system (eyes, retinas, early vision) seems to increase with the simplicity of the internal layers of the brain.

Scientists have developed a dynamic definition of a model, to explain both experimental results, and to describe the system under study. A good model should suggests new experiments in order to test its validity. Discrepancies suggest an update of the old model or its refutation. *Is a never end history!* I like better such view because it fits with both anthropomorphic and pragmatic view of computer vision.

Moreover, the last definition of model allows to relate strictly the hardware of the system under study to the paradigm of computation. In fact, it works for natural-hardware which is carbon-based (at least in our world), and it seems to show adaptive behaviour acquired during millions of years of learning and optimization-algorithms. Moreover, it works also for artificial-hardware which is silicon-based (at least until now), it shows deterministic and less flexible behaviour, but it has been trained during about only 50 years of logic programming. Artificial hardware is still at the prehistory, optical and bio chips are now coming up, and new paradigm of computation will probably be developed in the near future.

In 1946 Alan Turing suggested an *ideal* experiment in order to compare an artificial-brain with an human one. The experiment can be shortly described as follows: *a blind human investigator makes a set of questions to an human brain and a machine brain in order to discover who is the human and who is the machine. The machine will pass Turing's test (the machine is intelligent!!) if after the list of questions the blind human can not make a decision.*

Turing's test has some weak points: is not well formalized, and the results of the test is not definite, in fact we never will know if a question exists, that allows to discriminate the human from the artificial. Moreover, even if the behaviour of the two brains are indistinguishable, is the artificial system conscious of its behaviour?

Nevertheless, Turing's test suggests some remarks on the concept of visual models: *Given a visual system two or more models can describe the same behaviour, moreover one model can be adopted to describe mathematically more then one visual system.*

Nature May Suggest Practical Solutions

Natural neural nets suggest several architecture topologics and computation paradigms. Most of them are based on processing units, which are interconnected with a static topology (the links, connecting the PU's are physically and logically defined at the design stage).

Image analysis algorithms, implemented on multi-processors machines, require navigation of data among PU's, the mapping of the data on the PU's should minimize such overhead. Static topologics are tuned only for a given class of problem. Full connectionism could be the solution for data communication, however the actual limits of the VLSI technology do not allows high links density (for example, thousands links per processor).

The reconfigurability of the network topology seems to be more realistic, and efforts are now done in this direction. In fact, run time network reconfiguration may allow the tuning of the network path-ways with the kind of data exchange foreseen by given algorithms.

Another overhead factor is related to the I/O and the mapping of the data in the network. There exists the need to design smart artificial *retinas* and faster acquisition systems.

Visual perception is characterized by different levels of abstraction, and this implies a kind of heterogeneous computation. The machine vision design should consider this element by including the co-operation between specialized workers. Examples of dedicated modules are: banks of digital signal processors (e.g. for standard low level vision), reconfigurable networks of powerful processors and neural networks (e.g. for intermediate level vision), hosted in a main frame computer (e.g. to accomplish the interpretation phase).

The I/O of the visual information represents still a bottleneck. For this reason, efforts are done to design new chips for a fast acquisition. For example, smart retinas, as input for the vision system of a robot, seem to be one of the step toward the solution of the gap existing between the acquisition and the analysis phases. The main idea is to realize an opto-electronic layer based on a non-standard analogic/digital converter, that includes some pre-processing facilities (filtering, edge detection, thresholding, light adaptation, selection of *relevant* image zone).

Finally, vision systems whether human or artificial:
- *interact with the environment (collecting data from, and acting on the external word);*
- *perform computation (by using algorithms and internal models);*
- *modify internal status (algorithms, and models).*

Information (data, algorithms, methods, models) are inside a visual system like particles in a gas. The interactions are represented by the relations between different kind of information. The machine-architecture represents the constraints, and the "goal" the external fields of such "particle gas". The design of an artificial visual system should by guided by such factors in order to decrease its "entropy", and to reach a final status, which represents a satisfactory solution.

THE PSYCHOLOGISTS POINT OF VIEW (M. Zanforlin)

The problem so clearly illustrated by Vito Di Gesù from the point of view of A.I. research are quite complex and arise fundamental questions for psychologists too. As a comparative psychologist I have different perspective of the problem, but I will touch few points only.

As regard the distinction between anthropomorphic and practical approaches to visual perception, it does not appear to me a fruitful distinction. If by *anthropomorphic* we are referring to a general purpose visual system and to the human visual system as its model, then we have to take into account that we do not have a single model of the human visual system. At the moment there are various theories of visual perception; e.g. Gelstat, Helmholtz, Cognitivism, Gibson, etc., just to mention the main ones. And every theory imply also different approaches and methods to its study.

Moreover, while for some theories the term *model* can properly be used because they postulate a general working principle more or less formalized, others subdivide the perceptual process in stages with different working principles intervening at different stages. For example, Ramachandran maintains that the visual system is a *collection of tricks* that give workable solutions that may not be mathematically correct. To survive is enough. If by *practical model* of the visual system we imply a specific or limited function system then the Ramachandran conception of the human visual system may be considered a *collection of practical models*.

But the human visual system is just one of a large variety of biological visual systems with large differences in complexity. These systems are often very specialized for particular and specific functions.

It appears to me that the crucial distinction between biological and practical or artificial systems, lies in the way they solve the visual problems, posed by particular tasks, regardless of whether they are *simple* and specialized, or *complex* and general purpose.

I am convinced that the human visual system is not a collection of tricks but follows a general principle in resolving the various perceptual tasks. If this is the case, then this general principle can be mathematically described and artificially implemented. It is also possible that the same general principle may work even in simple specialized systems.

It may turn out, that even the various *tricks* that the human visual system appears to use in dealing with different tasks, are just various aspects of a general principle. For example Marr's rules for the *primal sketch*, which are very similar to Wertheimer's laws can be derived from a single general principle of *minimum differences*.

On the basis of these considerations a question arise: *do all the biological systems, whether simple or complex, of general purpose or specialized, work on the same principle?*

I think that a possibility of answering this question is given by the visual illusions. Visual illusions can be considered as non correspondences between the physical reality and what we perceive. In the same way that in some old pocket calculators when one pressed 2x2 obtained 3.999 instead of 4, and one could say that, although this is a good approximation, the answer is not correct. By analogy, the approximation or the correspondence wed the exact value, may be considered an *arithmetical illusion*. From my point of view the approximated answer of the calculator is more interesting than the correct one, because it suggest *haw the calculator works*. In the same way visual illusions may be used to formulate hypothesis on how the visual system works. But still more interesting is the consideration that if two different perceptual systems present the same visual illusion there is a high probability that they work in the same way.

For example, it is well known that in evaluating the speed of various objects the human subjects tend to perceive as more rapid small objects than large ones that move at the same physical speed.

The perceiving speed does not depend only on the size of the moving object but also on the level of illumination and the type of background against which the object moves. It is possible to test the visual perception of speed in the fly by measuring the threshold of its *landing reaction*. The fly shows this reaction when the expansion pattern of an approaching surface reaches a certain speed; the fly pushes forward the front legs, that are retracted during flight, and modifies the wing beat. In this way it has been possible to demonstrate that the landing reaction threshold (i.e. the perceived speed of the fly) depends not only on the physical speed of the pattern but also on the figural characteristics of the pattern and its level of illumination, in the same way as in the human subjects.

Many other similar examples from other animals can be quoted concerning the perception of form, depth, etc., in all cases *visual illusions* or discrepancies between physical and perceived values can be observed.

Now a further problem may be raised: *are these visual illusions really due to the way the visual information is elaborated by the system, i.e. the way it works, or are they due to the fact that the biological material (the neurons) have properties that are different from the electronic gadgets in which the same working principle can be implemented? In other words, given a biological system and an electronic one that works on the same principle, will they show the same visual illusions?*

My answer is yes. I think that the discrepancies, or visual illusions, depend on the working principles and not on the material used for its implementation. My conviction derives from the fact that many perceptual problems, as for example the extraction depth from motion, are of the type called *hill posed*, because they have no unique solution. But it is possible to obtain an unique solution if some *constraints* (usually these are mini-max methods or some *assumption* on extremal values) are imposed.

These constraints give solutions that are not always exactly correspondent to the form or the metric values of the real object. But the evidence that I have collected so far is rather scanty and the problem is still open.

THE POINT OF VIEW OF THE BIOPHYSICIST (S. Vallerga)

Importance of Hierarchy in Natural Visual Systems

From the study of the phylogeny of the visual system we learn that the circuitry of visual systems has developed in responses to the need of processing visual features of increasing complexity, from light intensity to complex patters motion.

The elements of the network sub serve different purposes. At each phase one feature of the visual scene is extracted. A single photoreceptor is enough to tell light intensity, diverse photoreceptors are needed to detect colours, image forming requires second order neurones and motion detection is performed in parallel as well as sequentially, and feedback circuitry tune the whole process. Modules add up to increase the general efficiency of the system and adapt the new habitats. Thus modularity is apparently more important than hierarchy in a natural visual system.

The diversity of natural visual systems reflects the extraordinary diversity of animal life. There is the single scanning light-detector of tiny marine copepod; spiders and scorpions have up to four different systems co-operating to produce the visual image; the eyes of insects see colours and movements inaccessible to the human eye; in snakes infrared sensing organs form heat-sensitive images to help visual perception; lizards have both foveated retina and ancestral third eye on the top of their head; human eyes scan the details of visual world. Differences are qualitative, in terms of how many parameters are extracted from the visual scene. Thus there is difference in design as well as difference in complexity for a given system. There is no a unique model for vision. All that is common to all visual systems is that they are light perceivers.

An Example of a Non-Human Visual System

Artificial vision seems to suffer from too much anthropomorphism and too often artificial vision and artificial intelligence synonym. Machine might have to perform visual tasks for which human vision is not always the best system. Nature has provided a large number of examples of visual systems fitted for different tasks.

Spiders have far better vision than thought by old vision theories. Nature worked on that system for over three hundred millions of years, while it got the chance to improve human vision for only about one million of years.

Spiders have a form of colour vision, detect UV and polarized light and can see the whole visual scene around them with four pairs of eyes. The frontal eye shave the direct retina (photoreceptors facing incoming light) of invertebrates and the lateral, medio-dorsal and dorsal eyes have the reverse retina of vertebrates. Each pair of eyes is devoted to a specific task: the frontal pair to image formation and detection of polarized light, the lateral pairs to peripheral vision, distance estimation and motion detection, extremely interesting and almost unknown in the integration of such signal.

The spider's eyes have fixed lenses, but the image is perceived clearly moving the retina back and forth. The retina can also be moved sideways increasing greatly the field of vision.

A spider can see clearly up to 40 cm, the best image is at 8-10 cm. The lens of the frontal eyes of a nocturnal species of net casting spiders has a f-stop of 0.58 and is totally free of spherical and chromatic aberration. It is composed of concentric material whose refractive index decreases from centre to periphery with such a precise gradient that all rays entering the lens are brought into focus at one point. This eye can absorb 2000 times as many photons than a human eye. On the whole it is quite a remarkable visual system and possibly also a good model for machine vision.

Optical Illusions and Camouflage

Visual illusions, quoted as possible proof for a unique working principle for vision, are not important for the physiology of visual systems. Vision was devised to support survival, namely feeding, escape from predators, reproduction. Images perceived ambiguously are not important for survivals. Visual illusions are far from being common in nature (I cannot think of a single case) they are mainly man-made tricks to test the visual system in extreme situations. Escher's drawings are not terribly important for our life. It is most likely that if ambiguous images were important for survival of creatures, nature would have designed tools for disentangle uncertainty.

Camouflage has been presented as an example of visual illusion, but in such case the visual system sees perfectly what is there: a dress to help the animal to melt with its habitat or to pretend to be another creature. There is no alternate perception between two possible choices. Humans commonly use camouflage, they change dresses, may use make-up, pretend to be some body else but this does not result in visual illusions.

THE POINT OF VIEW OF THE COMPUTER SCIENTIST (S. Castan)

My presentation has two parts, in the first I discuss about the anthropomorphic approach in computer vision, and in the second part I examine the problem of the choice of a specific architecture.

Has Computer Vision an Anthropomorphic or Pragmatic Approach?

To try to answer at this question , I will examine two main problems in Computer Vision: the 3D primitives and motion detection, and image understanding approach. In 3D primitives extraction we can select three main approaches:
• *passive stereo vision;*
• *structured light;*
• *active vision.*
In passive stereo vision, the most popular approach is binocular. In such way the choice of two cameras is anthropomorphic, but the choice of features to be matched, and the strategies of matching are purely pragmatic. We can notice that some researchers try to match with neural networks, and in that particular case could we say that is an anthropomorphic approach? In axial stereo vision, the use of only one camera with a multi focal len (zoom), is pragmatic. And in the same way the use of three cameras for avoiding the false target problem is pragmatic too.

The structured light approach consists of projecting some laser beams or plans onto the surface to be analysed, or projecting some grid having a specific design in order to parametrise directly the surface from few 3D curves. This approach is pragmatic too.

Active vision uses an anthropomorphic head, provided of ocular reflexes system such as Iris, of focus and vergence for gaze control, to command a robot arm by visual servant, without any calibration. To perform such task the anthropomorphic head has to have pan, tilt, vergence possibilities with lens motorisation. In that approach the head itself and the result of ocular reflexes are anthropomorphic, but algorithms to perform them are pragmatic.

In motion analysis, a token traker using the temporal covariance model to detect moving points can be seen as an anthropomorphic approach, but features used such as edges or regions detection are obtained in a pragmatic way.

In image understanding approach, it is interesting to point the difference between human and computer vision respect to the two following problems: the first concerns classical optical illusion, difficult to solve for human, becoming an obvious binary image processing job for computer. Conversely the second task consisting to interpret the well known Esher painting, is very easy for human and almost impossible in computer vision.

In conclusion we cannot say if in computer vision the anthropomorphic is better than the pragmatic approach, or vice versa, but these two approaches can be used in a complementary way.

Secondly we can say that computer vision systems are more efficient than human vision each time we have to extract some measurements with a good accuracy. But human vision is much more effective each time we have to extract relative information in size, position, velocity, or, when we have to make any complicated deduction like in the Esher fountain.

How to Choose an Architecture?

The first question could be: why a specific architecture? The two main constraints are the time to obtain a given result, and the volume of data associated to their structure (iconic or symbolic). Volume of data is application area dependant, from 16 KBytes in some very simple robotics applications, up to 83x106 Bytes/sec. in High Definition TV image sequence.

In other way, structure is problem dependant: we have iconic treatment process at pixel level or region level, and symbolic treatment describing the image.

So we can distinguish pixels processing, region processing and symbolic processing, each of them having a specific architecture more suitable than an other.

The pixel level treatment are devoted to SIMD architecture, such architecture able to take into account the spatial position of a given pixel and its fast access to a more or less large neighbourhood.

MIMD approach seems more suitable for region processing; segmentation algorithms runs faster in MIMD mode than in SIMD mode.

To reduce the volume of data, a multi-resolution structure could be suitable. Different organisation of the pyramid have been proposed, combining SIMD and MIMD approaches between the stages of the pyramid.

Let us notice that specialised modules can be used in very specific repetitive works . The high level processing is generally devoted to the High Level Language machine or more conventional processor. Nevertheless we can remark the capability to use SIMD line processor to extract information from a very large data base, using the image memory as a buffer for the data base and a line or column for the knowledge to be extracted. The problem becoming a very simple and very fast algorithmic comparison of a line or column to the image, added with a new organization of data.

So a general vision system could have these different computation paradigms: SIMD, MIMD, specialised, and general purpose processors sharing more or less the same data.

In conclusion to respect the price performance ratio constraint, the choice could be a reconfigurable machine able to switch, partially or completely as well, from one architecture to another, algorithmically driven to obtain a complex processing result in a given short time.

COMMENTS ON VISUAL SYSTEMS (B. Zavidovique)

About pragmatism, to start: nature has got some pragmatism too, although it is not the same as in engineering communities. On the one hand nature does not care about accuracy or precision, it reaches it most often in cross correlating a great enough number of independent and - rather - inaccurately measured phenomena. On the other hand, machines that we build don't need to survive themselves and anyway they don't reproduce at least can we infer from that, they won't have all the human functions, so whatever complex they could be, they are unlikely to reach any leaving complexity. And this is not new: cars and planes are extremely complex and automatized systems today, yet they are not as complex as an animal leg that have the same mobility function and certainly less complex - say difficult to model - than a six-neuron crab.

So, even if there is always some pragmatic choice in any design, pragmatism is not meant to serve ignorance with an alibi. And that probably is what misunderstanding might be all about: we really need to know what the problem is we are trying to solve. We work at understanding this (new?) part of physics, dealing with systems that transform INFORMATION into ACTION (in that respect retrieving some previously stored data or fact from a knowledge base is considered action here too). This is science, and would we have to build devices (dedicated for sake of achievement) it would be only to experiment and know better what systems to build. We shall be too happy if they can be used before we have understood all, but this is not the main goal.

As this is science hopefully there can be several models, and accordingly several theories, one very unlikely being comparable to the other if they are correct. For instance would we say that *rigid wings and body plus speed* is a "better" method than *flexible body and waving wings*. At least if we are to compare should we know the measure which we are trying to optimize and we, engineering people, would better not choose "self reproducibility and survival over centuries" as a measure.

Another difficulty here again was to define the system which will be modelled; and vision cannot be studied in isolation. The questions arose to know whether more sensors would be simplifying the process. One possible answer is that, by all means, we have to consider that the "system" would have a finite power supply. If it were ever autonomous, then adding a sensor should have consequences elsewhere and the final balance is not sure. For instance would human beings have 3 eyes, they probably would have a stiff neck! And probably would one eye be different from the other pair, say less complicated in some way.

We have models from control theory which allow to prove theorems like: for a given scene (distribution of events) one rough sensor adjunct of a fairly general sensor (in terms of the number of variables it can instanciate among all those defining the scene) makes a pair which gathers more information, than a pair of two identical average sensors, respective global generalities of the two pairs and other things being equal. This would not be true if instead of maximizing the information we maximize some event likelihood.

There are many other issues that could be dealt with here and make a difference. For instance, would we say that biological systems are continuous time-ones? If yes, that is a major difference with artificial systems, and there are definitely several models. When systems get complex are there mandatory hierarchies? Then different kinds of interaction between levels make again different models.

Finally, whether it is already existing (nature) or it is to be built (robots), it is worth building models to understand, and that is why even this "visual front end" trial is worth being experimented on as said by Cantoni.

COMMENTS ON ACTIVE VISION (H. Yeshurun)

I would like to comment on the relevance of biological (and specifically human) vision to machine vision. I think that visual psycho physics, that is the functionality of biological systems and their input-output relations is extremely important for computer vision, while their physiology and anatomy is usually less important.

The rationale for this position is based on the assumption that vision is a highly complicated task, and at this stage, there is no point in trying to outperform it by artificial systems. For example, it is becoming more evident that the human visual system is not calculating a complete 3D representation of the environment, and thus, postulating the role of computer vision as a hierarchical system that is supposed to produce such a representation is pointless, and probably impossible to achieve. In most models of stereo vision, there is an effort to overcome the problem of matching of similar texture in the left and right eye. However, humans are probably not solving this problem at all, as is easily experienced by fixating at a textured cloth (which usually results in losing its depth). Thus, it is important to draw the functional borders of biological vision and use them as the ultimate target of computer vision.

Physiology and anatomy of biological systems might be important, but they are in the level of implementation, and it is not sure that the same design principles apply for machines. Foveation, for example, is important, since it means that even the mighty human visual system can not concurrently process all the field of view, and who are we, computer scientists, to even try. But, another popular example, the log polar cortical representation of retinal images, might be misleading. We have adopted it in computer vision due to its scale and rotation invariance. But it might be in the cortex only due to developmental reasons, as it is more economical to specify cortical connections as conformal mappings (less parameters required), and then we are using it for the wrong reasons.

STRUCTURE AND FUNCTION IN THE RETINA

Gian Michele Ratto[1] and Paolo Martini[2]

[1] Istituto di Neurofisiologia del CNR
   Via San Zeno 51, I - 56127 Pisa, Italy
[2] Technobiochip
   Marciana, I - 57100 Livorno, Italy

1. INTRODUCTION

What makes the retina such an appealing object of investigation to neuroscientists? Obviously every student has his own answers, but still we believe all would agree on one point: the retina is comparable in complexity to the cortex itself, and yet it differs in one fundamental aspect - it is an isolated computational unit. We mean by this that its input and output can be known to a good level of accuracy, the input being an image projected by the lens onto the photoreceptors, and the output the electrical activity that can be recorded from the optic nerve. It is therefore possible, at least in principle, to determine precisely its input-output function and study the mechanisms by which this computation is achieved. Indeed this condition very seldom occurs for any later stage of cerebral processing.

Here we shall try to illustrate some key features of retinal architecture that bear some interest in the field of artificial vision, with the hope of clarifying the strategies employed by the natural system in dealing with the world of light and images. We shall first briefly describe the structure of the retina, summarising in a few paragraphs knowledge collected in years of work by many scientists who spent most of their scientific life in the field. To convey some of the ideas that are growing up about the processing that occurs in such structure we will also deal with recent studies that help to illustrate some points we intend to deliver. In doing so we shall mainly focalize on matters nearer to our personal field of research; the references cited will however be wide enough to satisfy those interested in more general topics.

2. A QUICK REVIEW OF RETINAL STRUCTURE

The retina is a sheet of nervous tissue covering the posterior part of the eye (see figure 1), serving the function of interface between the image projected by the lens and the centres devoted to vision.

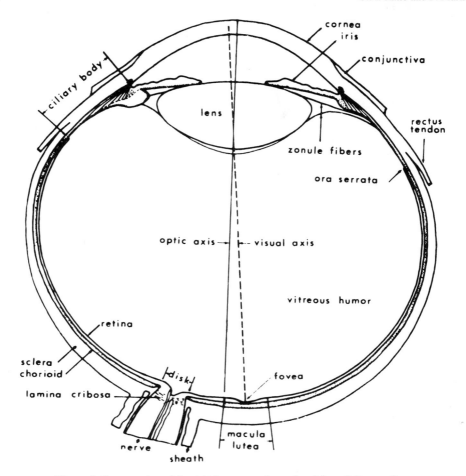

Figure 1. Cross-section of the right human eye (reproduced from Saltzmann[1]).

A closer inspection reveals its very fine structure. The drawing in figure 2 shows how it looks when we stain only a few neurons, but in fact the packing of cells is really extremely dense. You can see in figure 9 how close are the various cellular elements: indeed there is almost no extra cellular space. That the cells are ordered in layers is the first fundamental observation we gain by this simple anatomical analysis. It turns out that the general structure and composition of these layers is substantially identical throughout all the retina. In transversal sections we meet different neurons with different morphological and functional properties, whereas in longitudinal sections no substantial architectonic variation can be appreciated. This structural characteristic suggests that the same processing occurs in every portion of the retina, so that each element of the image undergoes a similar computational analysis. The retina thus reveals itself as an assembly of parallel modules, each composed of a pipeline of processors that deals with a small fraction of the image.

On a strictly qualitative basis intraretinal cells can be divided in two broad classes of elements: narrow cells extending perpendicularly across different retinal layers (bipolar and

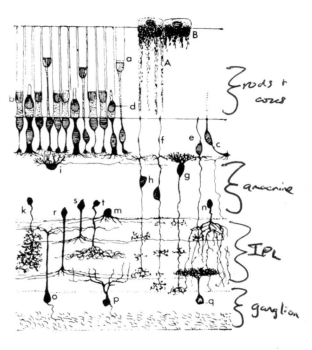

Figure 2. Typical retinal cells of the frog retina. Light enters the retina from below and crosses its entire thickness to be finally absorbed by the photoreceptors (**b** and **a** are rods, **d** is a cone). Other cell types are labelled as follows. **i**: horizontal cell. **h, f, g, e, c**: bipolar cells. **r, s, t, m**: stratified amacrine cells. **k, n**: diffused amacrine cells. **q**: displaced amacrine cell. **o, p**: ganglion cell. The layer between the amacrine and ganglion cells bodies is virtually devoid of cells bodies and it only hosts the bipolar cells terminals and the dendritic trees of amacrine and ganglion cells: it is named the inner plexiform layer (IPL). The scale can be guessed by considering that the total thickness of the retina is about 200 microns (reproduced from Ramón y Cajal[2]).

diffused amacrine cells), and very wide neurons confined in a single layer (horizontal, ganglion and stratified amacrine cells). While the first class of elements is responsible for establishing connections between different layers, the second class instead combines, within a single layer, information originating from different regions of the image. To the extent that performing a computation on an image amounts to applying an operator on spatially localised regions of that image, it is tempting to assume that the narrowly stratified but spatially wide elements are responsible for implementing some sort of computation on their input. The layers

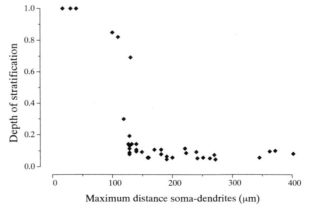

Figure 3. Relationship between stratification depth and dendritic tree width for a sample of amacrine cells in the retina of the sparid fish *Boops boops*. The maximum distance between the cell body and the farthest terminations is plotted against the depth of the stratification expressed as percentage of the IPL total thickness.

in which the arborisation of these neurons are confined thus act as stacks of processing devices. Visual information is fed to these layers and transferred between them by the vertical elements acting as communication units, whose small size would indeed help in maintaining the spatial resolution. A quantitative description of the correlation between lateral spread and vertical extent of dendrites can be found in figure 3. A sample of 38 amacrines fully stained with the Golgi method[3] has been studied. We measured (Ratto and Vallerga, unpublished data) the vertical extent of the processes in the inner plexiform layer (IPL) and plotted it as a function of the size of the dendritic field. There is a clear inverse relationship between field size and vertical extent of the dendrites.

A model in which "narrow" neurons simply transfer information between layers of processing devices (the "wide" neurons) is appealing, but great care should be taken in dividing neural elements in communication and computational devices. We shall see that this division is not at all clear-cut in natural systems: in fact each structure transmitting information also executes a certain amount of computation on it. This is indeed a major difference between the natural and the artificial. In artificial systems communication lines carry information without performing computations of any sort: in other words communication has no semantic

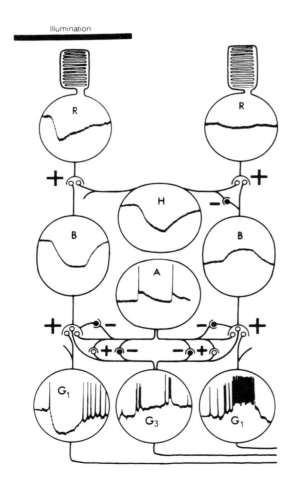

Figure 4. Scheme of synaptic organisation in the mudpuppy retina. Open circles represent excitatory synapses, filled circles are inhibitory synapses. **B** are two hyper polarising bipolar cells, G1 is an off-centre ganglion cell, **A** is an amacrine cell, **R** are two rods (only the left one receives illumination), **H** is a horizontal cell (reproduced from Dowling, 1990, originally published in Dowling[4], with kind permission from J.B. Lippincott Company).

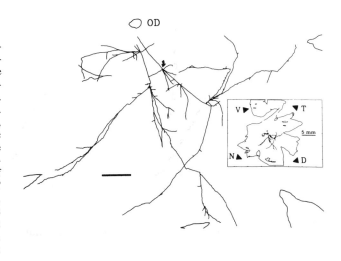

Figure 5. Computer reconstruction of two efferent fibres in a monkey retina. The arrow indicates the bifurcation reproduced in figure 6. The calibration bar is 1 mm. **OD** indicates the optic disk, the site at which the optic nerve leaves the retina. In the inset the two fibres are reproduced within the profile of the retina; **V**, **N**, **D** and **T** indicate the ventral, nasal, dorsal and temporal regions of the retina (reproduced from Usai et al.[12], Copyright (1991), with kind permission from John Wiley & Sons Inc.).

value. In natural systems, however, information carried over a communication line is always subject to some degree of computational processing (see section 4).

The electrical activity of retinal neurons becomes progressively more elaborate as we move from the photoreceptors (the entry point to the retina) towards the ganglion cells (the output), reflecting the ongoing computations that take place along the path. Figure 4 gives a highly simplified scheme of the electrical activity of some retinal neurons.

Photoreceptors respond to light with a hyper-polarisation whose amplitude depends mainly on the intensity of the light absorbed, and in a much lesser extent on the activity of neighbouring photoreceptors through a certain degree of electrical coupling. Horizontal cells respond to light with a sustained hyper-polarisation and their receptive field is much larger than their dendritic tree size. Bipolar cells respond to light projected in the centre of their receptive field either with a depolarisation (ON-centre) or a hyper-polarisation (OFF-centre). Amacrine cells often display a spiking behaviour; the computation they perform on the signal is complex enough to give rise to a class of amacrine cells exhibiting orientation-selective responses[5]. Ganglion cells can be roughly divided in sustained and transient types. The former type functions as a kind of simple integrator, thus acting as a sort of luminance detector, while the latter is very good in returning the temporal derivative of the light stimulus. This scheme is obviously very simplified and incomplete (for reviews see Kaplan, Lee, and Shapley[6]; Sakai and Naka[7]) and is given here with the sole scope of conveying the idea that the functional properties of retinal neurons, as seen when we move vertically from the photoreceptors to the ganglion cells, become progressively more elaborate as the extent of the computation performed on the visual signal increases.

3. FEEDBACK TO THE RETINA

The retina receives a retrograde input from cortical centres through a system of fibres that spread widely within its layers. This system has been well described in avian, fish and amphibian retinae[8-11] but its role and relevance in mammals are still somewhat obscure. We have recently offered evidences for the existence of such a system in the monkey retina[12].

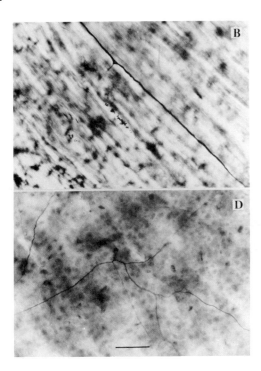

Figure 6. Microphotograph showing the morphology of the bifurcation indicated by the arrow in figure 5. The main process of the fibre runs horizontally over the ganglion cells body close to the surface of the retina (panel B). The very thin process detaches at the bifurcation and plunges in the retina, where it reaches the IPL and branches in a narrow plane (panel D). It is worth mentioning that at this magnification (400X) the depth of focus of the photographs is just a few microns. Bar = 50 μm (reproduced from Usai et al.[12], Copyright (1991), with kind permission from John Wiley & Sons Inc.).

Figure 5 is a computer reconstruction[13] of two such fibres. They enter the eye together with the optic nerve and cover a large fraction of the retinal surface. The main process runs for about 4 mm and generates at intervals secondary branches that eventually bifurcate within the IPL. These centrifugal fibres do satisfy the layering principle we have outlined above: being very wide they stratify narrowly. Figure 6 shows an enlargement of the field indicated by the arrow in figure 5.

The root branch of the fibre is very thick and runs horizontally just over the ganglion cells layer. The thin processes, originating at the bifurcation points, drop within the retinal layers giving rise to extended arborisations very narrowly stratified in the IPL (lower panel). The role of these efferent fibres in mammals is still unknown; in avian and amphibian species it has been observed that their activity modulates the responsivity of ganglion cells[14, 15].

4. ON THE FUNCTION-FORM RELATIONSHIP IN RETINAL NEURONS

It can easily be seen from figure 6 that the diameter of the secondary processes is much smaller than the diameter of the main branch (2.1 μm *vs.* 0.3 μm). This calibre difference has functional consequences, since the electrical resistance met by incoming signals depends on the process diameter. It is therefore reasonable to think that the distribution of an incoming spike train beyond the bifurcation point is going to be influenced by the different impedance of the two branches. This fact has important consequences in terms of signal conduction. For a long time it has been recognised that the nervous impulses are not diffused equally in an axonal termination, but there are factors that can modulate the spatial and temporal propagation of

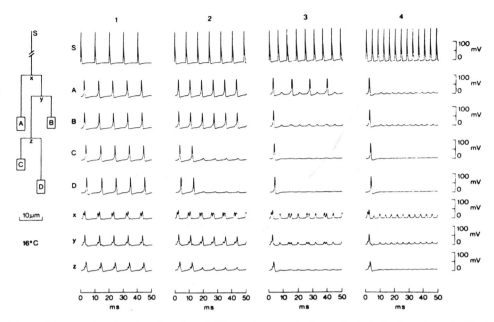

Figure 7. Propagation of a train of impulses of different frequency in the terminal arborisation schematised in the leftmost diagram. S indicates the train as fed at the entrance. A, B, C and D are four synaptic boutons. The potentials in S, A, B, C, and D are plotted in the diagrams labelled 1, 2, 3, 4. The inter- spike period decreases from 10 ms (column 1) down to 8, 6 and 4 ms. At different frequencies different boutons receive different patterns of spike activity (reproduced from Lüsher and Shiner[19], with kind permission from Biophysical Society).

action potentials[16]. In a number of papers it has been shown that the capability of a train of action potentials to propagate in an axonal arborisation depends critically on its geometry, on the electrophysiological properties of the membrane and on the temporal characteristics of the spike train. In other words the arborisation does not behave as a loyal messenger, distributing everywhere a trustworthy copy of the signal: rather different *versions* of the same signal do arrive at synapses located in different parts of the arborisation[17, 18]. Bifurcation points thus act as frequency dependent spatial filters, and this property could be used to control the distribution of information in large arborisations connecting many neurons. Figure 7 shows how the spike propagation to the synaptic boutons depends both on the pulses frequency and on the morphology of the termination[19].

Most of these studies are based on simulations performed on equivalent circuits. An equivalent circuit simulates the functioning of a small compartment of the neural component under study; many of such elementary compartments assembled together constitute the entire neuron. Figure 8 shows how a complex dendritic tree can be reconstructed by connecting several such elements. Each element represents a cylindrical segment of the fibre, and its membrane is modelled as a leaky isolator with a certain capacity. This model correctly describes the passive potential propagation along a cylindrical fibre[20].

Using the same model a current (the synaptic input) can be injected at a given location and the resulting post-synaptic potential be propagated within the neuron. This computation is usually achieved by means of a general-purpose tool for the study of circuits (SPICE is the best known of such programs[22, 23]). The study in figure 7 was performed in this way, solving with SPICE an equivalent circuit modelling the axon active membrane.

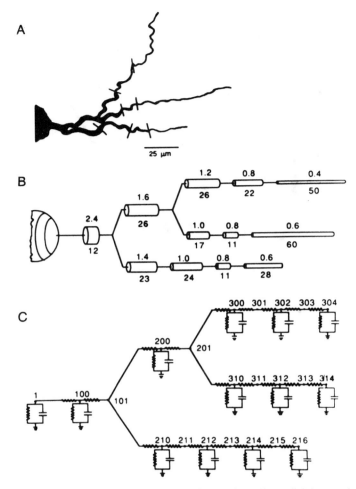

Figure 8. A complete dendritic tree is reconstructed by connecting in series and in parallel elements that emulate simple cylinders of uniform diameter. The upper panel shows the actual morphology; the middle panel is the conversion in compartments of uniform diameter; the lower panel is the equivalent circuit (reproduced from Segev et al.[21], with kind permission from MIT Press).

We now come to a specific example showing the strict interaction between function and form in retinal neurons. Figure 9 shows synapses in the IPL among four different amacrine cells.

Two of these cells form a reciprocal synapse, showing that amacrines can generate a signal that is processed locally and reverberated in the network; such a circuit is localised in a very small region of space. Since amacrine cells are axonless, all their output is generated locally in the dendrites. A typical amacrine cell can implement many of such circuits and an interesting question is to what extent these local circuits are functionally independent one from the other. Independence is achieved if there is a limited propagation of post-synaptic potentials between different circuits.

The amacrine cell shown in figure 10 is an interesting case. These cells are characterised by having long and very thin (0.3 μm or less) dendrites, with frequent varicosities 1 to 2 μm

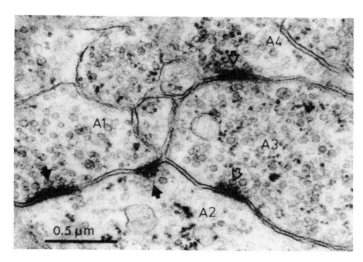

Figure 9. Electron micrograph showing serial and reciprocal synapses (indicated by arrows) between different amacrine cells. Filled arrows indicate reciprocal synapses: A_1 talks to A_2 with the synapse on the left and it receives a feedback through the synapse on the right. A_4, A_3, A_2, A_1 are serially connected (modified from Dowling[24]).

in diameter. The modelling of such morphology shows (see figure 11) that the varicosities contribute to decreasing the spread of the signal along the dendrite[25]. Furthermore, the huge load associated with the cell body effectively isolates the various dendrites, which are therefore capable of operating as separate units. It is very difficult to evaluate the amount of uncoupling present on the dendrite itself without a better knowledge of the membrane properties, and without an actual measure of the amount of leakage existing on the dendritic membrane. However, other things being equal varicose dendrites are certainly much less equipotential than a smooth dendrite. It is of course possible to obtain a similar attenuation by using a much thinner fibre (with possible problems arising from the necessity of storing intracellular structures), but this result would be obtained at the expense of a much higher input impedance.

The peculiar architecture of these amacrine cells thus accomplishes some very interesting goals. A great number of independent functional units can be densely packed in a single cell, saving up space and metabolic energy; this can be achieved because the large varicosities decrease the input impedance, and together with the thin dendrites make for a large electrotonic attenuation.

5. THE SIMPLEST RETINAL NEURON: THE PHOTORECEPTOR

We conclude our fragmentary exploration of the retina by giving now a closer look at the physiology of the best understood neuron of all: the photoreceptor. This cell is less clever at hiding its secrets than its more elusive colleagues: being easily accessible it is relatively easy to record its electrical activity, but most important, it offers the extraordinary opportunity of studying the electrical output of a system whose input (the light stimuli) is perfectly defined.

Figure 12 depicts a number of photoreceptors taken from the retinae of various vertebrates specie: you can see that all share a common structure. We can roughly divide it into

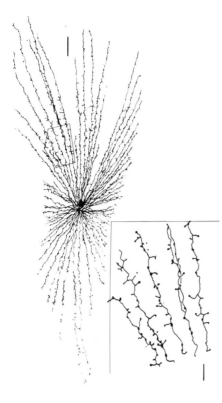

Figure 10. Radiate amacrine cell showing the characteristic radiate morphology and the dendritic varicosities. The inset shows, on a larger scale, some of the many varicosities populating the dendrites. The calibration bars are 50 and 10 μm (modified from Vallerga and Deplano[3]).

three compartments: the outer segment, the inner segment and the synaptic terminal. The outer segment contains the pigment molecules and the enzymatic machinery used for generating the photo-response. The inner segment provides the engine responsible for maintaining the homeostasis of the entire cell. Finally, the synaptic ending is the interface to the rest of the retina. A photoreceptor represents at most a single pixel of the image and the computation performed by this neuron precedes any further retinal elaboration. Given the amount of processing occurring between the photoreceptors and the retinal output one could speculate that its computational properties would somehow be hidden by elaboration performed within other retinal layers. Instead in a number of occasions it has been shown that the photoreceptors play a clear role in shaping the overall performance of the visual system, often acting as the limiting factor of such performance. In the remaining of this chapter we shall illustrate how the sensitivity of a particular class of photoreceptors, the rods, affects the overall visual sensitivity. Rods are best suited to operate in dim light and are responsible for most of our visual performance in conditions of low illumination. Their morphological and physiological organisation shows a remarkable degree of dedication to the task of detecting feeble light stimuli. All rods have the same spectral sensitivity and night vision is therefore strictly monochromatic. Since colour vision depends finally on the activity of subtractive mechanisms that reduce the efficiency of the system as a light detector, the loss of colour discrimination in rod vision is one of the prices paid in the quest for a high sensitivity. As a matter of fact rod sensitivity is really exquisite. In very dim light rods show sizeable responses to the absorption of a single photon. Figure 13 shows the responses of a rat photoreceptor

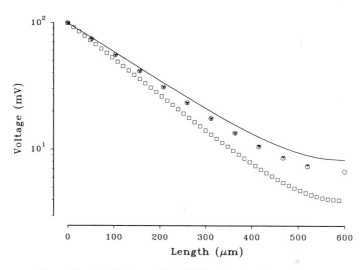

Figure 11. Voltage spread in a uniform dendrite (smooth line) of 600 μm length and 0.2 μm diameter. A post synaptic voltage of 100 mV was applied at the left extreme. Voltage spread in a dendrite carrying 11 varicosities of 1.2 μm diameter and 1.8 μm length (circles and triangles). Voltage spread in a dendrite carrying 50 varicosities (squares). Varicosities increase the voltage attenuation; similar computations show that they also decrease the input impedance.

elicited by flashes of light that determined on average the absorption of a single photon[27].

In these conditions the behaviour of the rod is dominated by shot noise: the number of photons absorbed varies around an average value following a Poisson statistic[28,29]. The response size is quantified: the photoreceptor gives a stereotyped response of fixed size to every single photon absorbed, and different sizes correspond to different numbers of absorbed quanta. As the light intensity grows, so does the response amplitude, eventually reaching a saturation level (see the two largest traces in figure 15). The intensity-response relationship follows an exponential law[30].

A striking demonstration of the limitations imposed by the photoreceptor sensitivity on the visual system performance can be seen during the mammalian post-natal development. Sensitivity to light is low in neonatal mammals compared with the adult: it gradually reaches adult levels within a few weeks or months from birth according to the specie. Figure 14 shows behavioural data collected on a human adult and an infant[31].

In this experiment a stimulus covering a large part of the visual field is presented to the subject for a given amount of time (plotted on the horizontal axis). The stimulus luminance is reduced until the threshold luminance (at which the stimulus is just detectable) is found. The plot shows that dimmer and dimmer light intensities become detectable as the duration of the stimulus is increased, until an absolute threshold is met (the flat part of the curve). There are many interesting questions opened by the results of this experiment, but suffice to notice that for every stimulus duration the threshold light intensity is always much higher in the infant subject compared to the adult. This implies a lower light sensitivity of the new-born visual system: at the age of the subject (2 months) the difference in sensitivity is about a factor of 50. We shall see that the neonatal low light-sensitivity can be attributed to a lower light sensitivity of the immature photoreceptors. Figure 15 shows light responses from rat rod photoreceptors obtained from two animals at different ages.

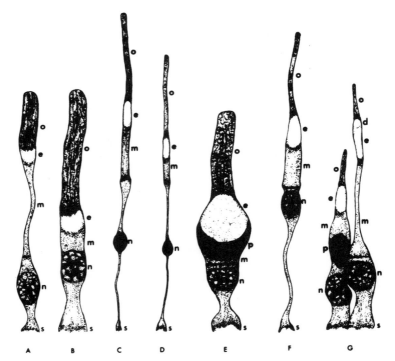

Figure 12. Photoreceptors from different vertebrate species. **A, B** green and red rod from the frog; **C**, human rod; **D**, rat rod; **E**, lizard rod (*Gecko gecko*); **F**, human cone; **G**, frog double cone. **o** is the outer segment; **e** and **m** together constitute the inner segment; **s**, is the synaptic body. Cells are drawn to scale; for comparison the human rod diameter is about 2 μm (reproduced from Young[26], with kind permission from University of California Press).

Both responses have the same shape, but the brightness of the stimuli employed to elicit the response is very different in the two cases: the saturating flashes in the older and younger receptors are respectively 7600 and 44000 photons per flash μm^{-2}. As a further example the two responses marked with a diamond are elicited by stimuli of the same intensity of 1150 photons per flash μm^2. In this case the younger photoreceptor is about 80 times less sensitive than the average adult rod.

6. CONCLUSIONS

What we have said in this chapter can be summarised in the following way. The retina is an assembly of parallel modules, each calculating some possibly complex computation on a small fragment of the image; this computational analysis is performed by neurons narrowly stratified across the retinal layers; the image characteristics they extract become progressively more elaborate as we move from the receptors layer towards the ganglion cells layer; at the output level (ganglion cells) different channels exist for each module that convey diverse image quantities to further visual centres. A feedback control from cortical centres reaches the retina and probably modulates the responsivity of ganglion cells. The membrane properties and the form of each cellular element are crucial determinants of its function[33], the amacrine cells being

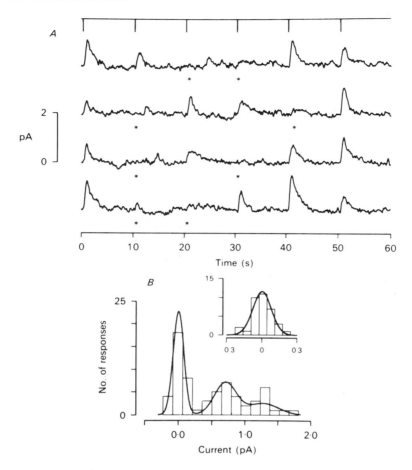

Figure 13. Current responses from an albino rat rod to 24 presentations of a dim flash (3.5 photons μm^{-2}; panel A). The uppermost trace indicates the light flashes. The histogram (panel B) shows the amplitude distribution. The continuous line has been computed assuming a Poisson distribution typical of shot noise (modified from Robinson et al.[27]).

a striking example of such concept. In natural systems like the retina information transport always has a semantic value: this can be seen by considering how the form and diameter of the fibre affect signal conduction in neuronal arborisations. Every stage of computational process-ing in the retina can shape the performance of the entire visual system: this is true also in the case of the photoreceptor, that is the simplest retinal neuron and where everything originates.

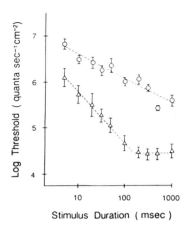

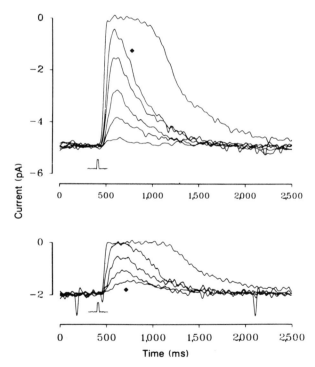

Figure 14. Psychophysical temporal summation functions for a 10 weeks old infant (open circles) and for an adult (open triangles) (reproduced from Fulton et al.[31], Copyright (1991), with kind permission from Elsevier Science Ltd).

Figure 15. Light responses collected from rat photoreceptors of 35 and 13 days of age (upper and lower panel, respectively). The light flash (20 ms. long) is delivered at 400 ms. The traces indicated by a diamond are elicited by flashes of identical brightness (reproduced from Ratto et al.[32], Copyright (1991), with kind permission from Macmillan Magazines Ltd).

REFERENCES

1. M. Saltzmann, *The Anatomy and Histology of the Human Eyeball in the Normal State*, Chicago University Press, Chicago, IL (1912).
2. S. Ramón y Cajal, *Histologie du System Nerveux de l'Homme et des Vertébrés*, A. Malsine, Paris, F (1911).
3. S. Vallerga and S. Deplano, Differentiation, extent and layering of amacrine cell dendrites in the retina of a sparid fish, *Proc. R. Soc. Lond. B,* The Royal Society, Vol.221, pp.465-477 (1984).
4. J.E. Dowling, Organisation of vertebrate retinas, *Invest. Ophthalmol.,* J.B. Lippincott Company, Philadelphia, PA, Vol.9, pp.655-680 (1970).
5. S.A. Bloomfield, Two types of orientation-sensitive responses of amacrine cells in the mammalian retina, *Nature,* Vol.350, pp.347-350 (1991).
6. E. Kaplan, B.B. Lee, and R.M. Shapley, New views of primate retinal function, *Progress in Retinal Research,* Vol.9, pp.273-336 (1990).
7. H.M. Sakai and K.I. Naka, Neuron network in catfish retina: 1968-1987, *Progress in Retinal Research,* Vol.7, pp.149-209 (1988).
8. J. Reperant, N.P. Vesselkin, J.P. Rio, T.V. Ermakova, D. Miceli, J. Peyrichoux, and C. Weidner, La voie visuelle centrifuge n'existe-t-elle que chez les oiseaux?, *Rev. Can. Biol.,* Vol.40, pp.29-46 (1981).
9. A.D. Springer, Centrifugal innervation of goldfish retina from ganglion cells of the nervus terminalis, *J. Comp. Neurol.,* Vol.214, pp.404-415 (1983).
10. C.L. Zucker and J.E. Dowling, Centrifugal fibres synapse on dopaminergic interplexiform cells in the teleost retina, *Nature,* Vol.300, pp.166-168 (1987).
11. J.E. Dowling, *The retina: an approchable part of the brain*, Harvard University Press, Cambridge, MA (1987).
12. C. Usai, G.M. Ratto, and S. Bisti, Two systems of branching axons in monkey's retina, *Journal of Comp. Neurology*, John Wiley & Sons Inc., Vol.308, pp.149-161 (1991).
13. G.M. Ratto and C. Usai, Computer aided tracing and encoding of axonal arborisations, *J. Neurosci. Methods,* Vol.36, pp.33-43 (1991).
14. F.A. Miles, Centrifugal control of the avian retina. III Effects of electrical stimulation of the isthmo-optic tract on the receptive field properties of retinal ganglion cells, *Brain Res.*, Vol.48, pp.115-129 (1972).
15. L. Cervetto, P.L. Marchiafava, and E. Pasino, Influence of efferent retinal fibres on responsiveness of ganglion cells to light, *Nature,* Vol.260, pp.56-57 (1976).
16. D.H. Barron and B.H.C. Matthews, Intermittent conduction in the spinal cord, *J. Physiol. (Lond.),* Vol.85, pp.73-103 (1935).
17. N. Stockbridge, Differential conduction at axonal bifurcations. II. Theoretical basis, *J. Neurophysiol. (Bethesda),* Vol.59, pp.1286-1295 (1988).
18. N. Stockbridge and L.L. Stockbridge, Differential conduction at axonal bifurcations. I. Effect of electrotonic length, *J. Neurophysiol. (Bethesda),* Vol.59, pp.1277-1285 (1988).
19. H.R. Lüsher and J.S. Shiner, Simulation of axon potential propagation in complex terminal arborisations, *Biophys. J.,* Biophysical Society, Vol.58, pp.1389-1399 (1990).
20. W. Rall, Core conductor theory and cable properties of neurons, in *Handbook of Physiology: the Nervous System,* Kandel, Brookhardt and Mountcastle eds., Williams and Wilkins Co. Baltimore, Vol.1, pp.39-98 (1977).
21. I. Segev, J.W. Fleshman, and R.E. Burke, Compartmental models of complex neurons, in *Methods in Neuronal Modeling,* Koch and Segev eds., The MIT Press, Cambridge, MA, pp.63-96 (1989).
22. I. Segev, J.W. Fleshman, J.P. Miller, and B. Bunow, Modeling the electrical behavior of anatomically complex neurons using a network analysis program: passive membrane, *Biol. Cybern.,* Vol.53, pp.27-40 (1985).
23. B. Bunow, I. Segev, and J.W. Fleshman, Modeling the electrical properties of anatomically complex neurons using a network analysis program: excitable membrane, *Biol. Cyber.,* Vol.53, pp.41-56 (1985).
24. J.E. Dowling, Synaptic organisation of the frog retina: an electron microscopic analysis comparing the retinas of frog and primates, *Proc. R. Soc. Lond. B,* The Royal Society, Vol.170, pp.205-228 (1968).
25. S.A. Elias and J.K. Stevens, *Brain. Res.,* Vol.196, pp.365-372 (1980).
26. R.W. Young, The organisation of vertebrate photoreceptor cells, in *The Retina: Morphology, Function and Clinical Characteristics,* Straatsma, Hall, Allen and Crescitelli eds., Forum in Medical Sciences 8, University of California Press, Berkeley, CA, pp.177-210 (1969).
27. D.W. Robinson, G.M. Ratto, L. Lagnado, and P.A. McNaughton, Temperature dependence of the light response in rat rods, *J. Physiol. (Lond.),* Vol.462, pp.465-481 (1993).
28. D.A. Baylor, T.D. Lamb, and K.W. Yau, Responses of retinal rods to single photons, *J. Physiol. (Lond.),* Vol.288, pp.613-634 (1979).
29. D.A. Baylor, B.J. Nunn, and J.L. Schnapf, The photocurrent, noise and spectral sensitivity of rods of the monkey Macaca fascicularis, *J. Physiol. (Lond.),* Vol.357, pp.575-607 (1984).

30. T.D. Lamb, P.A. McNaughton, and K.W. Yau, Spatial spread of activation and background desensitization in toad rod outer segments, *J. Physiol. (Lond.),* Vol.263, pp.257-286 (1981).
31. A.B. Fulton, R.M. Hansen, Yuan-Lih Yeh, and C.W. Tyler, Temporal summation in dark-adapted 10-week old infant, *Vision Research*, Elsevier Science Ltd, The Boulevard, Langford Lane, Kidlington 0X5 1GB, UK, Vol.31, pp.1259-1269 (1991).
32. G.M. Ratto, D.W. Robinson, B. Yan, and P.A. McNaughton, Development of the light response in neonatal mammalian rods, *Nature,* Macmillan Magazines Ltd, Vol.351, pp.654-657 (1991).
33. L. Cervetto and G.M. Ratto, Neuronal circuits in living organisms, in *Towards Biochips*, ed. Nicolini, World Scientific Publishing Co., Singapore (1990).

ON-CHIP VISION

Marco Ferretti

Dipartimento di Informatica e Sistemistica, Università di Pavia
Via Abbiategrasso 209, I - 27100 Pavia, Italy

ABSTRACT

This paper reviews the ever increasing activity in smart sensors, initiated in the 80's by Mead's silicon retina design. The gap between the sensors and the processors in a traditional image processing system can be partially reduced by embedding into the same silicon die both the transduction of incoming illumination and its interpretation through some processing function. If we consider the silicon die a "real estate" on which to build the smart sensor, many possible alternatives arise, which either privilege the sensing function or the processing one. On-chip-vision is the domain of integration that tries to balance the available silicon resources between these two extreme approaches. Analog processing is a viable solution for many tasks, though it leads to implementations that lack a general programmability. Application specific solutions, according to the ASIC paradigm, take a more flexible approach in mixing both analog and digital processing, according to the single task embedded in the silicon die. General purpose digital retinas, which combine an array of sensors with a programmable digital cellular array of simple processors, have also been proposed and built; the area trade-off between the sensing function and the near-neighbour connectivity suited to low level image processing is a major issue for further research.

1. INTRODUCTION

The usual architectural framework for image processing and computer vision consists of a set of distinct, though co-operating sub-units: an image acquisition device, a processing sub-system and an interconnection pathway linking the two. A rich variety of solutions exist for each such component. For instance, the acquisition device can be realised with standard Vidicon cameras, featuring interlaced or not-interlaced frame acquisition with serial scanning of the scene, or with CCD based digital cameras, which acquire the image in parallel through an array of photo detectors. As to the processing sub-system, the alternatives are even richer and more distinct in design: at one extreme, a single specialised chip embeds into silicon the sole function to be carried out on the incoming image (detection and localisation of an object for industrial inspection purposes); at the opposite extreme, a huge massively parallel processor, such as a Thinking Machines CM-5 or an INTEL Paragon system, is set up to carry out extremely complex, general purpose computations. The interconnection unit can range anywhere between a high-speed video bus, typical of image processing systems composed of a frame grabber and dedicated processing boards, and the ad-hoc parallel pathways designed for arrays processors such as the well-known CLIP4, MPP, and the other massively parallel systems designed and built in the 80's.

This rough characterization of the architectural framework highlights a feature common to all systems: the acquisition and the processing of images are kept separate and comply with the distinct technology limits of the respective sub-systems. The term "on-chip vision" denotes an architecture that, to a certain degree, abolishes this gap and that embeds onto the same substrate, a silicon die, both the functions of a tradition camera and the facilities of one of the targeted processing sub-units. This paper analyses the factors that constrain such an integrated solution and reviews some of the realisations so far obtained. The approach followed in this analysis is predominantly an engineering one: silicon technology and VLSI techniques are the major actors which set the ground for what is feasible and worth doing. Still, a brief excursus in the domain of natural vision systems helps in clarifying, by contrast rather than by analogy, some of the problems to be solved in designing and fabricating "on-chip vision" devices.

The periphery of the visual pathways in primates consists of a double set of photoreceptors, organised into two retinas (details on the organisation of the visual system can be found in other chapters of this book). In humans, each retina consists of four types of photoreceptors[1]: rods, for scotopic vision, and three kinds of cones, for photopic vision and colour detection. The number of rods is in the order of 125×10^6, that of cones 5.5×10^6. The layers of retinal cells reduce the complexity of the overall signals by a factor of 10^2, since the optic nerve consists of roughly 10^6 axons from the ganglion cells. This organisation highlights two important aspects: processing of the visual signals starts immediately in the visual pathway, right after signal transduction, and furthermore in a highly parallel sub-system (from this point of view, the irregular, foveated lay-out of photoreceptors is irrelevant); the pathway towards "higher level" processing areas (lateral geniculate body and the cortex) is itself highly parallel.

It is likely that the striate cortex supports many different forms of parallel processing (multiple functions carried out concurrently on extensive sets of signal). Continuous results from neuro-physiology increase the knowledge on the functional specialisations of this part of the visual systems, and the variety of "hardware solutions" in the brain is astonishing. In any case, the overall architecture of the visual systems does not exhibit the tremendous mismatch between sensing and processing that is typical of artificial systems.

Indeed, on the sensing side, the latest advances in artificial light sensors have led to cameras capable of 10^6 photoreceptors on a single silicon substrate[2,3]. On the processing side, the fine-grained parallel systems for vision have been designed for a maximum of 10^5 simple processing elements (PEs in the following), arranged into chips containing each at the most 10^2 such PEs. The interconnection between the two sub-system is however a dramatic bottleneck (see figure 1): for one thing, the light signal, though acquired in parallel, is extracted from the camera serially, or at the most in column (row) parallel way; the buses for the connection, even in the specially designed system (such as the MPP, CLIP4, and CM-2), can sustain roughly 10^2 links. Overall, the usual image processing system is completely un-balanced.

On-chip" vision is the attempt to bridge this gap. If we consider a silicon die of roughly one inch diagonal as the "real estate" to be used, two extreme approaches are possible (see figure 2): the first exploits this resource for the sensing function only, leading to silicon cameras of the above mentioned capacity; the second allocates silicon to processing only, resulting in chips with a highly packed array of PEs. All intermediate approaches try to balance the use of the

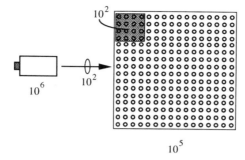

Figure 1. The bandwith bottleneck of standard computer vision systems.

resource for both functions. This is the domain of on-chip vision. We can further detail the design choices according to many criteria: technology (CDD vs. CMOS), processing mode (analog vs. digital), and finally processing scope (a single function embedded together with the sensing, in a so-called ASIC approach, or a general purpose solution, that allows one to embed many simple operations with a software approach). This paper covers the various alternatives and does not commit to a definition of on-chip-vision as the all-analog (or all-digital) approach to image processing within a chip. An analogous review is available in Zavidovique and Bernard[4], with an extensive bibliography on many detailed solutions not covered here.

2. THE TECHNOLOGY

The physical principles of light transduction through semiconductors materials are long well known. Electron-hole pairs are generated by incident photons with an efficiency that depends on the energy band of the semiconductor and on its crystalline structure.

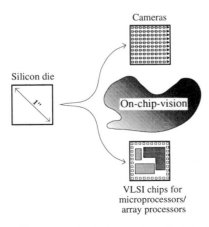

Figure 2. The silicon die as the design paradigm for "On-chip-vision".

The primary mechanism for detecting photons[5] is the use of a bi-polar arrangement of differently doped silicon bars: an electric field across a reverse-biased junction collects the photo-generated charge. On this scheme, the photo-diode and the photo-transistor have been properly used: the increase in the reverse saturation current is linear in the number of incident photons, that is in the incident energy. The gain in current in the photo-transistor is at the expense of larger silicon area and of a slightly longer response time; it is not a mandatory requirement for photo-detection.

The increase in carriers due to photons in a uniform silicon substrate causes a decrease in resistivity: the photoconductor is a device whose conductivity is modulated by incident light. Area and response time are critical, and for this reason photoconductors are never used for building arrays of photodetectors. Yet, there are proposals that hint to specific applications for which this approach could be useful; one of these proposals[6] will be described briefly later.

The range of incident light for which the electronic devices should be capable of a response, to be comparable with the performance of photoreceptors in humans, is very wide: at a wavelength of $0.55\mu m$, direct light from the sun produces 10^5 lux, while a cloudy night only 10^{-4}. The usual linearity assumption typical of small signal systems cannot be pursued. The visual system in humans compresses this enormous dynamic range through a non linear response, which begins with a chemical saturation effect in the "whitening" of pigmented proteins in photoreceptors, and is re-inforced by the centre-surround antagonism of the receptive fields of the first layers in the retina (contiguous photoreceptors measure contrast rather than absolute difference). A non linear response is thus required in silicon transducers as well: the photodiode and the phototransistor linear response to photo-generation must be followed by a controlled non-linearity, to produce usable signals, either for sub-sequent analog processing, or for straight analog-to-digital conversion. The "sub-threshold" operation of MOS transistors[7] offers such a non-linearity: when operated with gate voltage within a half volt range, the transistor yields a saturation current which is exponential on the gate voltage itself. We will analyse later the role of analog processing in setting up both linear and non linear functions for signal conditioning and processing. Figure 3a shows the physical set-up of a phototransistor [8]. Incident photons generate carriers at the junction formed by the P-doped substrate and the N-doped source. These electrons cross the transistor channel and are drained out. The effect of the connection between the gate and the source is to place the transistor in the sub-threshold region, thus realising the logarithmic dependence of the drain-to-source voltage on the photogenerated current. The schematic in figure 3b shows the correspondence between the physical arrangement and the circuit symbols. The reverse biased junction is described as a photo-diode. This scheme differs from Mead's photodetector, since it merges in a single device three effects: light transduction, current gain due to the transistor, and sub-threshold non linearity. Solutions based on distinct devices are more common, as they optimize the three effects separately.

The principle of using a reverse biased junction for light detection is common to both families of silicon technologies: CCD and CMOS. After this stage, silicon area is used in remarkably different ways in either case: CCD has long established itself as the main technique for building very large arrays of photo-detectors for cameras. Charge storage and transfer along row-oriented registers, with proper sequences of clock signals, is best suited to the row-major scanning scheme of commercial TV. The chips containing a CCD area sensor are based on the same principle: transduced signal is extracted serially. Nethertheless, the advances in fabrication have been impressive. Recent disclosures from companies in consumer electronics[2,3] revealed chips with an 1-inch image format (14.0 mm height x 7.9 mm width), assembling up to 1920 x 1036 pixels (7.3 μm x 7.6 μm each); both sensitivity (60-80 nA/lux) and dynamic range (70-90 dB) are remarkable. There are even more compact 2D-sensors[9], that cram in a 1/3-inch area 4×10^5 pixels, each 6.4 x 7.5 μm^2 large.

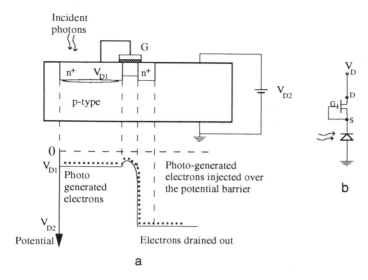

Figure 3. The phototransistor. a) A cross-section view of the silicon layers that make up the reverse-biased junction responsible for the separation of photogenerated carriers and the channel for injecting electrons over the potential barrier. b) The equivalent circuit.

In CCD device, charge transfer efficiency[10] is a critical parameter: charge due to transduced light must be preserved during its movement along the shifting register. Values in the range of 99.997% per stage are usual. A more critic issue is "smear reduction". At the small geometry of the sensing area, light diffraction at the air-silicon oxide transition, and leakage due to oblique incidence cannot be ignored. On-chip lens have been successfully integrated on top of the sensing array, so that incident light is concentrated on the sensing element. Finally, we quote a third physical parameter: charge storage capacity is limited. Under high incident light density, a blooming effect ensues, which saturates a larger area of the sensor. Antiblooming protection is best realised in CMOS technology[11].

CCD cameras can be built for colour image acquisition directly at the level of the sensor. On-chip micro filters can be laid out on top of the sensing array with various topologies. Green, red and blue filters are arranged in 2-D sensors according to interline, Bayer and more specialized (GSFS) geometries[12]; linear sensors use either three linear arrays one for each colour with three registers, or a single array with micro colour filters alternating along the sensors and a single register. There are also proposals for reducing colour mixing[13].

We conclude this brief analysis of silicon devices optimised for the sensing function by noting that the lay-out of the pixels and of colour filters in the sensor is highly regular. This due to the structure of row-oriented (interlaced or not interlaced) signal required by commercial TV. The disposition of photoreceptors in the retina of primates is instead far from the 2D-mesh. The density of cones in the fovea and of both rods and cones in the lateral areas changes in non-linear way with eccentricity. The same holds true for the dimension of the receptive fields of retinal cells, so that the sampling of incident signal obeys a more advanced principle. There have been proposals to build silicon sensors that emulate at least the sampling geometry at the photo-receptors level. A foveated retinal-like CCD device[14] uses the silicon area with two distinct topologies: a regularly sampled 2-D array of a hundred of pixels for the fovea and a set of 30 circular arrangements of pixels (64 along each circumference) for the lateral areas. The density and area of the sensors in this zone change with eccentricity: a logarithmic mapping of the coordinates of the sensor elements produces an image plane which is invariant to rotations and scaling. This scheme has advantages in many pattern recognition tasks, and it also embeds in the same device facilities to implement focus-of-attention and peripheral analysis. From a technological point of view, the devices embeds the sensor in a 11x11 mm² silicon die. Read-out of the acquired signal remains serial, and requires a complex clocking scheme (up to 18 different clocks), because of the different length of CCD registers that make up the circumferences. It is also worth noting that the non-linear space sampling function of this device is very different from the analogous effect in the retina of humans: in this one, the aggregation of signal does not happen at the level of transducers, but rather at the sub-sequent stages of processing, carried out by the layers of retinal cells.

3. THE ROLE OF ANALOG PROCESSING

Embedding linear and non linear functions in silicon to perform computations on the signal in its native representation (charge, current, voltage) is the rationale of analog processing. It seems quite useful to process the signal in its native representation, so as to avoid the area and time consuming process of conversion to a digital form. Among the advantages of an all-analog processing mode we can also quote a time-continuous approach to computation: by properly choosing the time constants of the elements in a network, one can guarantee a input-to-output time that suites the requirements of the application.

The efficiency of analog processing in silicon depends primarily on the format of the signal. As we have seen, the primary output of light-matter interaction is charge. CCD technology is very well suited to temporarily storing and to moving charge. There are a few more operations that can be carried out on charge representing signal.

Charge splitting is the most elementary operation[15]. A multiplication by a fraction smaller than one is physically implemented by separating charge packets as they move along the shifting pathway. Field oxide barriers are placed along the CCD channel, thus creating separate pathways: the split ratio determines the multiplicative coefficient, and it depends on the position of the field oxide barrier within the channel. If the barrier is placed within the transfer gate, the split ratio depends on the distance from the channel boundaries; if it is placed in the storage gate, it depends on the fraction of area available to charge on either side of the barrier. The precision of this splitting mechanism is very important for setting up filtering operations.

Charge multiplication is a way to increase the signal transduced by the reverse biased junctions of photo-diodes. It is a relevant function in large CCD area sensors, where the pixel of the photo-receptor is made to shrink to smaller and smaller areas. In multi-phase CCD channels[16], high electric fields are established at proper gates to induce impact ionisation on the photo-generated charges. The electric field is suitably controlled, so as to obtain the desired multiplication ratio.

A more difficult operation in charge domain is charge differencing. Charge packet summation is straightforward, since it consists of physically grouping distinct charge packets into a single area, thus effectively summing charges. A common technique for differencing has been a preliminary charge-to-voltage conversion, with a subsequent voltage differencing operation and voltage-to-charge back conversion. This procedure does change the representation of the signal and is demanding in silicon area. Other techniques[17] are based on direct charge manipulation, that involve the use of pre-charged electrodes for establishing surface potential directly proportional to the required output packet.

The storage of charge for subsequent re-use is a difficult operation in analog processing, since accessing the signal involves a physical (usually destructive) operation. It is questionable whether an analog storage is indeed required for low-level image processing. The neural network paradigm is one case in which such a problem has been tackled. The use of weights for connections requires a structure that is able both of a permanent storage capability and of a flexible updating scheme. EEPROMS[18] have been shown to be a feasible approach to this requirements.

A much wider variety of analog processing modes is available when the light signal is brought to a current/voltage representation. Mead's pioneering work on signal manipulation in the analog domain remains unparalleled. In his book "Analog VLSI and Neural systems"[7], one finds a set of linear and non-linear functions that manipulate currents and voltages according to the required transformations. The main building block for such operations is the transconductance amplifier, that exhibits a current-to-voltage transfer function useful both in linear and non linear processing. Its transconductance in the linear region of operation depends on the bias current; this allows one to built a variable dependence of the output on the input differential voltage and leads to the realisation of the four-quadrant multiplier. Its enhancement to support wide-input signals meanwhile yielding wide-output variations, along with its relative high voltage gain, makes it a "operational amplifier" with an extremely simple control on the transconductance.

Simple current and voltage transforms, such as addition and subtraction, multiplication, absolute value, exponentiation, logarithms and square root are a library of analog circuits for the designer of advanced circuits, embedding in time continuous mode the processing functions on the signal in its native representation. Collecting signals and combining them into aggregated values is fundamental in many signal processing tasks, and all the more in the vision domain. The computation of a mean value is one important case. The voltage follower configuration of the transconductance amplifier allows one to inject voltages into a single aggregating node: the output voltage at such a node is driven towards the mean of the input voltages through the transconductances at each input node.

Electronic spread of currents injected into a ladder of horizontal resistors (R) that are grounded through vertical resistors (1/G) leads to spatial 1-D averaging of signals. The decay of the voltage at each junction between a horizontal and a vertical resistor resulting from a single driving input is fairly well approximated through an exponential function, whose space constant depends only on the ratio between the horizontal and vertical resistance (RG). Such a network allows to compute a spatial convolution by applying the superposition principle: input voltages drive through their transconductance G the horizontal network to output voltages that are the convolution of the inputs with a kernel that depends only on RG.

The extension of such resistive networks to 2D topology is very convenient for image processing operations. Approximate solutions in the case of the hexagonal topology (see figure 6a) are easily computed, and show an exponential decay similar to that of the 1D case. The aggregating property of such a resistive network has been described by Mead as producing "an average that is a nearly ideal way to derive a reference with which local signals can be compared". The silicon implementation of this structure is based on the circuit of the transconductance amplifier for the vertical resistors, and on a active device for the horizontal resistor. A series of two pass transistors, whose gate voltages are biased by the same voltage reference, produces a current-to-voltage dependency that exhibits a linearity region in the range of 200 millivolts: the slope of the curve in this region is set by the bias voltage. The convolution kernel of the 2D hexagon resistive network realised with these circuits is therefore controlled by two voltages, that must be available at each node of the network.

The non-linear behaviour of the horizontal resistor outside the 200 millivolt range introduces a saturating effect that comes handy for signal segmentation. The "resistive fuse"[19] enhances the behaviour of the resistive horizontal arrangements of pass transistors of Mead's implementation by adding a saturating absolute value circuit that turns off the resistor at large voltage differences, hence the "fuse" operation. The importance of this in segmentation is clear. If the signal distribution over a resistive network exhibits abrupt changes, at the discontinuity the horizontal connections (resistive fuses) do not drain current and the voltages are pulled towards the averaged value in the respective regions.

4. PHOTORECEPTOR ARRAYS

While CCD technology yields optimal results when one tries to optimize the area for the photo-transduction function (as is required for TV cameras), it poses tight constraints on the processing modality. The serial mode of extracting the information from the sensor can be coupled with a few image processing operations, such as filtering[20], but it is not suited for general purpose operations, especially those involving near-neighbour pixel oriented operations.

Photoreceptors arrays have been designed in CMOS technology in an attempt to overcome these limitations. Such proposals are an intermediate stage towards true "general purpose retinas", where we can distinguish between the sensing function and an advanced processing modality at the pixel level. In the sequel, we analyse some of these intermediate solutions.

The serial access constraint of CCD sensor was the first problem tackled. An arrangement of 2D photodiodes can be easily fabricated in a standard MOS process. The widespread availability of MOS design and fabrication facilities is one of the reasons for the trend towards such photodetector arrays. Pixel addressability can be obtained at reduced silicon cost. At the system level, such sensors consist of a 2D mesh of photodiodes, with an associated pixel storage device, plus a read-out pathway. Overall, we can describe them as special purpose RAMs: the sensing function embedded into the photodiode at each cell is separate for the reading function. While the sensor acquires and converts incoming light in a single "integration period", the reading out can be executed many times and with a scanning law that obeys the requirements of the computation (which is still outside the sensor). A design that matches this scheme[21] consists of an array of 80x80 cells, each containing a photodiode, a sample and hold stage, a memory capacitor and a read-out buffer. The voltage stored at the memory capacitor can be read-out in multiple phases. Pixel addressability is obtained through row and column selectors. This solution outputs the addressed accessed pixel and reads out the associated voltage. A rather conservative 3 μm p-well CMOS process lead to the integration of the 80x80 array into a chip, with a single cell consuming a maximum of 72x75 μm^2 of silicon area.

Another sensor that allows direct addressing of each sample data is the Multi-port Access photo-Receptor (MAR) designed at Naval University, Canada[22]. The array of photo-receptors is laid out in hexagonal connectivity. The novel feature of this project is the availability of the analog near-neighbourhood of the addressed photoreceptor at chip periphery. Such outputs can be processed in parallel and in the analog domain, albeit outside the sensor chip.

A first step towards a more flexible sensor is the provision for analog-to-digital conversion. The rationale for this step is the following: on one side, one can use such addressable sensors with a coupled external digital processor (such as a general purpose processor, or a dedicated board), which demands for proper interfacing; on another side, one can conceive to integrate into the sensor chip a minimal digital processing capability, for instance a row of simple processors associated to each column of sensors. In either case, the conversion function must be the least area-consuming, so that silicon usage for the sensing is not impaired by the additional capability. This is why analog-to-digital conversion is carried out with simple techniques, usually the lengthy, ramp method (as an instance, see figure 4, adapted from Gottardi et al.[23]).

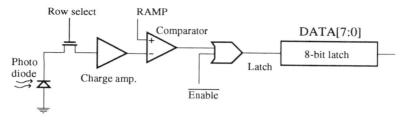

Figure 4. A scheme for analog-to-digital conversion.

There have been a number of 2D photodetector arrays[23-26] proposed and built from research institutions as intermediate experiments towards commercial products. The number of sensors integrated on the silicon areas varies quite considerably, depending on the technology available to the research groups. We shall not discuss the efficiency of such solutions from the point of view of area optimization, since we consider them more as architectural alternatives.

The GIB1 chip[24] was designed by an Italian group as a test for mixed analog/digital design and as a step towards the more advanced GIB2 photodetector array. It integrates an imager of 64x64 pixels, each 36 μm in both directions, into a die of 44 mm^2, along with a row of sense amplifier/comparators and peripheral circuitry for outputting the digitized image in serial form. The chip allows to alter the sensitivity of the imager either by varying a reference voltage, or by setting the integration time. The same research group is currently pursuing the design and construction of an enlarged sensor chip, hosting an array of 128x128 photo-diodes[23]. This design has a provision for a windowing read-out operation.

The provision for a processing capability is pursued at the level of a single array of simple processors. The GIB2 chip[24] embeds into the silicon die a 32x32 sensor array, a bank of 32x32x4 memory bits used as on-chip RAM, and 32 PEs (see figure 5). The row of sense amplifiers executes a simple thresholding operation, a very simple analog-to-digital conversion yielding a bit value for each pixel of the addressed row of sensors. The processing elements operate in SIMD mode under the control of an external device, which sends in instruction codes that are decoded by the proper circuitry on-chip. The processing elements can execute all 16 Boolean functions of two operands and can access neighbouring values through shift operations, along with on chip memory. The interface with off-chip circuitry is obtained through a 16-bit bus and a 12-bit address bus. FPGAs and external RAM are used to set-up a complete, though small, "digital camera". The chip die is 66 mm^2. This figure, if compared to the sensor-only GIB1 design, gives a rough estimate of the cost of embedding the processing function as a series of simple Boolean processors (both the GIB1 and the GIB2 chips were fabricated with a double-metal 3 μm analog CMOS process by Invomec).

The University of Linköping has contributed many designs in the photodetector array domain. A recent doctorate thesis[27] reviews them. We will hint briefly to the major steps in this long-term achievement.

The PASIC smart sensor chip[25] integrates an array of 128x128 photodiodes along with a A/D conversion row that converts in parallel the values addressed in a row of sensors. The size of each photo-diode is 48x48 μm^2, with an inter-diode distance in both directions of 60 μm. The chip contains furthermore a bank of 128 processing elements, each having an 8-bit shift register (used also for storing the value converted from a sensor), three single-bit Boolean registers, a three input ALU, and 127 bits of on-chip-memory (realised as dynamic, not refreshed RAM)

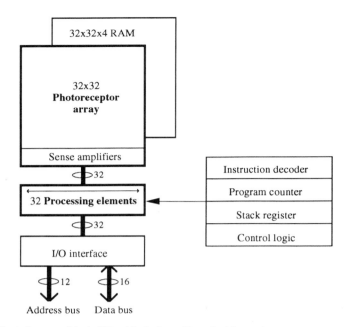

Figure 5. The block diagram of the building blocks for a chip embedding a photoreceptor array and a linear array of processing elements.

with 3 transistors per bit, per processing element. Although the datapath and the bus of the processing element are 1 bit wide, the interconnection of the processors is obtained by the parallel connection of the 8-bit registers, which are loaded serially through the bus from the A/D latches. This parallel distributed shift register can connect the border processors of the row in 1-D torus connectivity. An important feature of the array of processing elements is the Global-OR signal. This is the wired-OR of one of the Boolean registers of all 128 processors: it allows to build a global signal for control, and can also be used in many low-level algorithms. Its provision is typical of many cellular arrays, and its usefulness has been long recognized. The PASIC chip is 7.7 mm wide and 10 mm high. The sensor array takes 77% of the silicon area.

A 256x256 photo-diode image sensor array with an 8-bit digital output[26] has been designed, fabricated and tested as test chip for the commercially available MAPP2200 photodetector array and picture processor[28, 29]. The technology used in these projects is a 1.6 μm CMOS process. The area consumed by the light sensor is 20x28 μm^2; the overall image sensor area is 68 mm^2. The preliminary sensor chip outputs the 8-bit converted image data in row-serial mode at a frame rate of 150 frames/s. Analog-to-digital conversion is carried out with the ramp approach in parallel on the 256 sensed column data. A stepping counter is loaded with an initial offset value; variable increments allow to change the required precision (in the first prototype, 7-bit precision was obtained as a maximum) for a more flexible subsequent processing of the acquired images. If the stepping counter is not incremented at all, binary conversion is obtained, using the loaded offset as a threshold.

The commercially available MAPP2200 chip[28, 29] inherits the sensor array previously described, along with many features of the PASIC prototype. The 256 row wide sensor is coupled with 256 processors through the usual A/D conversion layer. With respect to PASIC,

MAPP2200 has a smaller memory (96 bits per processor) but a more advanced processing modality. Indeed the ALU of the processors can execute three types of instructions: i) pointwise local operations (no interaction among processors); ii) near-neighbour operations, on a 3x1 window of 1-bit data collected from the two neighbouring processors and from one local register; iii) a global operation that identifies connected segments along the row of processors, on the basis of a set of seeds stored in one of the Boolean registers. The chip is designed in the same technology of the above described sensor array, and embeds 500.000 transistors in a silicon area of 10x15 mm^2.

The photodetector array approach is a very interesting solution in the "on-chip-paradigm". It offers a viable compromise between the processing functionality required for general purpose operations on images and the area-demanding requirements of area image sensing.

5. ANALOG RETINAS

Modeling the behaviour of the retinal layers in biological systems, notably those of primates, in a silicon device is the primarily goal of Mead's "silicon retina". Extensive neurophysiology analysis of the functions carried out by the neural cells that make up the retina have long proved the existence of a space-temporal computation carried out on the incoming light through sets of highly specialized cells. Cones and rods transduce the light into potentials that activate synaptic junctions with bipolar and horizontal cells. Bipolar cells carry signals in traverse direction towards the ganglion cells, whose axons make up the optic nerve. Horizontal and amacrine cells distribute signals across space in lateral spread. The receptive fields of ganglion cells have a centre surround behaviour: the signal at the axon of ON/OFF ganglion carries an information that integrates positive responses from the receptors that project into the centre by contrasting it with inhibiting signals from the receptors of the surround. Among the other effects, one result of the arrangement of the retinal layers is to produce an edge detection capability that makes ganglion cells fire strongly frequency modulated sequences of impulses along their axons when their receptive field is stimulated by a non uniform distribution of illumination. In terms of signal processing operation, the mathematical model that best describes this behaviour is the so-called "difference-of-Gaussian" operator.

VLSI circuits, such as Mead's, that embed into silicon a computation that emulates this function of the retina draw on the analog processing facilities described in previous section: non-linear, logarithmic response to photodetector output and local averaging through resistive network for building a "receptive field" of adjustable dimensions.

Mead's proposal (see figure 6a and 6b) consists of two elements: an hexagonal resistive network, built with voltage adjustable horizontal resistors, and a node processor, containing the photodetector (a floating gate photo-transistor and the associated diode connected transistors for the logarithmic dependence of transduced voltage on the photo-generated inverse saturation current) and the circuitry for driving the resistive network and for sensing its computed local average.

The silicon photodetector exhibits a remarkably wide region of logarithmic response (four to five decades). The transconductance amplifier drives the node of the resistive network towards the input voltage: as explained above, the horizontal network stabilises so that the voltage at each node gives a "local" average of the inputs, with the weighting function being set by the product between the horizontal resistance and the vertical transconductance (both adjustable, at least in principle, at each node differently with the proper bias voltages). Such a local average emulates the inhibition from the "surround" within the "silicon receptive field". The second amplifier measures the voltage difference between the local input ("the ON centre of the silicon receptive field") and local average, thus producing a "difference" signal.

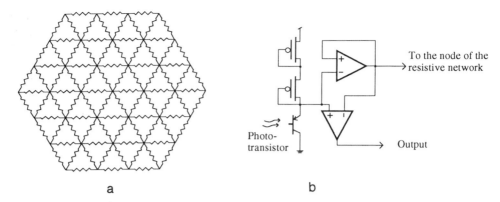

a   b

Figure 6. a) A bidimensional resistive network with hexagonal connectivity. b) Mead's schematic for the processing element associated to each node of the resistive network.

Mead's device shows a remarkable similarity to biological retinas with respect to sensitivity to incoming illumination; to the time response to abrupt changes in illumination; to edges in space distribution of light. Obviously, the uniform lay-out of the silicon photoreceptors and of the horizontal network does not compare with the distribution of cones and rods and with the dependency on eccentricity of receptive fields in human retinas. Refinements to the initial design[30] made the circuit more sensible to small differences between the centre and the surround responses, though an adaptation is required to prevent high-gain saturation effects: without such compensation, most pixel outputs are driven either to high or to low extreme under uniform illumination. Stability issues of the resistive network feedback effect have been also analysed[31]. The 48x48 pixels of this retina and the computed voltages are accessed through vertical and horizontal scan lines and are output in serial form for off-chip current sensing transformation. Overall, within the paradigm of this paper, Mead's silicon retina is the embodiment of a single spatial and temporal differentiation function carried out directly on the transduced signal, along with the sensing operation.

Among the many off-springs of the Mead's pioneering work we can quote a certain number of silicon analog retinas[32-34] aimed at image smoothing and segmentation.

The network of resistors in hexagonal topology used for averaging signals through electron spread exhibits an impulse response that can be approximated by an exponentiation. A more common response, and a much more used one for image processing, is Gaussian shaped. To obtain such a spatial response one is required to augment the network of horizontal resistors with an additional layer of resistor-like elements, so as to build a smoother decay in space. One such network is derived for the mono-dimensional case by applying regularization constraints to the equation of a curve that fits a set of discrete samples with a least-mean-square error criterion[32]. The resulting equation can be interpreted as the node equation of a network consisting of: i) the usual resistors R_0 (leakage resistors) connecting nodes to ground; ii) the horizontal resistors R_1 connecting contiguous nodes; iii) and *negative* resistors R_2 connecting every other node, being $R_2=-4R_1$. In the bidimensional case, the best approximation to a Gaussian smoothing function is obtained on the hexagonal grid by connecting closest neighbours with resistors R_1 and neighbours at a distance of 2 with negative resistors R_2. The kernel of the Gaussian convolution was shown to depend on $(R_0/R_1)^{1/4}$: if one makes R_1 depend on R_0, a single signal at a node of the network sets the convolution kernel. A prototype chip was fabricated through MOSIS foundry facilities. In a 2 μm CMOS technology, it embeds a network of 40x45 photoreceptors interconnected by this special resistive network in a silicon die of 7.9x9.2 mm², for a total complexity of 100.000 transistors. The output of the bi-dimensional convolution carried out within the chip can be accessed through the pins in row-parallel way.

Another approach[34] to quasi-Gaussian convolution applies the principle of "difference-of-exponentiation" (DOE) rather than usual "difference-of-Gaussian" (DOG). The resulting filter approximates well the $\nabla^2 G$ Laplacian-of-Gaussian kernel. By using two networks of horizontal positive resistors, interconnected only at the closest neighbour level, the network implements an image contrast enhancement (instead of the simple Gaussian smoothing filter of the previous case) without resorting to the potentially unstable negative resistors. The proposed architecture has been integrated with CMOS technology in a chip of 7.9x9.2 mm², containing an array of 53x52 photoreceptors.

6. ASIC SOLUTIONS

Analog retinas have been mainly motivated by the scientific search for electronic circuits that exhibit a processing modality similar to that of the periphery of biological visual systems. In a more engineering approach, the area of a silicon die can be used to solve a single task for a specific application. The design objective in such a situation is to produce a device that, while embedding a limited sensing capability, does solve the task by specialising the rest of the silicon area to the processing function(s) required. In this context, the technology used for the sensor (CCD or CMOS), and the processing modality (all-analog, mixed analog-digital, or even all-digital) are chosen on the basis of cost criteria imposed by the application. We can thus distinguish these ASIC chips from analog retinas because of the application driven specification of the processing carried out in silicon (Mead's retina and related projects are, in a sense, dedicated devices, since they only embed spatio-temporal differentiation or image smoothing, but do so without a driving application) and from photoreceptor arrays and "digital retinas", described shortly in the next section, because of the lack of a general purpose programmability.

There is a huge number of specific designs that suit the above definition of ASIC image sensors. We shall not give an exhaustive review (see Zavidovique and Bernard[4] for more examples), but shall rather highlight the main applications and discuss briefly some significant cases.

By large, motion detection and motion estimation is the application that has the largest number of proposed circuits (a sub-set is available in references[35-41]). A very general approach to motion is to compute the optical flow field. The all-analog chip by Tanner and Mead[35] solves the known aperture problem by distributing to each cell a global estimate of the overall velocity in a uniformly changing velocity field. A constraint circuitry minimises the error due to multiple possible local interpretations of the equation that relates intensity derivatives in time and in space (Horn's and Schunk's approach).

A rather straightforward approach to motion detection with no effort to estimate local velocities, is the design goal of the chip described in Chong et al.[37]. The array of photoreceptor is coupled with an array of current-mirror differentiators, that measure the variation in time of the incoming signal. These measures, obtained point-wise without interaction among neighbouring sensors, are aggregated in the form of currents summed up into a single signal available at chip periphery: the on-set of this signal shows that the sensors is detecting a global movement in the scene.

A novel mixed technology device is proposed in Gottardi and Yang[38]. This unique device uses both CMOS and CCD processes to set up a liner motion sensor, able to compute 1-D motion correlation between frames separated in space by 10 pixels at the most, meanwhile estimating motion velocities in the range $\pm 1 \div \pm 5000$ pixels/s. It embeds an array of 115 sensors (photodiodes, each 200x52 μm^2, in a 2 μm, standard double poly-double metal CMOS/CCD process), a row of CMOS dual sample-and-hold memories, a CCD image processor that computes the correlation between image frames and a winner-take-all circuit, that selects the strongest output from the ten correlations computed by the CCD processor. Overall, the chip takes 6.8x4.6 mm^2 in silicon area.

The visual system of insects has also been used as a driving biological example for obtaining motion estimates[41]. A linear arrangements of cells, consisting of on-centre-off surround spatial filters, interfaces the photodetectors with a linear array of current differentiators: these estimate local velocities by using the outputs from nearest neighbouring cell filters (at displacements +1 and -1). A global velocity is then obtained by summing up the local estimates for both positive and negative displacements. A chip[40] for tracking objects has been designed with an analogous approach, though with more complex computations carried out in silicon. A detection circuitry comprises 60 photodiodes, followed by a differentiation stage that detects changes in time through a feedback operated amplifier: the approximated transfer function of this stage is $H(s)=-RCs$ and has a time constant in the order of microseconds, well matched to the required sampling in time for the changes to be detected (approximately 10 ms). The rest of the chip contains an all-digital circuitry for storing templates that represent local changes in velocity, 12 special purpose processing units and RAM for intermediate results. The final result of the processing unit is the computation of velocities in the scanned visual field by template matching; 360° vision is possible by mounting 6 chips onto a rotation rod, since the visual angle of the array of sensor is 60°. The chip was integrated in 2 μm p-well, double poly, double metal CMOS technology; the die size is 4.5x4.6 mm^2.

Other solutions have been proposed for stereo matching[42] and for range estimation[43]. The first chip, although designed for an important image processing task, does not include a sensor, and is therefore outside the main discussion of this paper. The second, instead, contains an array of 28x32 circular shaped sensors and is suited to range finding techniques based on the light-stripe approach. The array of sensors replaces a conventional camera and detects peaks in light intensity projected onto the sensor by the swept stripe of light reflected from the scene. Each sensor converts incoming light into voltage and remembers the time at which a voltage peak occurs: this is, by definition, the instant when the voltage overcomes a chosen threshold. Storage is analog and read-out of voltages is performed in serial mode, on a row by row basis. For uniform response to stripe-line orientation, the sensors are circular.

A very interesting design and chip for fingerprint verification is reported in Anderson[44] and Renshaw[45]. As in the previously quoted chip[40] for motion estimation, this realisation combines the analog sensing with a digital processing function, a further example of the flexible approach to be followed in ASIC design. On a die of 11.5x11.5 mm^2, in a 1.2 μm CMOS process by European Silicon Structure foundry, the chip integrates 272.000 transistors that embed the following: a 258x258 array of photodiodes, a set of static RAMs (2x512x16, 1x16x16 bits) for intermediate results and ROMs (2x704x12 bits), a specialized processing unit with a 3x3 smoothing filter, a 9-bit DAC and comparator module, an adaptive thresholding unit, an 8x8 correlator cell and a rank-value filter. In this solution, the array of sensors is scanned serially as in a standard CCD camera, but the subsequent processing is carried out on the same silicon substrate: the reference fingerprint data is stored in the local RAM and is used for correlation with acquired image. The integrated sensor-processor realises the fingerprint verification system by substituting a board containing 16 VLSI/LSI devices with only 3 ICs and by cutting down the response time from 1.5 s to 0.5 s (average times).

Image acquisition and subsequent object position and orientation detection are realised with an all-analog implementation[46]. The computation of the position of the object and of its orientation are based on the techniques of moments: the chip acquires a scene with an array of 29x29 photoreceptors that are interconnected with a network of resistors. Currents injected into this resistive network by the incoming light are collected at the periphery of the sensor array into currents that drive distributed resistive lines: properly injected into both linear and quadratic resistive lines, the currents build up signals that are directly proportional to the centroid of the illumination and to the principal axis of the shape. The accuracy of such measures is considerable: the position of a 25x25 square over an 100x100 image field is within ±0.3%, and orientation is within ±2° for elongated and moderately sized objects.

A further implementation suited for robot movement control is described in Friedman and Clark[47]. A small array of photodiodes is integrated in a chip along with analog and digital circuitry, whose final goal is to produce a small set of digital signals to be forwarded to the control system of an autonomous robot. The task carried out by the chip is to compute an estimate of apparent motion and to encode the occupancy in space of a step edge projected onto the sensor. Beside the sensing of incoming light, the functional units realised in analog way within the circuitry are: i) intensity averaging; ii) spatial convolutions to detect discrete approximations to the first order derivative in X and Y directions, together with a discrete approximation to the Laplacian operator. Digital processing uses the three approximations to establish the octant in which a light-dark edge crosses the array, if any. Three bits are output at chip periphery. The sensing capabilities are minimal, since the prototype circuits contains only 25 phototransistors: the layout of the complete circuitry is enclosed in a small die of 2.25 mm on a side. The analog processing subunits take up 87% of the area. This small sensor is designed to be part of a more sophisticated sensory system for the robot, and its almost miniature sensing capabilities will be compensated for by other signals.

CCD technology is not completely absent in the ASIC paradigm. In addition to the motion correlator chip described in Hakkarainen and Lee[42], other proposals have been put forward for tasks such as lossless image compression[48]. The usual frame-based image sensor typical of this technology is augmented with a CCD-based neighbourhood reconstructor at chip periphery. The purpose of this neighbourhood circuitry is to encode pixel differences in pyramidal way, thus producing a compact image suitable for subsequent transmission. The IC is implemented in a triple-poly, double metal n-channel CCD process, for realising cross-channel transits. The size of each pixel is 15 μm on both sides, and the 256x256 array of sensors occupies 3.9x7.74 mm² of the die size; the near-neighbourhood circuitry takes only 2% of the chip area.

A specially designed edge detector sensor[49] has been fabricated with amorphous silicon layers on top of a usual crystalline silicon Metal-Oxide-Semiconductor (MOS). Detection of contrast in local illumination is obtained by two p-i-n cells connected back-to-back. The shape of the two photoreceptors is an ellipse, for better directional detection. The back-to-back arrangement produces no output when both photodiodes are illuminated uniformly, or when illumination varies in step edge perpendicular to the major axis of the ellipse. In any other situation, the two opposite currents are un-balanced and the detector outputs a voltage proportional to the difference of illumination. This is a rather crude, but fairly simple step-like edge detector.

The principle of variation of conductivity in response to illumination is used in a design proposed for general shape recognition[6]. In a circularly arranged array of photo sensible material, an illumination caused by a single shape causes the on-set of a field of currents. It can be shown that the potential measured at the periphery of the device has a distribution tightly linked to the Fourier transform of the shape. The proposed device would be used to discriminate a known shape against non-pertinent ones, and could be made insensible to shape rotations around the centre of the sensor array. Simple translations cannot be easily handled. The availability of proper materials for building the sensor is still questionable, since silicon does not exhibit the required sensitivity to light in a feasible area.

7. DIGITAL RETINAS

Within the "on-chip-vision" paradigm highlighted in figure 2, we can identify analog retinas as the point where silicon is allocated to a single, though complex low level image processing function. Electronic spread and non linear, analog processing build a network of wired-in near-neighbour processors, which convert incoming light and process it in a time-continuous form. The analog layer is the only processing layer, and it embeds all of the interactions within neighbours.

At another extreme in silicon usage we find massively parallel Boolean processors, that have been so popular in the 80's. In the chips designed for these systems, the silicon "real estate" is used for setting up a network of locally interconnected serial processors. Access to the near-neighbours is often accomplished by wiring an 8-connected set of links, that allow each processor to read-in data from the neighbours and to compute a Boolean logic function on the contents of the neighbours and on a selected local Boolean register. In some designs[50,51], an asynchronous propagation modality enhances the architecture and allows one to use connected components in binary images as atomic data structure. Cellular logic and shift-invariant operations have opened up the road to effective low-level image processing.

We can define a "digital retina" as the architecture that extends such Boolean cellular processors to the sensing function, by stacking a layer of photoreceptors on top of the network of locally interconnected Boolean processors. As highlighted in Zavidovique and Bernard[4], both technological and image processing related factors influence the design of such retinas. The use of silicon for the highly interconnected mesh of processors reduces the role of analog processing to the sensing function only (thus justifying the digital retina denomination). The Boolean processing modality advocates for a careful approach to analog-to-digital conversion: although bit-serial processing is quite feasible in massively parallel Boolean processor, this is much harder in a digital retina, where the number of memory cells per processor is bound by the tight constraint on silicon area.

The strongest motivation for building a digital retina is the need for a general-purpose engine that allows one to program the "smart sensor". Rather than committing to a single, optimised function obtained by clever analog techniques, one tries to reach a compromise in silicon usage for setting up a programmable device. In the following, we report on a design[52,53] that achieves such a compromise, by casting in the silicon die a 2D array of 65x67 sensors and Boolean processors.

The NCP retina (the acronym stands for Neighbourhood Combinatorial Processing) consists of a 2D arrangement of sensors/processors (see figure 7), linked with 4-connectivity. Each processor has three bits of memory (M1-M3), a photo-diode and a minimal Boolean processing capability, consisting of the four operations: i) M2←M1; ii) M2←NOT(M1); iii) M2←M1 AND M2; iv) M2←M2 AND not(M1). The interconnection between adjacent processor is obtained through mono-directional pathways (denoted with X and Y in figure 7). The design has adopted a minimal, though complete, computation facility and a communication pathway that minimises control signals, meanwhile requiring some detours to shift data in the orthogonal directions (a simple local operation such as a conjunction between two bits can take up as much as ten clock cycles). With reference to the alternative typical design of cellular Boolean processor, the mesh of NCP processors does fit neither the gating approach, nor the multiplexed solution. A single near-neighbour function requires multiple steps of shifts in proper directions to bring the data of the neighbouring processors into the local storage, and a sequence of intermediate computations as the single data are latched in. The procedure is therefore lengthy, if compared to the analogous cases of cellular processors. Yet, the system supports a complete set of cellular logic operations, and can emulate in time complex pattern matching operations.

We can appreciate the compromise thus achieved by considering the hardware of a single

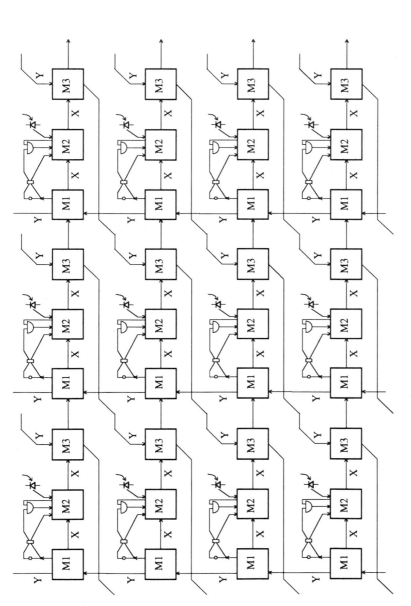

Figure 7. The arrangement of processing elements and photodetectors in the "digital retina". Each processing elements consists of three memory bits (M1, M2, M3), of a photodiode and of a very simple boolean processing logic.

processor. The storage elements have been designed as semi-static one-bit memories; overall, the NCP processor contains only 28 CMOS transistors, including the silicon for the photo-diode; 6 clocking signals are routed throughout the device for implementing the various shifting operations required for moving data within a processor and among the processors. With a 2 μm CMOS technology, the basic cell in the chip takes 80x100 μm^2 area; the 65x67 retina is thus embedded into a 50 mm^2 die. The main clock signal has been tested to drive correctly the chip at 20 MHz. A simple binary image edge detection is executed in 1.5 μs, since it requires 30 cycles of the main clock.

In the process of bridging the gap between the sensing and the processing function (see figure 1), the NCP retina has gone a considerable step forwards. A host of low-level image processing function can be carried out on the same device that acquires the image. The designers introduce the notion of "rough vision" to describe an image processing modality that matches the required general programming capability with extremely tight constraints in the analog-to-digital conversion phase. Indeed, one can easily note that the Boolean processing mode of the device is better developed than the light transduction. The acquired signal can only be translated to a binary value. While dithering scheme[54] are suited to the neighbouring access structure of the NCP retina, they produce admittedly poor halftone renditions of the analog acquired image. Nevertheless, even with the extremely simple analog-to-binary conversion, this design has a notable breath in conception and is the closest to be found in the "on-chip-vision" paradigm to a programmable retina.

At a much coarser scale of integration, one more proposal aims at the eventual construction of a digital retina[55]. In this case, a prototype chip contains 8 Processing Elements (PE), arranged in a row, but equipped with 4-connected links to neighbours. Each processor contains both a photoreceptor for image acquisition and a LED, for image generation, a unique solution among the proposals for smart sensor. Although designed for bit-serial processing, the processing element contains three eight-bit shift register, a two-input four-bit multiplier and a ALU for Boolean operations. The present implementation is obtained in a rather minimal gate-array technology, with a PE made up by 337 gates: the whole chip contains 2696 gates for the processor, plus an extra 247 gates for common circuitry. A prototype system comprising 64x64 processors has been built. Admittedly, the current physical dimensions of the system (height 1010 mm, width 880 mm, depth 350 mm) are enormous and serve as a gross scale comparison to the NCP design: a much richer processing capability at the level of the processor, along with an integration technique scarcely suited to the design, make the digital retina concept a more effective one.

The NSIP architecture[56, 27] is at an early stage of development. It also draws much from previous designs at the University of Linköping. It features a four connected mesh of simple processors, each embedding a photodiode, a simple ALU and some local memory. This design is unique in its approach to the sensing function, which is tightly coupled to the processing: the NSIP programming mode consists of a loop that tests in a non-destructive way the output from the photodiode and iterates same basic local operation until a global condition is met in the whole array. As with the PASIC project, this architecture supports a global propagation mode by using a gated access to the four neighbours. It is not yet clear how much of this conceptual design can be integrated into silicon. Though much more flexible than the NCP retina, the NSIP architecture still lacks a VLSI implementation, even a prototype one. The closest device available is the commercial one-dimensional sensor LAPP1100[57].

8. CONCLUSIONS

In this paper we have analysed the many alternatives that are available for building a smart sensor and for reducing the gap of traditional image processing systems, which separate the image acquisition phase from the subsequent processing. We believe that the on-chip-vision paradigm advocates for a mixed mode of processing the transduced signal: the clear advantages of analog processing are somewhat offset by the lack of general programmable capability, typical of digital retinas. These, in turn, are as yet minimal in their analog-to-digital conversion layer. Some photodetector arrays offer an intermediate solution, by allowing a more advanced processing modality on rows of the transduced image. The ASIC approach is always a feasible way to tackle a specific engineering problem, when a single task (or a relatively small set of tasks) is to be carried out.

REFERENCES

1. J.E. Dowling, *The Retina: An Approachable Part of the Brain*, Belknap Press of Harvard University, Cambridge, MA (1987).
2. K. Sakakibara, H. Yamamoto, S. Maegawa, H. Kawashima, Y. Nishioka, M. Yamawaki, S. Asai, N. Tsubouchi, T. Okumura, and J. Fujino, A 1 Format 1.5M Pixel IT-CCD Image Sensor for and HDTV Camera System, *IEEE Trans. on Consumer Electronics*, Vol.37, No.3, pp.487-493 (1991).
3. M. Negishi, H. Yamada, K. Harada, M. Yamagishi, and K. Yonemoto, A Low Smear Structure for 2M-Pixel CCD Image Sensors, *IEEE Trans. on Consumer Electronics*, Vol.37, No.3, pp.494-499 (1991).
4. B.Y. Zavidovique and T. M. Bernard, Generic Functions for On-Chip Vision, *Proc. IEEE ICPR-92*, The Hague, The Netherlands, NL, pp.1-10 (1992).
5. S.M. Sze, *Semiconductor Devices: Physics and Technology*, John Wiley and Sons, New York, NY, pp.24, 523 (1985).
6. J.E. Hershey, R. Liberati, and D. S. Hammer, New Focal Plane Architecture and Transform for Fast Target Recognition, *Applied Optics*, Vol.28, No.18, pp.3810-3813 (1989).
7. C. Mead, *Analog VLSI and Neural Systems*, Addison-Wesley, Reading, MA (1989).
8. S.G. Chamberlain and J.P.Y. Lee, A Novel Wide Dynamic Range Silicon Photodetector and Linear Imaging Array, *IEEE Journal of Solid-State Circuits*, Vol.SC-19, No.1, pp.41-48 (1984).
9. H. Akimoto, H. Ando, H. Nakagawa, Y. Nakahara, M. Hikiba, and H. Ohta, A 1/3-in 410000-Pixel CCD Image Sensor with Feedback Field-Plate Amplifier, *IEEE Journal of Solid-State Circuits*, Vol.SC-26, No.12, pp.1907-1913 (1991).
10. E.R. Fossum, Wire Transfer of Charge Packets Using a CCD-BBD Structure for Charge-Domain Signal Processing, *IEEE Trans. on Electron Devices*, Vol.ED-38, No.2, pp.291-298 (1991).
11. E.G. Stevens, Photoresponse Nonlinearity of Solid-State Image Sensors with Antiblooming Protection, *IEEE Trans. on Electron Devices*, Vol.ED-38, No.2, pp.299-302 (1991).
12. T. Watanabe, K. Hashiguchi, T. Yamano, J. Nakai, S. Miyatake, O. Matsui, and K. Awane, A CCD Color Signal Separation IC for Single-Chip Color Imagers, *IEEE Journal of Solid-State Circuits*, Vol.SC-19, No.1, pp.49-54 (1984).
13. S. Kawamoto, Y. Watanabe, Y. Otsuka, and T. Narabu, A CCD Color Linear Image Sensor Employing New Transfer Method, *IEEE Trans. on Consumer Electronics*, Vol.37, No.3, pp.481-486 (1991).
14. J. Van der Spiegel, G. Kreider, C. Claeys, I. Debusschere, G. Sandini, P. Dario, F. Fantini, P. Bellutti, and G. Soncini, A Foveated-Like Sensor Using CCD Technology, in *Analog VLSI Implementation of Neural Systems*, C. Mead and M. Ismail eds., Kluwer Academic Publisher, Norwell, MA, pp.189-211 (1989).
15. S.S. Bencuya and A.J. Steckl, Charge-Packet Splitting in Charge-Domain Devices, *IEEE Trans. on Electron Devices*, Vol.ED-31, No.10, pp.1494-1501 (1984).
16. J. Hynecek, CCM-A New Low-Noise Charge Carrier Multiplier Suitable for Detection of Charge in Small Pixel CCD Image Sensors, *IEEE Trans. on Electron Devices*, Vol.ED-39, No.8, pp.1972-1975 (1992).
17. E.R. Fossum and R.C. Barker, A Linear and Compact Charge-Coupled Charge Packet Differencer/Replicator, *IEEE Trans. on Electron Devices*, Vol.ED-31, No.12, pp.1784-1789 (1984).
18. C. Sin, A. Kramer, V. Hu, R.R. Chu, and P. K. Ko, EEPROM as an Analog Storage Device, with Particular Applications in Neural Networks, *IEEE Trans. on Electron Devices*, Vol.ED-39, No.6, pp.1410-1419 (1992).
19. J. Harris, C. Koch, J. Luo, and J. Wyatt, Resitive Fuses: Analog Hardware for Detecting Discontinuities in Early Vision, in *Analog VLSI Implementation of Neural Systems*, C. Mead and M. Ismail eds., Kluwer Academic Publisher, Norwell, MA, pp.27-56 (1989).
20. A.M. Chiang and M.L. Chuang, A CCD Programmable Image Processor and its Neural Network Applications, *IEEE Journal of Solid-State Circuits*, Vol.SC-26, No.12, pp.1894-1901 (1991).
21. O. Yadid-Pecht, R. Ginosar, and Y. Shacham Diamand, A Random Access Photodiode Array for Intelligent Image Capture, *IEEE Trans. on Electron Devices*, Vol.ED-38, No.8, pp.1772-1780 (1991).

22. M. Tremblay, D. Larendeau, and D. Poussart, Multi module focal plane processing sensor with parallel analog support for computer vision, *Proc. Symposium on Integrated Systema*, Seattle, WA (1993).

23. M. Gottardi, A. Sartori, and A. Simoni, POLIFEMO-An Addressable CMOS 128x128-Pixel Image Sensor with Digital Interface, *IRST TR* (1992).

24. A. Sartori, A Smart Camera, in *FPGAs*, Moore and Luks eds., Abingdon EE&CS Books, Abingdon, pp.353-362 (1991).

25. K. Chen, A. Aström, and P.E. Danielsson, PASIC a smart sensor for computer vision, *Proc. ICPR*, Atlantic City, NJ, pp.286-291 (1990).

26. C. Jansson, P. Ingelhag, C. Svensson, and R. Forchheimer, An Addressable 256x256 Photodiode Image Sensor Array with an 8-bit Digital Output, *Proc. ESSCIRC-92*, pp.151-154 (1992).

27. A. Aström, *Smart Image Sensors*, Linköping Studies in Science and Technology, Dissertation No.319, Linköping, S (1993).

28. R. Forchheimer, P. Ingelhag, and C. Jansson, MAPP2200: a second generation smart optical sensor, *SPIE 92*, Vol.1659, pp.2-11 (1992).

29. IVP Integrated Vision Products AB, *MAPP2200 Technical Description*, Linköping, S (1991).

30. C. Mead, Adaptive retina, in *Analog VLSI Implementation of Neural Systems*, C. Mead and M. Ismail eds., Kluwer Academic Publisher, Norwell, MA, pp.239-246 (1989).

31. D.L. Standley and J.L.Jr. Wyatt, Stability Criterion for Lateral Inhibition and Related Networks that is Robust in the Presence of Integrated Circuits Parasitics, *IEEE Trans. on Circuits and Systems*, Vol.CAS-36, No.5, pp.675-681 (1989).

32. H. Kobayashi, J. White, and A.A. Abidi, An active resistor network for gaussian filtering of images, *IEEE Journal of Solid-State Circuits*, Vol.26, No.5, pp.738-748 (1991).

33. P.C. Yu and H. Lee, A CMOS resitive-fuse processor for 2-D image acquisition, smoothing and segmentation, *Proc. ESSCIRC 92*, pp.147-150 (1992).

34. T. Shimmi, H. Kobayashi, T. Yagi, T. Sawaji, T. Matsumoto, and A.A. Abidi, A parallel analog CMOS signal processor for image contrast enhancement, *Proc. ESSCIRC 92*, pp.163-166 (1992).

35. J. Tanner and C. Mead, Optical motion sensor, in *VLSI Signal Processing*, Kung, Owen and Nash eds., IEEE Press, pp.59-76 (1987).

36. J.L. Wyatt Jr., D.L. Standley, and W. Yang, The MIT vision chip project: analog VSLI systems for fast image acquisition and early vision processing, *Proc. IEEE Intern. Conference on Robotics and Automation*, Sacramento, CA, pp.1330-1335 (1991).

37. C. Chong, C. Andre, T. Salama, and K.C. Smith, Image-Motion Detection Using Analog VLSI, *IEEE Journal of Solid-State Circuits*, Vol.27, No.1, pp.93-96 (1992).

38. M. Gottardi and W. Yang, A CCD/CMOS Image Motion Sensor, *Proc. 1993 IEEE ISSCC*, pp.194-195 (1993).

39. R. Etienne-Cummings, S. Fernando, N. Takahashi, V. Shtonov, J. Van der Spiegel, and P. Mueller, A new temporal domain optical flow measurement technique for focal plane VLSI implementation, *Proc. CAMP93*, New Orleans, LA, pp.241-250 (1993).

40. A. Yakovleff, A. Moini, A. Bouzerdoum, X.T. Nguyen, R.E. Bogner, K. Eshraghian, and D. Abbott, A microsensor based on insect vision, *Proc. CAMP93*, New Orleans, LA, pp.137-146 (1993).

41. A. Andreou, K. Strhben, and R.E. Jenkins, Silicon retina for motion computation, *Proc. ISCAS 91*, Singapore, pp.1373-1376 (1991).

42. J.M. Hakkarainen and H.S. Lee, A 40x40 CCD/CMOS AVD Processor for Use in a Stereo Vision System, *Proc. ESSCIRC-92*, pp.155-158 (1992).

43. A. Gruss, L.R. Carley, and T. Kanade, Integrated sensor and range-finding analog signal processor, *IEEE Journal of Solid-state Circuits*, Vol.26, No.3, pp.184-191 (1991).

44. S. Anderson, W.H. Bruce, P.B. Denyer, D. Renshaw, and G. Wang, A Single Chip Sensor and Image Processor for Fingerprint Verification, *Proc. IEEE 1991 ICCD*, pp.12.1.1-12.1.4 (1991).

45. D. Renshaw, P.B. Denyer, G. Wang, and M. Lu, ASIC Image Sensor, *Proc. ISCAS-90*, pp.3038-3041 (1990).

46. D. Standley, An Object Position and Orientation IC with Embedded Imager, *IEEE Journal of Solid-State Circuits*, Vol.26, No.12, pp.1853-1859 (1991).

47. D.J. Friedman and J.J. Clark, VLSI Sensor/processor Circuitry for Autonomous Robots, *Optics, Illumination, and Image Sensing for Machine Vision*, SPIE, Vol.1614, pp.99-110 (1991).

48. S. Kemeny, H. Torby, H. Meadows, R. Bredthauer, M. LaShell, and E.R. Fossum, CCD image sensor with differential pyramidal output for lossless image compression, *Proc. IEEE CICC 91*, pp.12.6.1-12.6.4 (1991).

49. H. Lin, W. Sah, and S. Lee, The Vertical Integration of Christalline NMOS and Amorphous Orientational Edge Detector, *IEEE Trans. on Electron Devices*, Vol.ED-39, No.12, pp.2810-2812 (1992).

50. V. Cantoni, V. Di Gesù, M. Ferretti, S. Levialdi, R. Negrini, and R. Stefanelli, The PAPIA System, *Journal of VLSI Signal Processing*, Vol.2, pp.195-217 (1991).

51. M.J.B. Duff, Propagation in cellular logic arrays, in *Proc. Workshop on Picture Data Description and Management*, pp.259-262 (1980).

52. T. Bernard, B. Zavidovique, and F. Devos, A programmable artificial retina, *IEEE Journal of Solid-State Circuits*, Vol.28, pp.789-798 (1993).

53. T.M. Bernard, B. Zavidovique, and F. Devos, The NCP retina: an imager, a halftoner and a micro-grained array processor on the same chip, *Proc. ESSCIRC 92*, Copenhagen, DK, pp.159-162 (1992).

54. B.E. Bayer, An optimum medium for two-level rendition of continuos-tone pictures, *IEEE Intern. Conf. on Communications*, pp.2611-2615 (1973).

55. M. Ishikawa, A. Morita, and N. Takayanagi, High speed vision system using massively parallel processing, *Proc. IEEE/RSJ Int. Conf. on Intelligent Robots and Systems*, Raleigh, NC, pp.373-377 (1992).

56. A. Aström, R. Forchheimer, and P. Ingelhag, An integrated sensor/processor architecture based on near-sensor image processing, *Proc. CAMP93*, New Orleans, LA, pp.147-154 (1993).

57. IVP Integrated Vision Products AB, *LAPP1100 Product Description*, Linköping, S (1989).

PANEL SUMMARY
FOVEATION, LOG-POLAR MAPPING AND MULTISCALE APPROACHES TO EARLY VISION

Hezy Yeshurun[1] *(chairman)*, Ivo De Lotto[2], and Concettina Guerra[3]

[1] Department of Computer Science
   Tel Aviv University, IL - 69978 Tel Aviv, Israel
[2] Dipartimento di Informatica e Sistemistica, Università di Pavia
   Via Abbiategrasso 209, I - 27100 Pavia, Italy
[3] Dipartimento di Elettronica e Informatica, Università di Padova
   Via Gradenigo 6A, I - 35131 Padova, Italy

ABSTRACT

Hot Points

One of the ways by which early human vision is clearly distinguished from current machine vision is the fact that human vision is strongly space variant, and build up a (multiscale) representation of the world from those space variant fixations. In this panel, we will ask how relevant is this principle of human vision to machine vision (Y. Yeshurun), and present the principles of multiscale (C. Guerra) and wavelets (I. DeLotto) approaches that are closely linked to the issue of representation.

This panel discussed some solutions to vision problems based on variable resolution approaches. The pre-defined basis of the discussion included the following hot points:
1. Why is foveated vision necessary (if at all..)?
   or: Should we blindly imitate biological vision?
2. What is the preferred type of foveated vision?
   (e.g. complex log, pyramid)
3. What are the effects of foveated vision on customary vision algorithms?
   (from edge detection to object recognition)
4. Effectiveness of multiscale and wavelet techniques.

FOVEATION AND LOG-POLAR MAPPING (H. Yeshurun)

Biological vision is foveated, highly goal oriented and task dependent. This observation, which is rather clear if we trace the behaviour of practically every vertebrate, is now being taken seriously into consideration by the computer vision community. This is evident from recent work on active vision systems and heads and general active vision concepts and algorithms.

One of the fundamental features of active vision is the use of space-variant vision and sensors that allow, in the case of the log-polar representation data reduction as well as a certain degree of size and rotation invariance[1,2].

A major question that arises is why should computer vision blindly imitate biological vision? is it really necessary to use foveation at all? Analysing human vision, we find out that the human visual system covers more than 200 degrees of field of view. Though the resolution of the visual system depends on the exact phrasing (visual acuity, hyperacuity etc.), it is probably safe to assume that the fact that the central 10x10 degrees are sampled by 1000x1000 ganglion cells imposes a lower bound on the resolution. The fact that the number of axons in the optic nerve is two order of magnitude less than the number of retinal photoreceptors can be an indication that visual information is passed by a relatively narrow band channel. Thus, describing biological parameters as engineering design principles, we can come to the conclusion that if one needs high resolution data in a large field of view, using limited channel (and probably limited computational resources), foveation is mandatory.

There are quite a few alternatives for implementing foveation. As already mentioned, the log-polar transformation was found to be a good fit to the retino-cortical mapping. From a computational point of view, this mapping offers the benefits of scale and rotation invariance. As can be easily demonstrated, an avoidance and alarm detection system can be efficiently implemented by a log-polar mapping for objects that move towards us along the line of sight. However, it is still worthwhile to consider the possibility that the main advantage of this mapping, which is conformal, is developmental and not computational, in the sense that only a few parameters should be specified in order to carry it out.

Thus, we might implement foveation by a multiscale approach based system, such as the pyramid. In this case we use a small (typically 4) levels of representation instead of the continuous log-polar one. The multiscale approach can be brought to its extreme by using just two cameras: a wide field of view for general acquisition of images, and a zoom, narrow field of view camera that is used to scan in details selected areas. This approach is actually being taken in a current implementation of an autonomous vehicle.

It is quite clear that the specific choice of foveation is task dependent, and as such, should be decided by cost/benefit analysis. Selecting the foveation scheme, it is usually possible to implement it either by hardware or by software. The current trend is hardware implementation, which will ultimately result in "Silicon Retina"[3]. Hardware foveation is fast, but possibly task customized and rigid. Current performance of general purpose chips and machines makes it possible now to implement foveation in software. While these implementations might be slower than the hardware ones, they offer the convenience of off-the-shelf equipment, and they are highly flexible and allow a wide scope of experiments.

The same argumentation applies also for the more general question of electronical vs. mechanical active vision system. While the current mechanical systems imitate biological systems by mounting a relatively narrow field cameras on 5 or 6 dof robots and "heads", it is possible, in principle, to scan the world with a high resolution and wide field of view camera, and implement foveation by software.

As many other points in this area, the decision is more economy driven than theory driven, in the sense that the relevant parameters are the cost of large CCDs as compared to high precision mechanical heads, and the ratio of the CPU power of commercially available machines to customized hardware.

MULTISCALE APPROACHES (C. Guerra)

Multiscale approaches are very popular in computer vision, especially for tasks that are high computation intensive. An important question is how to efficiently use the multiresolution data. For the matching problem that arises in object recognition, in the stereo problem, etc., the solution that is widely adopted in the literature is a hierarchical coarse-to-fine strategy, that is a procedure that tries to find correspondences at the coarsest level of resolution and to expand, at the finer levels, the hypothesized correspondences. Coarse-to-fine strategies have been successfully used in a variety of other image processing and vision applications, including optical flow computation, image reconstruction, texture determination, etc. However, the coarse-to-fine approach has generally the disadvantage that mis-matches at a coarse scale cause errors from which it is impossible to recover since the algorithms usually proceed by subdividing the corresponding coarse elements into parts. This limitation is heavy particularly for highly deformed shapes for which the matching is not very reliable at all levels. On the other hand, this method is fast because it allows to disregard large portions of the search space.

One can argue in favour of alternative strategies that avoid to incur into unrecoverable errors. A bottom-up search strategy has been suggested that overcomes the limitations of a coarse-to-fine strategy by processing the multiresolution data starting from the finest level, and using the coarser levels to impose constraints on the matching elements[4]. Unlike the other approach, this strategy is not likely to lead to a reduction in the computation time for the matching problem. However, it may give more reliable results. The best solution to this problem will probably depend on the formulation of the matching problem itself and on the application.

MULTISCALE ALGORITHMS AND WAVELET TRANSFORM (I. De Lotto)

Wavelet[5-7] theory represents a unified framework for many techniques of signal and image processing. Image analysis aims at extracting relevant informations on an image by trasforming it. Important operations such as parameter estimation, coding and pattern recognition can be performed on the transformed image where its relevant properties can be more evident. Multiscale wavelet algorithms have been designed to take advantage of the structural properties of images and of the way they are perceived by a human observer. For instance the separation of edges from the texture information of the image. The main characteristic of these techniques is connected with the compact support of the chosen wavelets, that is with their good localization in space and "frequency". In general the differentiation operator is not diagonal for a wavelet basis, that is "frequencies" are mixed up. But a good microlocalization leaves the differentiation operator nearly diagonal both for scale ("frequency") and space. The wavelet basis can be as well orthogonal, but with the price of irregular basis functions for the orthogonality with compact support. It must be noticed however[8] that orthogonality and even linear independence are not essential for representing functions.

It seems interesting to use this approach both to detect what part of an image can be meaningful to guide foveation, and to point out structural properties of an image. Algorithms for wavelet transforms are fast and often suitable for implementations on a chip. Many research groups are working on the subject, particularly for application fields different from artificial vision such as radar and sonar signal detection, image and acoustic signal coding, numerical analysis. The results so far obtained are considerable and in progress. They seem adaptable to the problems of early vision.

CLOSING REMARKS (H. Yeshurun)

It seems that the main point we can agree upon, is that the relevance of human vision to machine vision is strongly a cost/benefit issue. Thus, multi scale and wavelets approaches could be definitely used in the context of machine vision, even if they are not used exactly in the same manner by biological systems. Foveation by itself, is probably unavoidable, if we need to cover a large field of view having restricted computational resources. The specific type of foveation, however, could be task dependent, and not necessarily following the biological example. We have also discussed issue of software/hardware implementations of foveation in particular, and vision related function in general, and came again to the conclusion that this should be a pragmatic issue, rather than an ideological one.

REFERENCES

1. Y. Yeshurun and E.L. Schwartz, Shape description with a space variant sensor: algorithms for scan path, fusion, and convergence over multiple scans, *IEEE Trans.* PAMI 11, Vol.11, pp.1217-1222 (1989).
2. M. Tistarelli and G. Sandini, Estimation of depth from motion using an antropomorphic visual sensor, *Image and Vision Computing,* Vol.8, No.4, pp.271-278 (1990).
3. A. Rojer and E.L. Schwartz, Design consideration for a space variant visual sensor with complex logarithmic geometry, *Proc. 10th IAPR,* pp.278-285 (1990).
4. V. Cantoni, L. Cinque, C. Guerra, S. Levialdi, and L. Lombardi, Tree pattern matching for 2D multiresolution objects, *Proc. CAMP'93*, New Orleans, LA, pp.43-46 (1993).
5. S.G. Mallat, A theory for multiresolution signal decomposition: the wavelet representation, *IEEE Trans.* PAMI 11, Vol.7, pp.674-693 (1989).
6. I. Daubechies, The wavelet transform, time frequency localization and signal analysis, *IEEE Trans.* IT 36, Vol.5, pp.961-1005 (1990).
7. C.K. Chui, *Wavelets: a Tutorial in Theory and Applications*, Academic Press, NY (1992).
8. S.G. Mallat and Z. Zhang, Matching pursuit with time frequency dictionaries, *IEEE Trans.* SP (in press).

NEUROPHYSIOLOGY OF THE STRIATE CORTEX

Adriana Fiorentini

Istituto di Neurofisiologia del CNR
Via S. Zeno 51, I - 56127 Pisa, Italy

ABSTRACT

Recent findings on the structural and functional properties of the striate cortex of cats and primates are briefly reviewed. In particular these findings show that: i) different stimulus attributes are processed at least partially in parallel, ii) the responses of single neurones to a given visual stimulus are context-dependent, iii) some degree of neural plasticity is present even in the adult visual cortex, iv) multiplexed temporal codes may be used by visual neurons to transmit multiple messages about different stimulus qualities. Some properties of extrastriate visual areas are suggestive of further stages of visual processing that may be relevant for pattern recognition.

1. INTRODUCTION

The primary (striate) visual cortex analyses and transforms information coming from the retina through the Lateral Geniculate Nucleus (LGN) and transmits information about different stimulus properties to visual areas of higher order, to be further processed. The main properties of the primary visual cortex of cats and monkeys are the retinotopic representation, with a increased magnification of the more central portions of the visual field, the selectivity of single neurons for various stimulus parameters (for instance orientation, colour, velocity, etc.) and the cytoarchitectonic arrangement of cells with similar stimulus selectivity and ocular dominance[1].

While these properties are well known, the interpretation of their role in the processes that lead to visual perception is still a matter of debate. For instance, cortical neurons have been claimed to play a role in visual perception as *feature detectors*, or alternatively to act as spatial- and temporal-frequency *filters*[2].

A number of recent findings have improved our knowledge of the anatomical and neurophysiological properties of the striate cortex and have substantially modified previous models of its functional organization. Some of these recently discovered properties are particularly relevant to a discussion on visual primitives.

2. PARALLEL NEURAL PATHWAYS AND
   PROCESSING STREAMS

There is clear anatomical and physiological evidence for the existence in primates of parallel neural pathways that originate within the retina and remain largely (though not totally) segregated through the LGN and the striate (V1) and parastriate (V2) visual cortex[3-5]. These pathways ultimately project to different extrastriate areas in the inferotemporal and in the parietal cortex, respectively, that are known to subserve different tasks, namely object identification and spatial relationships[6].

It has been suggested that separate pathways deal with different types of visual information and that these contribute each to a single aspect of perception [4]. An alternative viewpoint assumes that various parameters of the visual stimuli are dealt with simultaneously by different anatomical pathways and each neural stream contributes to the perception of several attributes of objects in the external world[5]. The former view is that of a strict parallelism of visual processing, the second view favours redundancy of representation of stimulus properties and convergence and divergence of processing streams at subsequent stages of processing. The considerable anatomical interconnection within and between different visual areas would be in favour of the second view[7].

The cytoarchitectonic structure of the striate cortex imposes hardware constraints. The original scheme of the columnar arrangement of striate cortical cells according to their orientation preference and ocular dominance[8] has been replaced for the monkey by the more elaborate scheme that includes the so called "blobs" of layers 2 and 3, where non-oriented cells with spectrally opponent receptive fields prevail and which alternate with interblob regions where cells are orientational and many of them are not colour-coded[9].

The development of new optical methods has offered a more detailed account of the distribution of orientation and ocular dominance columns in the monkey striate cortex. Data on orientation preference and selectivity inferred from images of the cortical surface during visual stimulation show that the orientation columns are organized according to two competing schemes: *linear zones*, where orientation preference varies linearly along one direction and remains constant in the other, and discontinuities, where orientation preferences rotate continuously. The latter take the form of *singularities* if the orientation rotates through ±180 deg along a circular path. These are located near the centers of ocular dominance columns, where the cells are strongly dominated by one eye. The linear zones run perpendicularly to the edges of the ocular dominance columns: here the cells are highly selective for orientation and receive balanced inputs from the two eyes. In the frame of models of the striate cortex based on feature *detectors*, it may be speculated that the linear zones and the singularities could subserve detection of oriented edges and detection of textured surfaces, respectively[10, 11].

3. LONG RANGE INTERACTIONS

Most cells in the monkey striate cortex have small receptive fields. The receptive field of cortical cells, defined by the region where a visual stimulus can cause the cell to fire, was originally considered to have fixed properties (at least in adult animals) and, by definition, to be unaffected by stimuli located outside its boundary. In view of the modular structure of the striate cortex, and its subdivision in hypercolumns, each dealing with a small portion of the visual field, cells within a module were considered to be largely independent of the activity of cells of other modules having non-overlapping receptive fields. This view is not supported by recent anatomical and electrophysiological findings. The discovery of long range horizontal connections, and in particular those mediated by pyramidal neurons in the upper layers of the cat and monkey striate cortex, supplied anatomical evidence that cells having the same orientation preference, but widely separated receptive fields can interact with each other[12, 13]. Electrophysiological methods have shown that the response of a cell to a stimulus within the receptive field can be substantially modified by stimuli in the surrounding regions of the visual field, although the latter are unable to elicit a direct response from the cell[14, 15]. The response of a cell to a stimulus in the preferred orientation can be affected by stimuli presented simultaneously in the orthogonal orientation or in other orientations, even if the latter do not directly activate the cell[16]. Thus the response of a cell to a complex stimulus, comprising many elements and extending beyond the receptive field, cannot be predicted from its response to a simple stimulus. The filter properties of cortical cells are not static, but may change according to the visual context.

These findings are relevant to the understanding of some perceptual phenomena, like the tilt illusion and the preattentive detection (pop-out) of elements of a given orientation from a background of lines of different orientation[13, 17].

Simultaneous recordings with pairs of electrodes picking up the activity of cells from separate regions of the visual cortex showed that the discharges of cells located in different columns, but with the same orientation preference and the same ocular dominance are correlated, even for cells separated by several millimeters. This is indicated by a peak in the cross-correlogram of the discharges of the cells in response to a visual stimulus[18]. Similar findings have been reported for colour-coded non-oriented cells of monkey V1: correlated firing is found for cells of different blobs, provided they have the same type of color-opponency[19].

The horizontal connections may exert both excitatory and inhibitory influences onto the target cells, and their effect may depend upon the level of activation of other inputs converging onto the cell[13]. Thus they may have dynamic, context-dependent modulatory influences on the properties of the receptive fields.

4. SYNCHRONOUS NEURONAL OSCILLATIONS IN
THE CAT STRIATE CORTEX

The possibility of functional interactions between cortical cells with non-overlapping receptive fields, but similar tuning properties, is confirmed by the synchronization of high frequency firing in cells of the cat striate cortex[20]. Many cortical cells, mainly complex cells of the supra- and infra-granular layers, fire rythmically at frequencies between 30 and 50 Hz during presentation of visual stimuli. Synchronization of the rythmic activity during suitable stimulation has been observed either for cells having overlapping receptive fields, irrespective of their orientational tuning, or for cells having the same orientation preference, but located in different orientation columns[21].

These findings are relevant to the problem of the representation of visual objects in terms of neural activity[22]. It has been suggested that visual objects are represented by the synchronous activity of assemblies of neurons at a relatively early stage in the neural pathway, rather than by single "grand-mother cells" at a much higher level. The synchronization of the activity of cells with similar stimulus specificity may be interpreted as a means to bind together similar features from the visual scene in order to represent an object in terms of the activity of assemblies of cells. This suggests a functional role in visual processing for the rythmic oscillations observed in neurons of the primary visual cortex. On the other hand, the synchronous oscillations of cells of different orientation columns, but the same orientation preference, might be simply an epiphenomenon of their being connected with the same pyramidal cell[13]. Or may be a natural consequence of intrinsic cellular mechanisms combined with inhibitory feedback[23].

Forthy - fifty Hz oscillations seem to be a very general phenomenon in the nervous system: coherence of oscillatory activity in the 40 Hz range between subcortical and cortical sites has recently been found in the human brain by magnetic encephalography[24].

5. PLASTICITY IN THE ADULT VISUAL CORTEX

The neural plasticity of the striate cortex, revealed by deprivation experiments in cats and monkeys, is limited to an early postnatal developmental period of variable duration in different species: the so-called critical period[8]. In adult animals the structural and functional properties of the striate cortex have been believed not to be substantially modifiable by anomalies of visual experience. There is recent evidence, however, that functional and structural rearrangements may occur in the adult striate cortex, for instance a remapping of retinotopic projections and changes in receptive field size after localized retinal lesions[25].

An interesting question is whether repeated sensory stimulation may induce plastic changes in the adult cortex underlying learning processes. In the somatosensory system of the monkey, training in a tactile discrimination task can induce changes in cortical topography and in receptive field size of cortical neurons[26]. Similar evidence is not available so far for the primary visual cortex. However, examples of visual perceptual learning in man indicate that visual experience may improve considerably the performance in visual discrimination tasks, and that the improvement is limited to the area of the visual field that has been exposed to the training stimuli and does not transfer to adjacent areas. Moreover, the effects of training, that are very longlasting, are selective for properties of the stimuli as contour orientation, spatial frequency content, spectral composition, etc.[27]. These findings seem therefore to reflect changes occurring in the visual system at a cortical level where the visual field is represented retinotopically and where orientation and spatial frequency selectivity are preserved. Some of these findings might be consistent with plastic changes occurring in the receptive fields of striate cortical neurons. Others, however, are more likely to imply plasticity at an extrastriate cortical level, possibly in area V2. Recent findings on monkeys with ablations of visual area V4 suggest that this area may be crucial for transfer of visual learning from a trained to an untrained region of the visual field[28]. Thus, whether training in a visual discrimination task can induce plastic changes at the level of the striate visual cortex remains an open question.

6. NON-VISUAL INPUTS TO THE STRIATE VISUAL CORTEX

The visual sensory messages carried by LGN afferences are not the only input to the striate cortex. There is evidence in both cats and monkeys that the response to a visual stimulus of striate cortical neurons can be modified by information about the position of the eye in the orbit. In the cat, inflow information from proprioceptive receptors in eye muscles reach the primary cortex[29] and disruption of proprioceptive afferences impairs binocular steroacuity of the animal[30]. In behaving monkeys, the processing of binocular disparity leading to stereopsis is modulated by changes in ocular vergence and accommodation[31]. These findings indicate a considerable degree of interaction between sensory and oculomotor information in the striate cortex.

In extrastriate visual areas, the receptive field properties of single neurons can be substantially modified according to whether the animal is paying attention to the stimulus located in the receptive field[32]. This seems not to occur in the striate cortex.

7. NEURONAL CODE FOR CARRYING VISUAL INFORMATION

The intrinsic code used by visual neurons to carry visual information is largely unknown. The classical hypotesis is that visual information is encoded and transmitted in terms of the mean rate of discharge. However, the spike frequency of neurons at the various stations of the visual system, and in particular in the striate visual cortex, depends upon several stimulus parameters, like stimulus contrast, spatial and temporal frequency, contour orientation, etc. Therefore, the strength of the response to a visual stimulus cannot represent unambiguously the various stimulus parameters.

Another possibility is that information is encoded in terms of the temporal modulation of the rate of firing[33] and that multiplexed temporal codes are used to transmit multiple messages related to various properties of the stimulus. Evidence in favour of this hypothesis has been presented for neurons at four stations of the neural visual system, the retina, LGN, V1 and inferotemporal cortex (IT)[34]. The results obtained from behaving monkeys indicate a greater efficiency of temporal encoding with respect to mean-rate encoding in the transmission of visual information. The complexity of the temporal messages increases from the peripheral to the central stations of the visual system and the temporal codes at the various stations overlap in time. Thus concurrent processing may take place at different visual areas.

REFERENCES

1. D.H. Hubel and T. Wiesel, Receptive fields and functional architecture of monkey striate cortex, *J. Physiol. (Lond.)*, Vol.195, pp.215-243 (1968).
2. L. Maffei, Spatial frequency channels, in *Handbook of Sensory Physiology, Vol.VIII: Perception*, R. Held, H.W. Leibowitz, and H.L. Teuber eds., Springer-Verlag, Berlin, D, pp.39-63 (1978).
3. D.H. Hubel and M.S. Livingstone, Segregation of form, colour and stereopsis in primate area 18, *J. Neurosci.*, Vol.11, pp.3378-3415 (1987).
4. M.S. Livingstone and D.H. Hubel, Psychophysical evidence for separate channels for the perception of form, colour, movement and depth, *J. Neurosci.*, Vol.11, pp.3416-3468 (1987).
5. E.A. DeYoe and D.C. Van Essen, Concurrent processing streams in monkey visual cortex, *Trends Neurosci.*, Vol.11, pp.219-226 (1988).
6. M. Mishkin, L.G. Ungeleider, and K.A. Macko, Object vision and spatial vision. Two cortical pathways, *Trends Neurosci.*, Vol.6, pp.414-417 (1983).
7. D.C. Van Essen, C.H. Anderson, and D.J. Felleman, Information processing in the primate visual system: an integrated system perspective, *Science*, Vol.255, pp.419-423 (1992).
8. D.H. Hubel and T.N. Wiesel, Functional architecture of macaque monkey visual cortex, *Proc. Roy. Soc. Lond. B*, Vol.198, pp.1-59 (1977).
9. M.S. Livingstone and D.H. Hubel, Anatomy and physiology of a colour system in the primate visual cortex, *J. Neurosci.*, Vol.4, pp.309-356 (1984).
10. G.G. Blasdel, Differential imaging of ocular dominance and orientation selectivity in monkey striate cortex, *J. Neurosci.*, Vol.12, pp.3115-3138 (1992).
11. G.G. Blasdel, Orientation selectivity, preference and continuity in monkey striate cortex, *J. Neurosci.*, Vol.12, pp.3139-3161 (1992).
12. C.D. Gilbert and T.N. Wiesel, Columnar specificity of intrinsic horizontal and corticocortical connections in cat visual cortex, *J. Neurosci.*, Vol.9, pp.2432-2442 (1989).
13. C.D. Gilbert, Horizontal integration and cortical dynamics, *Neuron*, Vol.9, pp.1-13 (1992).
14. L. Maffei and A. Fiorentini, The unresponsive regions of visual cortical receptive fields, *Vision Res.*, Vol.16, pp.1131-1139 (1976).
15. C.D. Gilbert and T.N. Wiesel, The influence of contextual stimuli on the orientation selectivity of cells in primary visual cortex of the cat, *Vision Res.*, Vol.30, pp.1689-1701 (1990).
16. M.C. Morrone, D.C. Burr, and L. Maffei, Functional implications of cross-orientation inhibition of cortical visual cells. I: Neurophysiological evidence, *Proc. R. Soc. Lond. B*, Vol.216, pp.335-354 (1982).
17. D.C. Van Essen, E.A. De Yoe, J.F. Olivarria, J.J. Knierim, J.M. Fox, D. Sagi, and B. Julesz, Neural responses to static and moving texture patterns in visual cortex of the macaque monkey, in *Neural Mechanisms of Visual Perception*, D Man-Kit Lam and C.D. Gilbert eds., PPC-GPC, Houston, TX, pp.137-154 (1989).

18. D.Y. Ts'o, C.D. Gilbert, and T.N. Wiesel, Relationships between horizontal interactions and functional architecture in cat striate cortex as revealed by cross-correlation analysis, *J. Neurosci.*, Vol.8, pp.1160-1170 (1986).

19. D.Y.Ts'o and C.D. Gilbert, The organization of chromatic and spatial interactions in the primate striate cortex, *J. Neurosci.*, Vol.8, pp.1712-1727 (1988).

20. C.M. Gray and W. Singer, Stimulus-specific neuronal oscillations in orientation columns of cat visual cortex, *Proc. Natl. Acad. Sci.,* USA, Vol.86, pp.1698-1702 (1989).

21. C.M. Gray, P. Koenig, A.K. Engel, and W. Singer, Oscillatory responses in cat visual cortex exibit intercolumnar synchronization which reflects global stimulus properties, *Nature*, Vol.338, pp.334-337 (1989).

22. C.M. Gray, A.K. Engel, P. Koenig, and W. Singer, Synchronous neuronal oscillations in cat visual cortex: functional implications, in *Representations in Vision*, A.Gorea ed., Univ. Press, Cambridge (1990).

23. A.B. Bonds, R.K. Snider, J.F. Kabara, P. Bush, and T.J. Seinowski, On the origin of oscillations in cells of the cat striate cortex, *Inv. Ophth. Vis. Sci.*, Vol.34, p.909 (1993).

24. U. Ribary, R. Llinas, F. Lado, A. Mogilner, A. Ioannides, R. Jagow, M. Joliot, and J. Volkmann, Origin and characteristics of coherent thalamo-cortical 40-Hz oscillations in the human brain, *Soc. Neurosci. Abs.*, Vol.18, p.1420 (1992).

25. C.D. Gilbert and T.N. Wiesel, Receptive field dynamics in adult primary visual cortex, *Nature*, Vol.356, pp.150-152 (1992).

26. G.H. Recanzone, M.M. Merzenich, and W.M. Jenkins, Frequency discrimination training engaging a restricted skin surface results in an emergence of a cutaneous response zone in cortical area 3a, *J. Neurophysiol*, Vol.67, pp.1057-1070 (1992).

27. A. Fiorentini and N. Berardi, Limits in pattern discrimination: central and peripheral factors, in *Vision and Visual Dysfunction*, J.J. Kulikowsky, V. Valsh and J.J. Murray eds., MacMillan, London, UK, Vol.5 (1991).

28. P. Schiller, The effects of V4 and middle temporal (MT) area lesions on visual performance in the rhesus monkey, *Visual Neurosci.*, Vol.10, pp.717-746 (1993).

29. P. Buisseret and L. Maffei, Extraocular proprioceptive projections to the visual cortex, *Exp. Brain Res.*, Vol.28, pp.421-425 (1977).

30. A. Fiorentini, M.C. Cenni, and L. Maffei, Impairment of steroacuity in cats with oculomotor proprioceptive deafferentation, *Exp. Brain Res.*, Vol.63, pp.364-368 (1986).

31. Y. Trotter, S. Celebrini, B. Sticanne, S. Thorpe, and M. Imbert, Modulation of neural steroscopic processing in Primate area V1 by the viewing distance, *Science*, Vol.257, pp.1279-1281 (1992).

32. J. Moran and R. Desimone, Selective attention gates visual processing in the extrastriate cortex, *Science*, Vol.229, pp.782-784 (1985).

33. A. Cattaneo, L. Maffei, and M.C. Morrone, Patterns in the discharge of simple and complex visual cortical cells, *Proc. R. Soc. Lond.* B, Vol.212, pp.279-297 (1981).

34. J.W. McClurkin, L.M. Optican, B.J. Richmond, and T.J. Gawne, Concurrent processing and complexity of temporally encoded neuronal messages in visual perception, *Science*, Vol.253, pp.675-677 (1991).

VISUAL FORM REPRESENTATION

Gabriella Sanniti di Baja

Istituto di Cibernetica del CNR
Via Toiano 6, I - 80072 Arco Felice (NA), Italy

ABSTRACT

Schemes to compact spatial data into representations that facilitate the computation of geometrical properties and favor shape description are discussed in this paper. Any region, resulting after image segmentation, can be basically represented in terms of its external characteristics (the contour of the region), or of its internal characteristics (the pixels constituting the region). Hierarchical data structures, based on the principle of recursive decomposition, as well as approximate representations are also discussed.

1. INTRODUCTION

Picture processing necessarily involves the design of methods for both storing and manipulating images inside an electronic device. In particular, picture storing is a crucial task because pictorial data, as they appear after the acquisition and digitization process, require a very large number of bits. Moreover, the ability to efficiently process data greatly depends upon the suitability of the encoding technique used to represent the input image. Let us consider a simple task: the transmission of an alphanumeric character. A reasonably good reproduction of the form of a character requires a binary matrix having at least 15 columns and 20 rows, i.e., 300 bits. On the other hand, if the character is analysed and recognized, 8 bits would be enough to store and transmit its identifying code. This implies that memory occupation may diminish even by a reduction factor of 40.

Depending on the specific problem, different techniques can be adopted for pictorial data representation. When data compression is mainly motivated by the need of reducing memory occupation, the coding technique should be information preserving, in the sense that the input data should be perfectly reconstructed starting from the adopted representation scheme. When data compression is required in the framework of feature extraction applications, the coding technique should reduce memory occupation in such a way that enough information is preserved to automatically distinguish patterns belonging to different classes. In this case, the coding technique produces an approximate representation.

Images may appear in a variety of different forms, so that different coding methods can be devised. A monochrome grey level digital image I is an m x n array, whose pixels denote discrete light intensity values. It is a common practice in image processing to let m and n be integer powers of 2 and, at least when dealing with hierarchical data structure based on recursive decomposition, it is $m = n$. The number of permitted grey level values is generally at most 256, so that 8 bits are required to store each pixel, and 8 x m x n bits are necessary for the entire array. When colour, multispectral, or moving images are of interest, memory occupation increases proportionally to the number of colours, spectral components or time-ordered images constituting the sequence.

When the global effect produced by the silhouette of a pattern is sufficient for its recognition, the number of grey levels can be limited to two. This reduces memory occupation to m x n bits. Further compression of binary images can be achieved by means of contour-based and region-based representation systems. These representation systems are discussed in the following sections. An exhaustive study of this subject can be found in the literature (see for example[1-4]).

2. PRELIMINARY NOTIONS

Let B and W be a pattern and its complement in a 2^k x 2^k binary image I, digitized on the square grid. The sets B and W are also referred to as the sets of the black pixels and the white pixels. The 8-metric and the 4-metric are used for B and W, respectively. Without loss of generality, we suppose that B is constituted by a unique 8-connected component; no assumption is done on the number of 4-connected components constituting W. The frame of I is the set of all pixels in the first or last row of I, or the first or last column of I. We assume that all the pixels of the frame are white.

For every black pixel p, let $N(p) = \{n_k \mid 0 \le k \le 7\}$, where $n_0,..., n_7$ are the pixels successively encountered while going clockwise around p starting from the pixel to the north-east of p. The n_k are also called the neighbours of p.

The contour C of B includes every black pixel p for which at least one n_k, k odd, is white. Accordingly, C is an 8-connected curve, when B is simply connected, while is union of 8-connected curves, when B is multiply connected.

The Distance Transform DT of B is a multivalued replica of the pattern, where each pixel is labeled with its distance from W. In the DT, the label of a pixel measures the radius of the disc that can be centered on the pixel and fits the shape of the pattern. The shape of the disc depends on the distance function used to compute the DT. The discs are diamond shaped when using the city-block distance d_4, and square-shaped with the chessboard distance d_8. The non circular shape of these discs is due to the fact that both the city-block and the chessboard distance functions provide a quite rough approximation to the Euclidean distance. To get a closer approximation, one should suitably weigh the unit moves from a pixel p to any of its neighbours, so as to take into account that horizontal/vertical moves and diagonal moves have different Euclidean length. The resulting distance function is called a weighted distance function. To avoid resorting to real numbers, suitable integer weights are used. A thorough investigation of the weighted distance functions and weight selection can be found in Borgefors[5,6]. A good approximation is obtained by using the weights 3 and 4, for the horizontal/vertical and diagonal moves, respectively. The corresponding distance function is denoted by $d_{3,4}$. If this function is used to compute the DT, the disc associated with any pixel is octagon-shaped. An even better approximation of the Euclidean distance can be obtained by enlarging the set of neighbours of p, so as to include also the pixels that could be reached by the knight's move (in the game of chess). In this case, a third weight has to be introduced.

3. CONTOUR-BASED REPRESENTATION SYSTEMS

3.1. Information Preserving Representations

The contour of B can be stored as a list of planar coordinates of its pixels. Only two integer numbers have to be recorded for each contour pixel, which implies at most 16 bits per contour pixel, when the spatial resolution of I is 256 x 256.

The order in which the contour pixels are recorded influences the suitability of the obtained coding. If the contour pixels are identified and recorded while I is raster scan inspected, the obtained list does not facilitate pattern manipulation since pixels which are stored one after the other in the list are not necessarily neighbours along the contour. Contour tracing, in turn, allows one to keep track of the coordinates of the contour pixels according to the order in which the pixels are found along the contour.

The chain-code representation allows to reduce storage requirement and favours pattern processing[7]. When tracing the contour (clockwise) and moving from one pixel to the successive contour pixel, one of eight possible moves is done. Each move can be identified by one of the digits 0, 1,..., 7 and coded by 3 bits. Thus, the contour is completely defined by the coordinates of the pixel from which tracing is started (say the first black pixel found when raster scan inspecting I in forward fashion), and by a chain-code representing the sequence of moves done while entirely tracing the contour. The string "444334355433355557777700007777771121332111" is the chain-code representation for the pattern shown in figure 1.

As the obtained string generally contains sequences of repeated codes, a more compact representation is possible in terms of pairs of integers, where the first integer in the pair denotes the chain code, and the second is its occurrence number in the sequence.

The crack-code representation is also information preserving. If we follow the cracks between the pattern and its complement, one of four possible directions is taken for each move: left, top, right and bottom. Thus, only four different codes are employed, compared to eight used in the chain-code representation. However, the obtained string is generally longer, since more than one code may be necessary in correspondence with any contour pixel.

Geometric properties of B, as perimeter, area and higher order moments, can be directly computed from both the chain-code and the crack-code representations.

Computing the perimeter is a straightforward task, when using a contour-based represen-tation. The perimeter can be defined as the number of contour pixels. Alternatively, the

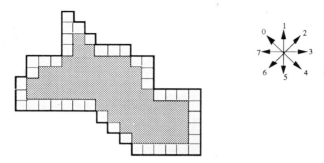

Figure 1. The string 444334355433355557777700007777771121332111 is the chain-code representation, obtained while clockwise tracing the contour starting from the topmost pixel.

perimeter can be defined as the number of unit moves done while completely tracing the contour. This number is larger than the number of contour pixels in presence of one-pixel wide pattern subsets, which are traced twice while following the contour. When the chain-code representation is used, a quasi Euclidean measure of the perimeter can be easily obtained, by counting any diagonal move as $\sqrt{2}$.

The area of each connected component of B is the number of black pixels constituting it. We limit ourselves to the case of the crack-code representation, and point out that a similar criterion allows area computation starting from the chain-code representation. Let us place the origin of the Cartesian coordinate system in the bottom left corner of the contour pixel from which clockwise contour following has been started when crack-coding B (see figure 2). The integral can be computed by considering B as made up of one-pixel wide columns, having different heights. Thus, the problem is that of computing the area of a single column. Let c^i_x and c^i_y be the projections on the x and y axes of the i-th crack-code c^i. As a crack-code corresponds to a unit horizontal/vertical move, any projection assumes one among three possible values: -1, 0, 1. Then, the ordinate y^k of any top corner of the k-th contour pixel, and the area of B can be respectively computed as:

$$y^k = \sum_{i=1}^{k} c^i_y \qquad\qquad A = \sum_{i=1}^{n} y^{k-1} c^i_x \qquad\qquad (1)$$

3.2. Approximate Representations: Polygonal Approximation

The contour can be approximated piecewise linearly, or by using higher order curves. Here, we limit ourselves to the polygonal approximation, and briefly describe two basic approaches (merging type and splitting type). A comprehensive description of contour approximation techniques can be found in Pavlidis[8].

Purpose of the polygonal approximation is to identify as vertices suitable pixels of C such that for each pair of successive vertices all the intermediate contour pixels are aligned in the limits of the adopted tolerance. Although, in principle, all the contour pixels are likely to be vertices of the polygonal approximation, only the extremes of any run of equally chain-coded pixels are considered. In fact, the intermediate pixels of the run are aligned along the direction denoted by the corresponding chain-code. The list of the extremes of the runs of equally chain-coded pixels defines the polygonal approximation with tolerance equal to 0, i.e., an information preserving representation, and provides the input data from which to extract the vertices of any

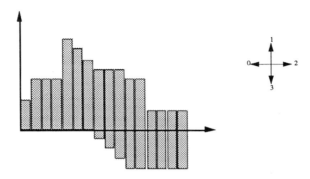

Figure 2. To compute the area, the pattern is interpreted as union of unit-wide rectangles.

polygonal approximation with tolerance greater than 0.

According to the merging type approach, starting from any contour arc delimited by two successive candidates, one keeps extending the arc as long as it remains a relatively good fit to a straight line. When straightness is violated, a new vertex is added to the list of vertices. The process terminates when the contour has been completely examined. An example is shown in figure 3, where dots denote the extremes of runs of equally chain-coded pixels, and black dots are the detected vertices. The weak point of this algorithm is its instability under isometric transformation of the pattern since the location of the vertices is conditioned by the location of the first vertex and by the order in which the contour pixels are examined.

According to the splitting type approach, the vertices are identified in a more "parallel" fashion. The input set of candidates is initially divided into two subsets, by suitably selecting two vertices. If the chord joining the selected vertices is not a good fit to any of the two arcs into which the contour is partitioned, each arc is subdivided into two sub-arcs at the point farthest from the chord. New sub-arcs are obtained and, if the corresponding chords do not provide a suitable approximation, the corresponding arcs are newly split into two sub-arcs. The process is iterated until any obtained chord provides a good fit to the corresponding arc. The weak point of this method is the selection of the two vertices necessary to accomplish the initial partition of the contour into two arcs. The two vertices should be identified by using some rotation invariant criterion, to have a stable representation. For instance, they could be found as the pixels farthest from the baricenter of the pattern.

A drawback common to both the above polygonal approximation approaches is that they require to *a priori* fix a value for a tolerance threshold on the distance of the candidates from a chord. Unfortunately, a unique threshold value is generally not adequate to correctly handle the entire contour, and some context dependency should be taken into account.

As the tolerance threshold increases, the number of vertices of the corresponding polygonal approximation generally decreases, and the representation becomes rougher and rougher.

3.3. Multiresolution and Hierarchical Representations

If the polygonal approximation is repeatedly accomplished by using different (increasing) threshold values, a multiresolution representation of the pattern becomes available. In particular, when the splitting type approach is followed, if the two vertices necessary to perform the initial partition of the contour into two arcs are fixed, the set of vertices found with a given threshold value, is a proper subset of the set of vertices found with the preceding (smaller) threshold. Thus, the multiresolution representation can be directly derived from the polygonal

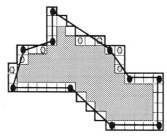

Figure 3. Black dots denote the vertices of a polygonal approximation of merging type; the first vertex is the topmost contour pixel and the contour is traced clockwise.

approximation performed with the smallest (positive) threshold, provided that for each vertex, together with the planar coordinates, also the distance from the chord is recorded. A vertex, found with the smallest positive threshold, is still a vertex in a resolution level corresponding to a higher threshold if its recorded distance is larger than the higher threshold. In this way, one can provide a compact representation, valid whichever is the adopted threshold, i.e., the resolution level.

A tree structure can be defined to provide a hierarchical representation in terms of successive approximations with increasing tolerance threshold. In the tree structure, each node of a given hierarchy level represents a polygon side, and the sons of the node are its immediate refinements, i.e., the corresponding sides at the next more accurate approximation. The sides of the coarsest polygon are the sons of the root node. The number of hierarchy levels depends on the number of adopted thresholds.

3.4. Approximate Representations: Curvature Maxima and Minima

A piecewise linear approximation of the contour can be obtained by joining the curvature maxima and minima. Curvature is ordinarily defined as the rate of change of slope, the slope being a multiple of 45°. A possible criterion to identify curvature maxima and minima is based on the use of the curvature codes. Each black pixel p of C is assigned a value counting the number of white pixels in $N(p)$. With reference to figure 4, one may note that pixels located on straight lines, aligned along any of the eight principal directions in the digital plane, are labeled 3; in turn, pixels placed in correspondence with contour concavities and convexities, i.e., the curvature minima and maxima, are labeled less than 3 and more than 3, respectively.

Curvature maxima and minima identified by using a small support (the contour arc joining the pixel to its preceding or following neighbour along the contour) provide a representation which is strongly conditioned by contour noise. Thus, one can define the "left" and "right" k-slope, as the slope of the straight line joining a pixel with a contour pixel k steps away along the contour in any of the two directions. Accordingly, the k-curvature at a pixel p is the difference between its left and right k-curvature. Selecting the proper value for k is the crucial point.

Starting from an initial set of vertices, iterative relaxation methods produce, at each iteration of the process, a refined selection based on the local context. The classification obtained at each stage of the process has to be interpreted as tentative, rather than firm, as it is when a polygonal approximation is performed.

To extract information on contour curvature from a larger support, one can proceed as suggested in Gallus and Neurath[9]. A curvature n-code c_i^n is assigned to every contour pixel by using the differential chain codes c_j [10], in a contour neighbourhood ($i-n+1 \leq j \leq i+n-1$) of size

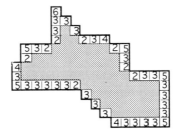

Figure 4. The contour pixels are labeled with the number of their white neighbours n_k (k=1, 8).

$(2n-1)$ pixels. Each c_i^n, $n>1$, is derived from the c_j's by means of the following expression:

$$c_i^n = n \times c_i + \sum_{k=1}^{n-1} (n - k) \times (c_{i-k} + c_{i+k})$$ (2)

The value of n need not be large; for instance, n=4 was sufficient in the quoted paper[9] to deal with chromosome contours.

One may generate various representations of the same contour, by using a suitable function of these codes, designed to enhance the curvature minima and maxima. The utility of this approach is twofold: it is possible to get rid of the contour noise in a small number of iterations; it is possible, by iterating the process until a stable configuration is obtained, and by introducing a threshold on the significance that curvature minima and maxima should have to be retained as vertices, to generate a set of different representations of the same contour at different resolutions. This multiresolution approach can be seen as convenient to facilitate recognition.

3.5. Approximate Representations: Dominant Points Detection

A piecewise approximation can be obtained by identifying the dominant points, i.e., the pixels that perceptually dominate contour arcs which delimit convex portions of B and whose curvatures are sufficiently high. Several procedures for detecting dominant points can be found in the literature (see for instance Teh and Chin[11], which discusses the performance of many of them). It is difficult to find a significant set of dominant points when the contour contains details of different sizes. A curvature-based definition of dominant points is satisfactory only when the resolution at which the contour has to be analyzed is known. To overcome this problem, one can extract sets of dominant points at various resolutions, by changing the value of n in the n-code according to some perceptual criterion.

In the quoted paper[11], the support on which the significance measure has to be accomplished is found for each contour pixel p_i, by using as parameters the length (l_{ik}) of the chord and the distance (d_{ik}) of the pixel from the chord. For p_i, the support is extended from k to $k+1$ pixels on both sides, until the following conditions are satisfied:

$$l_{ik} \geq l_{ik+1} \quad \text{or} \quad \begin{cases} d_{ik} / l_{ik} \geq d_{ik+1} / l_{ik+1} & \text{if} \quad d_{ik} \geq 0 \\ d_{ik} / l_{ik} \leq d_{ik+1} / l_{ik+1} & \text{if} \quad d_{ik} < 0 \end{cases}$$ (3)

Then a significance measure (for instance the k-curvature at p_i) can be done on the support properly identified.

In Arcelli and Ramella[12], the pixels having perceptual significance are directly identified, by means of an iterative process. The extremes of runs of pixels having the same chain-code constitute the starting set of candidates. For each candidate, perceptually significant parameters (e.g., the distance from the chord, or the area of the triangle whose vertices are the contour pixel and the extremes of the chord) are computed and the mean values of the parameters are determined. A comparison between the significance of a given candidate and the mean values is then used to establish whether the candidate has to be maintained in the set of vertices at that iteration. Also the past history of that vertex, suitably stored iteration after iteration, is taken into account. Removal of some vertices automatically causes the increase in size of the support of some of the remaining vertices. At the next iteration, the perceptual parameters and the mean values are newly computed. The process is iterated until it reaches a stable configuration. In this way a number of possible representations of the contour are available, from the one allowing complete recovery to the most roughly approximated one.

4. REGION-BASED REPRESENTATION SYSTEMS: BLOCKS

An image can be decomposed into a number of maximal blocks, each of which can be fully represented in a compact way. The run length coding and the quadtree representation are based on the decomposition into disjoint maximal blocks; the *MAT* representation originates a decomposition into partially overlapping maximal blocks.

4.1. Run Length Coding

Each row of a grey level image consists of a sequence of runs of pixels having the same grey level. These runs can be seen as blocks into which the image can be decomposed. In case of binary images, each row is completely specified by giving the length of the successive runs and the colour (black or white) of the first run only.

When *I* has a few rows, this representation is very economical and is often used for picture compression purpose. The number of bits required to store the length of the *r* runs in a row, which includes *n* pixels, is r $\log_2 n$ bits, since it takes $\log_2 n$ bits to specify the length of a single run (ranging from 1 to *n*). Since it is necessary to record the colour of the first run in the row, $1 + r \log_2 n$ bits are required for each row, in place of *n* bits necessary to store the row as it is.

The area of the pattern, when this consists of a unique connected component, is the sum of the lengths of the runs of black pixels. In case of more than one component of black pixels, a connected component labeling technique - still possible for a run length coded image - is necessary to identify the various components, prior to area computation. Perimeter can also be evaluated by taking into account the relations between black runs in a row with adjacent runs in the preceding and following rows. The maximal extent in the horizontal direction is defined by the run with the maximal length. Note that the maximal extent does not generally coincide with the maximal thickness, as this could occur in a direction different from the horizontal one. It is also possible to find the minimum rectangle enclosing the pattern, as well as the location of the centroid. This centroid could be useful for certain operations to be performed on the pattern, or as a reference point for the pattern itself.

4.2. Hierarchical Representations: the Binary Tree

A hierarchical representation of each row of the image is possible, in terms of a binary tree. See figure 5, referring to the row "0000110011100000" of an image *I* with size $2^k \times 2^k$, k=4. Since the pixels in the row have different color, the row is split in the middle, so as to divide it into two subsets. The splitting process is repeated, until black and white pixels still coexist. (The number of necessary iterations is at most k, when black and white pixels alternate.) In this

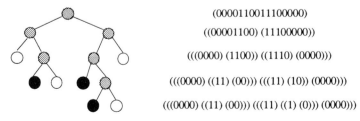

(0000110011100000)

((00001100) (11100000))

(((0000) (1100)) ((1110) (0000)))

((((0000) ((11) (00))) (((11) (10)) (0000)))

((((0000) ((11) (00))) (((11) ((1) (0))) (0000)))

Figure 5. Binary tree associated with the row: 0000110011100000.

way, a binary tree is built, where each node has at most two descendants, corresponding to the two halves in which it can be divided. The root is level 0 and coincide with the entire row. The root and the intermediate nodes are grey, as they include both black and white pixels. The leaf nodes represent the uniformly colored subsets, and the level at which they are found indicate the step of the splitting process at which the nodes satisfy the homogeneity criterion. Note that the obtained nodes are not necessarily runs of maximal length. In fact, a maximal run can be split into more than one node, depending on its length and position along the row, which could be not the power of 2 corresponding to the level at which the node is found.

The nodes at any level h<k, represent subsets of length 2^{k-h}, located at multiples of 2^{k-h}. To store the tree, the necessary amount of memory is proportional to the number of nodes. Generally speaking, this representation does not seem to be economical, nor has been really used for practical application.

4.3. Hierarchical Representations: the Quadtree

As the binary tree, also the quadtree representation is based on the principle of recursive decomposition and provides a hierarchical data structure. Hierarchy allows one to focus on the interesting subset of data, so that a hierarchical representation is more efficient and the computation time when processing hierarchical data structures is often more limited.

If I does not consist entirely of black pixels, or of white pixels, it is subdivided into four equal-sized quadrants. Each quadrant is divided into subquadrants and so on, until the block constituting the square region entirely consists of black or white pixels. The blocks are disjoint and have standard sizes, being the length of the side of the blocks a power of 2, and are placed in standard locations. Since the various blocks have different size, the quadtree can be interpreted as a variable resolution data structure. See figure 6.

The blocks are not necessarily the maximal homogeneous square blocks into which the image can be decomposed, in fact the representation is sensitive to pattern translation.

If blocks of arbitrary size are permitted, the quadtree representation can be conveniently adopted in a split-and-merge decomposition process leading to a subdivision of the pattern into maximal homogeneous blocks.

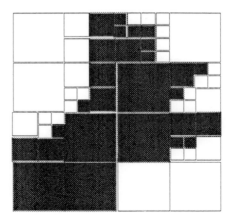

Figure 6. The image decomposition into quadrants originating the quadtree representation.

When dealing with grey level images, the uniformity criterion to stop the division process can be, for instance, based on the comparison between the standard deviation of the grey levels and a threshold. Each block, obtained in this way, may be uniformly labeled with the mean grey level pertaining to the pixels of the block. If this is done, the input image is piecewise approximated, but obviously exact reconstruction of the input image is no longer possible.

The quadtree representation is a close relative of a multiresolution representation system, known as the pyramid representation. A k pyramid is a sequence of arrays such that the array (i-1) is a version of the array i at half the scale of the array i. To build the pyramid, starting from the highest resolution level, groups of 2 x 2 pixels are merged to make a unique "parent" pixel. The root node consists of one pixel. The value of the parent depends on the four children. In an AND-pyramid, a pixel is black provided that all the 4 children are black. For such a pyramid, the root node is empty. See figure 7, referring to the AND-pyramid for an image I with spatial resolution $2^4 x\ 2^4$, where the root node is not illustrated. In an OR-pyramid, the color of a pixel is black provided that at least one of the 4 children is black.

Pyramids are useful for tasks like feature detection and extraction. In fact, one can limit the scope of the search: once some significant information is found at a coarse level, the finer resolution levels can be searched to get further refinements.

4.4. The MAT Representation

Any pattern can be interpreted as the envelope of its maximal discs. A maximal disc is a disc which fits the pattern and is not completely overlapped by any other disc fitting the pattern.

The centers of the maximal discs are located along the medial region of the pattern, in correspondence with the symmetry axes of the pattern. For this reason, the representation based

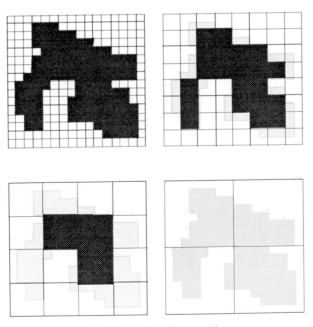

Figure 7. The AND-pyramid.

on the maximal discs of B is called the Medial Axis Transformation (MAT). The MAT is completely defined if the centers of all the maximal discs of B are found and recorded together with the radii of the corresponding maximal discs. Although a remarkable (information preserving) compression is obtained, we note that, since the maximal discs substantially overlap each other, the set of their centers is not the minimal set from which the pattern can be recovered. A minimal set of centers of maximal discs can be identified, at the expenses of some (non disregardable) extra computation to be accomplished after the centers of the maximal discs have been detected[13].

On the DT, the identification of the centers of the maximal discs of B is a straightforward task. In fact, since the size of the discs increases with the radius, a comparison between the label of p and the labels of its neighbours allows one to establish whether the disc centered on p is completely overlapped by the discs centered on its neighbours [14-17]. In figure 8, the centers of the maximal discs, found on the DT computed by using d_4, d_8 and $d_{3,4}$ respectively, are shown as bold labels.

The set of the centers of the maximal discs is thin, but is not a linear set. Moreover, it is not connected even if the pattern is connected. Thus, the MAT cannot be conveniently used for shape description purposes. However, the MAT is a very close relative of the skeleton, which is a region-based representation system having linear structure and reflecting topological and geometrical properties of the represented pattern.

5. REGION-BASED REPRESENTATION SYSTEMS: THE SKELETON

Thinning is a transformation which, applied to a binary image changes from black to white a number of pixels, without altering image topology, until the set of the black pixels is transformed into a subset T, unit wide and centered within B [18-21].

Elongated patterns, characterized by nearly constant width, can be thinned by repeatedly applying to the contour of the pattern a topological peeling process, aimed at removing the black pixels not necessary to preserve picture topology. When thinning patterns with non constant thickness, one should also preserve from deletion the end points, i.e., the pixels necessary to map in T the tips of the pattern protrusions. In this case, thinning is preferably named skeletonization and the set T is referred to as the skeleton (see for example Arcelli et al.[22]).

The skeleton T is a connected set, if the pattern is connected; has the same connectivity order as B, and each of its loops surrounds a hole of B ; is centered within B ; is unit wide; its pixels are labeled with their distance from W; includes all the centers of the maximal discs of B, except for those whose removal is indispensable to allow T to be a unit wide set.

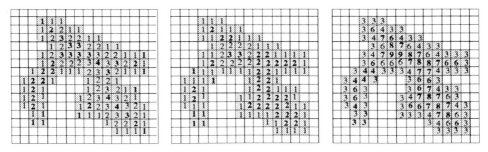

Figure 8. Bold labels denote the centers of the maximal discs on the DT computed by using d_4, d_8 and $d_{3,4}$ respectively.

Skeletonization is seldom completely information preserving, because thinness and inclusion of all the centers of the maximal discs are generally mutually exclusive, and thinness is preferred in most cases. Symmetric reduction of the set of the centers of maximal discs results in a recovered set whose shape is not significantly different from that of B, since the non-recovered pixels tend to be fairly uniformly distributed along the contour of B. Thus, it is still appropriate to talk of reversible (information preserving) skeletonization. This is the case also when a pruning process is accomplished on the skeleton, so as to rid it of branches originating from protrusions regarded as non meaningful in the problem domain.

Skeletonization can be performed by repeatedly applying topology preserving removal operations to the black pixels, which at any given iteration represent the current pattern contour; removable pixels are changed to white provided that they are not identified as end points. Selection of the contour pixels as the only candidates to removal is done to favor an isotropic compression of the pattern towards its innermost part; removal inhibition for the end points is done to avoid shortening excessively skeleton branches, which would otherwise cause lack in the skeleton representative power. The end point detection criterion is the weak point of many algorithms, and often compromises the isotropy of the transformation.

Due to the closeness between the MAT and the skeleton it is not surprising that skeletonization can be conveniently performed on the distance transform of the pattern [23-29]. As an example, the performance of the $d_{3,4}$ DT-based skeletonization algorithm [28] is illustrated in figure 9.

DT-based skeletonization allows to obtain the skeletal pixels directly labeled according to their distance from the complement, makes it easy to detect the centers of the maximal discs as well as the saddle points, and facilitates the detection of the remaining pixels necessary to gain skeleton connectedness. Moreover, using a DT-based skeletonization method, it is not necessary to check a condition specifically tailored to end point detection. In fact, end points are automatically identified when finding the centers of the maximal discs: the end points are centers of maximal discs, placed in correspondence with pattern protrusions. The general scheme of a DT-based skeletonization process can be outlined in a homogeneous way, whichever distance function is used to compute the DT. The process includes three steps: i) computation of the DT, ii) identification of the set ML of the skeletal pixels (i.e., the centers of maximal discs, saddle points and connecting pixels), and iii) reduction of ML to the unit-wide

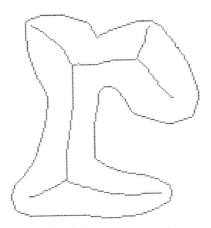

Figure 9. The skeleton of an r-shaped pattern.

skeleton T. On the DT, the identification of the ML can be accomplished at a limited computational cost. One inspection of the DT is sufficient to identify all the centers of maximal discs and the saddle points. (Saddle point detection is done by counting, for each pixel in the DT, the number of components of neighbours with smaller label and with higher label.) The connecting pixels are found by growing paths along the direction of the steepest gradient in the DT, starting from any already found skeletal pixel. In the third step of the process, ML is reduced to unit width, by employing topology-and-end-point preserving removal operations.

The type of distance function used to build the DT is important as concerns stability of skeletonization under pattern rotation. When rough approximations of the Euclidean distance are used, the structure of the skeleton is strongly modified under pattern rotation. In turn, the use of a weighted distance, or of the Euclidean distance, produces a skeleton quite stable under pattern rotation and is generally preferable for practical applications.

Algorithms to compute perimeter, area and moments of the pattern directly from its skeleton can be found in the literature[30-31].

A more compact approximated representation can be derived from the skeleton, by applying to each skeleton branch a polygonal approximation, so as to divide it into rectilinear portions. Since the skeletal pixels are labeled, one can represent the skeleton as a curve in the 3D space, where the three coordinates of any pixel are its planar coordinates and label. If a polygonal approximation is performed on this 3D curve, the skeleton results to be partitioned into a set of straight line segments, whose pixels are labeled with linearly changing values[32,33].

Each skeleton partition component can be interpreted as representative of a "simple region", i.e., a region characterized by two properties: i) the local thickness of the region linearly changes along its symmetry axis; ii) the contour arcs, with respect to which the skeleton component is symmetrically placed, are straight line segments.

The compression of the skeleton (and hence of the pattern) achieved in this way is very high. In fact, it is enough to keep track of coordinates and label of a few pixels, i.e., the vertices of the polygonal approximation, to be able to recover an approximated version of the pattern. Indeed, each simple region could be obtained by drawing the discs associated to the corresponding pair of vertices, and by joining the discs by means of a trapezium shaped strip. An example is illustrated in figure 10. The pixels of the $d_{3,4}$ DT-based skeleton, which have to be recorded for shape representation and description, are denoted by crosses in figure 10a; the simple

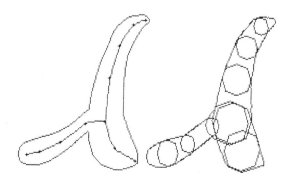

Figure 10. Decomposition into simple regions of a λ-shaped pattern originating from the partition of its skeleton by means of a polygonal approximation.

regions, recovered starting from the coordinates of the vertices found after the polygonalization are shown in figure 10b. The pattern is quite faithfully recovered, although the approximated representation system is very compact. Geometrical features (like area, perimeter, orientation, shape) of any simple region can be directly derived from the coordinates and labels of the corresponding pair of vertices, without resorting to pattern recovery.

By selecting different threshold during the polygonal approximation, different skeleton decompositions are possible, which become rougher and rougher representations of the input as the threshold increases. Each decomposition of the skeleton corresponds to a decomposition of the pattern into simple regions characterized by two properties on thickness and orientation.

REFERENCES

1. A. Rosenfeld and A.C. Kak, *Digital Picture Processing*, Academic Press, New York, NY (1982).
2. R.C. Gonzalez and P. Wintz, *Digital Image Processing*, Addison-Wesley, Reading, MA (1987).
3. B.G. Batchelor, D.A. Hill, and D.C. Hodgson eds., *Automated Visual Inspection*, IFS North-Holland, Bedford, NL (1985).
4. H. Samet, *The Design and Analysis of Spatial Data Structures*, Addison-Wesley, Reading, MA (1989).
5. G. Borgefors, Distance transformations in arbitrary dimensions, *Comput. Vision Graphics Image Process.*, Vol.27, pp.321-345 (1984).
6. G. Borgefors, Distance transformations in digital images, *Comput. Vision Graphics Image Process.*, Vol.34, pp.344-371 (1986).
7. H. Freeman, Computer processing of line drawings, *Comput. Surveys*, Vol.6, pp.57-97 (1974).
8. T. Pavlidis, *Structural pattern recognition*, Springer Verlag, New York, NY (1977).
9. G. Gallus and P. W. Neurath, Improved computer chromosome analysis incorporating preprocessing and boundary analysis, *Phys. Med. Biol.*, Vol.15, p.435 (1970).
10. H. Freeman, On the encoding of arbitrary geometric configurations, *IRE Trans. Electronic Computers*, Vol.10, pp.260-268 (1961).
11. C.H. Teh and R.T. Chin, On the detection of dominant points on digital curves, *IEEE Trans. Patt. Anal. Mach. Intell.*, Vol.11, p.859 (1989).
12. C. Arcelli and G. Ramella, Finding contour-based abstractions of planar patterns, *Pattern Recognition*, Vol.26, No.10, pp.1563-1577 (1993).
13. I. Ragnemalm and G. Borgefors, Towards a minimal shape representation using maximal disks, in *The Euclidean Distance Transform*, I. Ragnemalm, Dissertation No.304, Linköping University, S, p.245 (1993).
14. A. Rosenfeld and J.L. Pfaltz, Distance functions on digital pictures, *Pattern Recognition*, Vol.1, No.1, pp.33-61 (1968).
15. C. Arcelli and G. Sanniti di Baja, Weighted distance transforms: a caracterization, in *Image Analysis and Processing II*, V. Cantoni et al. eds., Plenum Press, New York, NY, pp.205-211 (1988).
16. G. Borgefors, I. Ragnemalm, and G. Sanniti di Baja, The Euclidean distance transform: finding the local maxima and reconstructing, *Proc. 7th Scandinavian Conf. on Image Analysis*, pp.974-981 (1991).
17. G. Borgefors, Centres of maximal discs in the 5-7-11 distance transform, *Proc. 8th Scandinavian Conf. on Image Analysis*, pp.105-111 (1993).
18. C.J. Hilditch, Comparison of thinning algorithms on a parallel processor, *Image Vision Comput.*, Vol.1, p.115 (1983).
19. N.J. Naccache and R. Shinghal, An investigation into the skeletonization approach of Hilditch, *Pattern Recogn.*, Vol.17, p.279 (1984).
20. F. Leymarie and M.D. Levine, Simulating the grassfire transform using an active contour model, *IEEE Trans. Patt. Anal. Mach. Intell.*, Vol.14, No.1, pp.56-75 (1992).
21. L. Lam, S.W. Lee, and C.Y. Suen, Thinning methodologies- A comprehensive survey, *IEEE Trans. Patt. Anal. Mach. Intell.*, Vol.14, p.869 (1992).
22. C. Arcelli, L.P. Cordella, and S. Levialdi, From local maxima to connected skeletons, *IEEE Trans. on PAMI*, Vol.3, pp.134-143 (1981).
23. C. Arcelli and G. Sanniti di Baja, A width-independent fast thinning algorithm, *IEEE Trans. on Pattern Analysis and Machine Intelligence*, Vol.7, pp.463-474 (1985).
24. C. Arcelli and G. Sanniti di Baja, A one- pass two-operations process to detect the skeletal pixels on the 4-distance transform, *IEEE Trans. on Pattern Analysis and Machine Intelligence*, Vol.11, pp.411-414 (1989).

25. L. Dorst, Pseudo-Euclidean skeletons, *Proc. 8th Int. Conf. on Pattern Recognition,* pp.286-288 (1986).

26. C. Arcelli and M. Frucci, Reversible skeletonization by (5, 7, 11)-erosion, in *Visual Form Analysis and Recognition,* C. Arcelli et al. eds., Plenum, New York, NY, pp.21-28 (1992).

27. C. Arcelli and G. Sanniti di Baja, Ridge Points in Euclidean Distance Maps, *Pattern Recognition Letters,* Vol.13, p.237 (1992).

28. G. Sanniti di Baja, Well-Shaped, stable and reversible skeletons from the (3, 4)-distance transform, *Journal of Visual Communication and Image Representation* (1993).

29. C. Arcelli and G. Sanniti di Baja, Euclidean skeleton via center-of-maximal-disc extraction, *Image and Vision Computing,* Vol.11, p.163 (1993).

30. L.P. Cordella and G. Sanniti di Baja, Geometric properties of the union of maximal neighbourhoods, *IEEE Trans. Patt. Anal. Mach. Intell.,* Vol.11, pp.214-217 (1989).

31. G. Sanniti di Baja, O (N) computation of projections and moments from the labelled skeleton, *Comput. Vision Graphics Image Process.,* Vol.49, p.369 (1990).

32. C. Arcelli, R. Colucci, and G. Sanniti di Baja, On the description of digital strips, *Proc. Int. Conf. on Artificial Intelligence Applications and Neural Networks,* pp.193-196 (1990).

33. G. Sanniti di Baja and E. Thiel, Shape Description via Weighted Skeleton Partition, *Proc. 7th Int. Conf. on Image Anal. and Process.,* Bari, I (in press).

LOOKING FOR VISUAL PRIMITIVES

Carlo Arcelli[1], Luigi P. Cordella[2] *(chairman)*, and Leila De Floriani[3]

[1] Istituto di Cibernetica del CNR
  Via Toiano 6, I - 80072 Arco Felice (NA), Italy
[2] Dipartimento di Informatica e Sistemistica, Università di Napoli "Federico II"
  Via Claudio 21, I - 80125 Napoli, Italy
[3] Dipartimento di Informatica e Scienze dell'Informazione, Università di Genova
  Viale Benedetto XV 3, I - 16132 Genova, Italy

ABSTRACT

Visual primitives can be considered as abstractions of those informative subsets of an image which are of interest in a given vision task. After discussing their nature and some problems related to their extraction, pattern description in terms of primitives is considered. Eventually, models relating 3-D visual primitives in high level vision are discussed.

INTRODUCTION

When referring to human vision, visual primitives may be defined as abstractions of some informative subset of what is seen. They originate at perceptual level as a response to visual stimuli detected at retinal level. In the early times of computer vision research, hints on the types of primitives to look for in the digital pictures to be analysed, came from the investigations on some of the nervous systems of various species (e.g., frog, squirrel, rabbit, cat and monkey). These investigations[1] showed that some specific triggers were events such as convex edges, oriented slits or bars (possibly moving), ends of lines, line segments and corners.

Machine vision is concerned with extracting information from visual sensors to enable machines to make "intelligent" decisions. In this respect, redundancy reduction is of basic importance in the design of information processing systems that can perceive and interact with the external world, since it gives evidence to the singularities of the images which will constitute the primitives to be utilized for higher level processing. In machine vision, visual primitives are the smallest image subsets with specific geometrical and/or structural properties, which are of interest in a given analysis and description task. They define the basic level of description, in a hierarchy of descriptions, used to identify a given pattern.

In the framework of computer vision, the term "visual primitive", or better "primitive component", has a clear meaning for people familiar with the structural approach to visual pattern recognition[2,3]. This approach assumes that a complex pattern can be decomposed into simpler subpatterns, possibly in a recursive way, and then characterized (i.e., described) in terms of simple components, which are called "primitives", and of their relations. In this way the structure of a pattern can be outlined. Primitive components should not be further structured, or better their structure should not be of interest for the considered purpose.

In the decision-theoretic approach to recognition[4,5], the term "feature" is certainly more familiar than the term "primitive". In this case, a pattern is characterized by a set of features (e.g., simply a set of measurements performed on the raw data) and a feature vector is assumed to represent the pattern to be recognized. Recognition implies the partition of a feature space into regions each pertaining to a different class. The elements of a feature vector should be such as to characterize a pattern so that it can be attributed to a specific class with high reliability. If this aim is achieved, it can be said that the used features effectively represent the pattern or that its description has been given in terms of essential features which include information about the primitives.

In pattern recognition, the most general meaning of the term feature could be: any pattern property that can be parameterized. The term could thus be applied uniformly to specify any type of entity assumed to characterize a pattern, from the simplest to the most structured one. However, a terminology for a feature hierarchy can be suggested: starting from the raw data, "local features" can be generated directly from them (absence of any structure); "intermediate features" are partial aggregates of the former ones, while a structured set of them, playing a special descriptive role for a pattern family, is a "primitive feature". Semantically significant assemblies of primitive features can be defined, and an object can be seen as a terminal (primitive) feature assembly[6]. This terminology may also suggest a strategy for achieving pattern description.

The determination of a set of primitives in terms of which the patterns of interest can be described is influenced by a number of circumstances, such as the nature of the data, the specific application considered and the technology available for implementing the recognition system.

Selecting primitive features is a complex problem that has no general solution. There isn't a "universal picture element" nor a simple alphabet of primitives generally valid. According to K.S.Fu[3], a general selection criterion is that the primitives should serve as basic pattern elements to provide a compact but adequate description of the data in terms of the specified structural relations; moreover they should be easily extracted.

In the next section, visual primitives and problems related to their extraction are discussed, while in the successive section, pattern description in terms of primitives is considered. 3-D visual primitives and models relating such primitives in high-level vision are discussed in the last section, where aspects regarding the use of CAD models and relational representations are emphasized.

VISUAL PRIMITIVES

Although features useful to form visual primitives can generally be evaluated from texture, shading, motion, depth, etc., for a large class of images they are mainly extracted from boundaries which, in their turn, consist of edges. Edges represent small areas of high local contrast in correspondence with discontinuities in intensity, colour, texture and so on. Accurate detection of edges is therefore crucial to automatic feature detection and object recognition.

In general, images may contain a number of edges occurring at any orientation and with sizes varying from very short (in this case the edges may be regarded as dots) to rather long.

Edges may constitute the boundaries of closed regions, or may originate other primitives such as crossings, junctions, corners and bends. The significance of these primitives was shown in an early paper[7], where they were taken as the lowest level units in a recognition procedure for line patterns. The usefulness of a set of similar features (horizontal, vertical and oblique straight lines; curves which are closed, open vertically, open horizontally) was also discussed[8].

The edge detectors usually form the basis for complex digital image processing and analysis. For instance, to extract arbitrary curves such as objects boundaries from a noisy background one can use line detection operations which characterize line-like features as a succession of short edges, aligned in a given direction, which are brighter (or darker) than the points on either side of them in the orthogonal direction.

Edges provide an indication of the shape of the objects in a picture. However, since in machine vision edge detection is concerned with ideal edges corrupted by various forms of noise, it is often difficult to design edge detectors which manifest good performance characteristics when the edges are incomplete and degraded.

It has been suggested that edges should be found by using a symbolic data representation rather than the original numerical output from an edge detector. To this end, a vector description is associated with each edge element in the two-dimensional image. This vector contains information about the type of edge, the degree of contrast as given by the gradient, the position of the edge, its orientation. After applying an elementary edge detector, the edges are aggregated into line segments by using rules based on the examination of the compatibility of the symbolic edge descriptions in a local region. The new data array results from a purely data driven procedure, not determined by the observer's knowledge of the semantic aspects of the visual input, and is termed the primal sketch[9].

According to Gestalt theory[10,11], human vision is designed to structure the spacial features of an object in such a way that they are perceived as properties of the object, all together in the same construction, and not merely as individual parts. Gestalt psychologists attempted to define the organizational principles by which the global aspects of a visual scene are abstracted from the details. For instance, while figures may differ considerably regarding the nature of their constituent parts, the proximity of the parts produces the same perceptual interpretation, and such figures are understood as equivalent patterns (see figure 1). Similarly, the role of good continuation is crucial to force the end points of the lines to be perceived as linked so as to form a boundary line (see figure 2).

Principles of good continuation, similarity and proximity play a major role in the perception of closed regions. Accordingly, in image analysis the closed regions may be identified i) by linking a suitable sequence of edges; ii) by grouping pixels on the basis of gray-level homogeneity (region growing).

Segmentation is the assignment of the pixels in a given picture to one of many disjoint sets such that the pixels in each set share a common property. The criterion used to assign each pixel to a set and the number of sets are largely dependent on the desired task, the desired description,

Figure 1. Examples of figures with equal global organization, but different constituent parts.

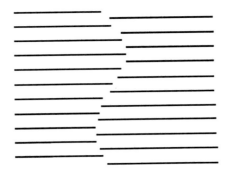

Figure 2. Boundary line defined by the good continuation of the end points of the right and left gratings.

and the scene. Thresholds are used to assess property significance. The processes of edge finding and region growing may cooperate to build a segmented image, producing better results than what would be obtained by either technique alone[12].

Texture analysis may also be considered as a problem of region segmentation, where the comparison has to be accomplished between neighbouring areas or patches of pixels. Comparison is still performed on the basis of similarity and proximity, but it takes into account local patterns, i.e., a whole set of primitives. A major source of difficulty in the study of texture has been how to describe these patterns. Features as coarseness, contrast, and edge orientation appear to be the primary factors which influence the aggregation processes originating textured regions.

From the applicative point of view, among the oldest attempts to primitive feature based description and recognition of 2-D patterns, are the works of Shaw[13] for bubble chamber tracks and of Ledley[14] for chromosome analysis (late 60's). Pieces of lines or edges of different curvature, orientation and size were typically considered as primitives in these cases and the relation between primitives was simply concatenation.

2-D primitives can be line configurations of the boundary or of the skeleton of a figure, but can also be two-dimensional parts of this latter. Figure decompositions into primary convex subsets have been quite common[2]. It has been claimed that the more the primitives are simple, the more inefficient and unreliable is the recognition process. However extracting more complex primitives implies the difficulty of grouping simple features to form more complex features. For instance, the variability of the descriptions possible for the P's shown in figure 3 depends on the way they are represented and on the primitives that are selected. If characters are approximated with polygonal lines, thus a possible primitive is the straight segment. The circular arc would be a more powerful primitive, under condition to have an effective way of approximating characters with circular arcs, i.e., in the specific case, of fitting circular arcs to digital lines (see figure 3).

As for 3-D objects, they are often represented by a single 2-D projection. In this case the problem is that of recognizing 3-D objects from 2-D images. The majority of the approaches uses simple 2-D primitives such as line segments, corners, inflections and 2-D perceptual structures.

3-D volumetric primitives, that assembled together in different ways can serve to build up objects, have been proposed and employed (polyhedra, generalized cylinders, geons)[15,16]. They are preferably specified in a qualitative way, so as their use can be more general. 3-D objects must be recognized independently of the particular view in which they may appear: this implies that a number of possible view per primitive has to be taken into account.

Most computer vision systems, which are designed to recognize three-dimensional

Figure 3. Some samples of handwritten characters (letter "P"); a) bit map and skeleton (superimposed), b) polygonal approximation of the skeleton (superimposed), c) approximation of the skeleton with circular arcs. For different representations (b or c), different primitives can be convenient (see text).

objects, compare a scene model, constructed by processing images obtained from one or more sensors, against entities in a model database containing a description of each object the system is expected to recognize[17]. The development of such model-based recognition techniques has occupied the attention of many researchers in the computer vision community for years[18-20]. Three-dimensional model-based computer vision uses geometric models of objects and sensor data to recognize objects in a scene. Likewise, CAD systems are used to interactively generate 3D models during the design process. Despite this similarity, there has been a dichotomy between the two fields. In recent years, the unification of CAD and vision systems has become the focus of research in the context of manufacturing automation[21]. The term CAD-based vision has been coined for research in vision employing CAD models for various visual tasks[20,21]. Other approaches to model-based recognition use object representations derived from a model of human vision[22].

IMAGE DESCRIPTION

Image description is a parameterization process according to which a suitable representation of an image is transformed into a data structure made up of features and of relations among them. In fact, a listing of features does not generally provide a sufficient description. One cannot distinguish between T and L if features such as "horizontal segment" and "vertical segment" are just listed. It is also necessary to know how the segments join together. Thus, in addition to a set of features, one must specify a set of relations, and a set of rules describing the patterns in terms of features and relations. Figure 4 exemplifies this problem. Examples of relational features are: right of, left of, above, below, inside, outside, at the center of, surrounded by, near, far, next to, attached to, visible from, overlapping, occluding, isolated, grouped, larger, smaller, longer, shorter.

Thus, for complex images, the most interesting approach seems to be the structural one according to which the image is divided into parts (high level primitives for that image) and described in terms of those primitives and their interrelationships; in turn, each part is seen as made of primitive components and analogously described. In principle, the process can be iterated till to reach the feature level most convenient for that type of image. In other words, given an image containing one or more objects, description requires to extract instances of the primitives and their connectivity relations. This can be made with a bottom up approach, combining simpler features to form complex primitives and then objects, although, as far as higher abstraction levels are reached, the need for model based knowledge is regarded unavoidable.

In early vision, information is processed without any prior knowledge of the viewed image. The goal is to extract relevant information useful for further analysis (e.g., sharp changes in image luminosity are a priori relevant information). The result of early processing is generally still an image, which is a (possible) representation of the original one. Further processing can lead, through successive stages, to obtain more convenient representations.

The aims of the representation phase are manyfold: it serves to reduce noise, to compact information, to decompose the pattern into "meaningful" primitive components, to represent components in such a way that not significant shape variations are hidden. In fact, features actually invariant within a class and peculiar of that class may not be directly detectable from the raw data. Suitable processing is generally needed. The representation phase should be intended as a first step of a process of abstraction that leads from the specimen to the prototype (model).

Although the decomposition process outlines specific features of an object, it leads to a

Figure 4. Schematic drawing of a face and some of its scrambled versions. When the relations among the constituent parts are changed, the figure may be no longer perceived as a face.

representation that is still susceptible of different descriptions depending on the scope, the wideness of the image class, and so on. It can be said[23] that a meaningful decomposition and description implies to single out primitives (features, component parts) as much as possible invariant with respect to the differences existing among the specimens of a same class, and to describe the components and the structure they form (i.e., what we have called the representation) in such an essential way as not to display the differences between members of the same class, but rather to put in evidence the similarities. Ideally, all the members of a same class should be described in the same way (still refer to figure 3).

Some generally accepted criteria for obtaining effective descriptions are: attention to the scope, conciseness, accessibility of the features to be used, uniqueness of the obtained descriptions. Still, a description must be robust to occlusion and noise, invariant over a range of viewpoints and scales, and computationally efficient. Moreover, descriptions should be stable, while remaining sensitive enough; however, to absorb variability within a same class, and not to destroy information necessary to discriminate between different classes, are generally conflicting aims.

Indeed, all the above issues are often competitive, in the sense that using certain features may maximally satisfy some criterion but preclude to maximize the others. This could suggest to introduce a redundancy of feature types.

As mentioned above, the number and the nature of the features to be used for describing a pattern depend on the task. If a description serves to the purpose of faithfully reproducing the described object, then it has to be quantitative and very detailed. On the contrary, for recognition purposes, few qualitative features are more convenient especially if, for the application at hand, the number of classes is restricted. In human vision, the dependency of the used features on the task, has been outlined since long time ago by recording eye movements during constrained and unconstrained observation of pictures[24]. These experiments also outline the role of knowledge when looking for primitive features.

One of the image feature that is most commonly used for description is shape. However, shape is generally a very complex feature, whose description is not trivial.

Recent studies on anatomical connections have shown that in non-human primates the cortex processing visual information can be divided into several areas, and that the visual pathways appear separated into two streams, running dorsally and ventrally[25]. Studies of the effects of brain lesions have also indicated that this ventral stream of processing is associated with shape perception and object recognition[26]. As for studying the human brain mechanism involved in recognition, a recent technique concentrated on monitoring local blood flow by using positron emission tomography within the brain during shape processing tasks[27]. All the previous studies, however, do not provide information about mechanisms by which shape is computed; for that information single cell recording techniques can provide useful insights. In this respect, key neurophysiological studies along the posterior anterior axis of the ventral pathway have recently been reviewed[28].

The term "shape" usually refers to a configuration consisting of several elements that somehow mutually relate. The elements are interchangeably called features, attributes, cues, or components. Examples of shape features are: boundary features (dominant points, curvature extrema, circular arcs,...), region features (convex and concave pattern components, maximal discs,...), boundary sinuosity, symmetry, compactness, angular variability, direction of the dominating axis, elongation. Each feature can assume values that may vary along a continuous or a discrete scale.

The quantitative study of object shape in terms of edges, angles, and contours began with Attneave[29]. Particularly, he suggested that a boundary curve can be segmented by means of critical points which coincide with the points of maximum inflection. An alternative study of object shape in terms of regions was carried out by Blum[30], who proposed a new geometry for shapes based on the primitive notion of symmetric point. In this case, the shape of the object can be described in terms of the symmetric points and of the maximal discs, obtained as growth of these points and completely contained inside the object.

Also suitable parts of the object may be regarded as shape primitives[31] (e.g., see figure 5). In this respect, a major problem in machine perception is to develop procedures that distinguish functional parts from purely nominal parts that lack psychological reality. In any case, a shape measure must be used to quantify the similarity or dissimilarity between shape descriptions.

MODEL-BASED OBJECT RECOGNITION

Three issues dominate the design of a modelling scheme for vision: (i) the method adopted for model acquisition; (ii) the choice of a reference frame for the model, and (iii) the choice of a representation[32].

Methods for acquiring models have received relatively little attention. Usually, the models used for recognition have been provided manually. Manual construction of object descriptions is obviously time consuming and requires a detailed knowledge of the object recognition system. This approach to model construction is impractical in applications where the set of objects to be recognized is large or changes frequently. An alternative is to construct models from examples using prototypical features extracted from images taken from a number of

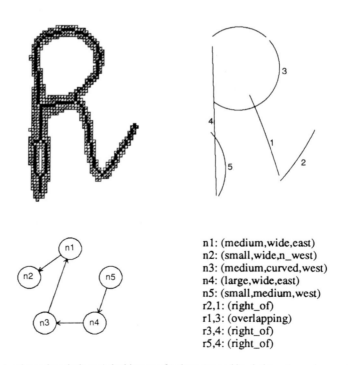

n1: (medium,wide,east)
n2: (small,wide,n_west)
n3: (medium,curved,west)
n4: (large,wide,east)
n5: (small,medium,west)
r2,1: (right_of)
r1,3: (overlapping)
r3,4: (right_of)
r5,4: (right_of)

Figure 5. Character shape description: a) the bit map of a character and its skeleton (superimposed); b) a possible representation and decomposition of the character in terms of circular arcs: c) description of the above representation with an Attributed Relational Graph whose nodes correspond to character components.

viewpoints. Models built by learning are often limited in their precision by the quality of the sensor. Much interest centers on using CAD models to construct models for object recognition. Object-centered CAD models provide a natural way to define an object and can be the source of the information necessary for its recognition, although they are often organized in ways that are not appropriate for vision.

The attributes and the geometric relations among component parts of an object must be defined with respect to a coordinate system, which in vision research is either object- or viewer-centered. Most work in object recognition has adopted object-centered models, since they provide a natural way to express objects independently of the view[33]. Recently, there has been a lot of interest on 3D multiview representations, which model objects by a finite set of viewer-centered descriptions[34]. Each member of the set models the object by its 2D projection as seen from one viewpoint on the view sphere. By using such representations, features extracted from images can be directly matched with features associated with each member of the multiview model set. The shift from object-centered models to viewer-centered ones shifts processing requirements from object recognition time to object modelling time.

Computer vision researchers have used a variety of model types, which can be broadly classified as either descriptive (or quantitative), i.e., the model can be used to generate a synthetic image of the object, or discriminatory (or qualitative), i.e., the model information can be used to distinguish between different objects, but not to generate synthetic imagery. Solid representations used in CAD are quantitative in nature: primitives (surfaces and volumes) are specified in terms of numerical parameters. If we are performing a visual recognition task and the objects to be recognized can be distinguished by examining some qualitative features of the segmented primitives, representations, which capture only those variations, might offer real advantages in processing[17].

The idea of qualitative representation was proposed by Biederman[22] as a model of human vision, but it also offers interesting properties for a computer model. In human vision, the retinal image is transformed at different levels of the visual pathway into various data representations as a precursor to possible object recognition. At the highest levels in this process, we have only a very sketchy knowledge of the exact details[16]. The main idea behind Biederman's approach is to coarsely reconstruct 3D objects using generic primitives, called geons.

While the geon representation has an intuitive appeal, the lack of quantitative information limits its usefulness in environments where discrimination between qualitatively similar, but quantitatively different objects is performed. If an assembly line is making a "family" of parts which differ only in scale, they would have identical geon representations, making discrimination between the different sized items impossible without additional (quantitative) information[17].

Quantitative modelling consists of using "classical" CAD models, i.e., a boundary model or a volumetric one. A boundary model describes a solid object as a collection of boundary entities (i.e., vertices, edges, faces) and their mutual adjacency/incidence relations. Volumetric models (like, Constructive Solid Geometry) describe a solid object as the boolean combination of predefined volumetric primitives. Thus, they are typical of a design environment, since they somehow simulate the process of object construction by a designer. Moreover, CAD systems based on a CSG representation always store a boundary description of the object to speed up rendering as well as analysis operations. On the other hand, a boundary representation of the solid should be augmented with geometrical information to speed up the matching process. CAD models contain information for the local design operations such as what shape to extrude or what is the profile curve for a sweep operation. Features used in construction of models are implicitly rather than explicitly used in the CAD representation. For example, a dihedral edge, comprised within its adjoining surfaces, is not modelled as an edge per se but as two surfaces with adjacency information.

Object recognition techniques are based for the most part on geometric features of the objects to be recognized. These include corners, edges and planar faces for polyhedra, as well as points, arcs of distinct curvature and regions of constant curvature for sculptured surfaces. Other features, such as axes of inertia, profile curves, surface textures properties, reflectance, etc. can also be used. Thus, several authors have recently proposed the use of a boundary model, produced by a commercial CAD system, augmented with additional features. For instance, in a boundary model, a line is characterized by its endpoints; an additional feature associated with a line is its length. Similarly, a planar surface is characterized in a boundary model by (i) coefficients of its plane equation and (ii) its bounding curve. Such description can be augmented, for instance, with a list of visible areas corresponding to the viewpoints on the uniformly sampled viewsphere.

Often, not only geometric features enhancing the object recognition task are added to boundary models, but relations other than the adjacency/incidence ones commonly encoded in a boundary data structures are stored in the representation. Examples of such relations are orientation, proximity, containment, covisibility, etc. Such relations are usually described in the form of a graph.

A graph representation has the flexibility of representing different types of attributes of an object model, and relations among their composing primitives[32,35]. Not only binary relations can be represented by a graph, but also higher-order relations can be encoded as hyperarcs connecting three or more nodes in a hypergraph[36]. Graph structures are the basis of classical boundary models. An example is given by the incidence graph, in which nodes describe the three basic topological entities (vertices, edges and faces) with their geometric attributes, while the arcs correspond to four incidence relations, i.e., Vertex-Edge, Edge-Vertex, Edge-Face and Face-Edge relations (see figure 6).

An example of the use of a graph structure as a view-independent representation of a solid is given by the work of Zhang, Sullivan and Baker[32]. A hypergraph model of an object is computed from its boundary description produced through a CAD system. In such hypergraph, nodes correspond to extended model features, including their shapes and the types of 2D image features that might match, while arcs and hyperarcs describe covisibility of model features. Covisibility of model features is generally view-dependent, in the sense that two features, that are covisible from one viewpoint, may not be covisible from another. Features are considered covisible if the probability of their co-occurrence in images is high.

Graphs are also used in vision as conceptual representation of the visibility structure of a scene. An example is provided by the aspect graph introduced by Koenderink and Van Doorn[37]. An aspect graph is a representation of an object's topology; thus, it captures all viewpoints of an object. The aspect is the topological appearance of the object from a particular viewpoint. Slight changes in the viewpoint change the size of features, edges and faces, but do not cause them to appear or disappear. When a slight change in viewpoint causes a feature to appear or disappear, an event takes place. An aspect graph, or visual potential graph, is obtained by representing aspects as nodes and events between aspects as paths between corresponding nodes[21].

We have seen that graph representation are the basis for both "classical" boundary representations used in CAD and for "enriched" boundary ones used in vision systems. Recently, a lot of attention in CAD/CAM has been devoted to the development of the so-called feature-based models. The term feature (shape or form feature) is used in the computer vision and in CAD/CAM literature with a different meaning, and sometimes a lot of confusion arises. Form features are a compact way of describing subparts of an object, which have a specific meaning in the context of the design or production process. There have been several attempts in giving a truly satisfactory general definition of form feature. Pratt[38] defines a form feature as "a related set of elements of a geometric model conforming to characteristic rules allowing

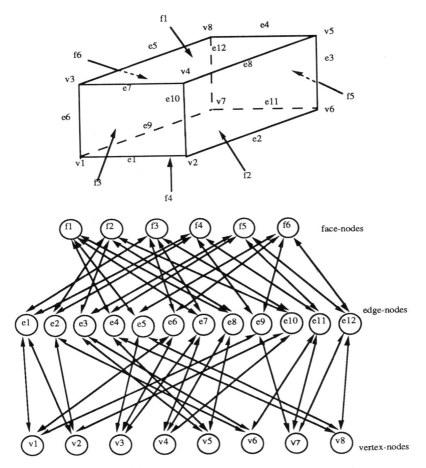

Figure 6. An example of incidence graph of an object.

its recognition and classification and that, regarded as an independent entity, has some significance during the life cycle of the modelled product". The elements might be volumetric primitives in a CSG representation or geometric and topological entities in a boundary model. The basic idea is that such elements occur in a recognizable pattern when we consider the model from the point of view of a specific application, such as process planning for machining, or assembly planning. Examples of form features are protrusions, blind holes, slots, pockets, through holes, etc.

Relational descriptions of the decomposition of a solid object into its form features have been developed in the form of hypergraphs, called feature graphs, which describe an object as a structured aggregation of face-adjacent object parts, called components (see figure 7). Such face-adjacencies are defined by subsets of the faces of each feature, shared by two or more components. Each node of the feature graph is again a graph describing the topology and the geometry of the feature boundary.

Although feature-based modelling is still a research issue in CAD/CAM, the development of feature-based models as well as of form feature recognition algorithms will make feature-

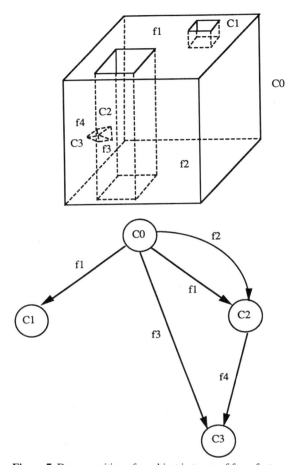

Figure 7. Decomposition of an object in terms of form features.

based models available as input representations for vision systems. In fact, feature-based models could be enhanced with geometric information as boundary models to allow the representation of features as geometric characteristics of a solid easily detectable from an image (i.e., visual features). Moreover, if an object is decomposed into shape features, we can use the aspects to represent a small set of primitives rather than the entire object. This could lead to a possible application of a "recognition-by-part" procedure. An open issue is whether the process knowledge embedded in a feature-based model can facilitate the object recognition process.

REFERENCES

1. H.B. Barlow, R. Narasimhan, and A. Rosenfeld, Visual pattern analysis in machines and animals, *Science*, Vol.177, pp.567-575 (1972).
2. T. Pavlidis, *Structural Pattern Recognition*, Springer-Verlag, Berlin, D (1977).
3. K.S. Fu, *Syntactic Methods in Pattern Recognition*, Academic Press, New York, NY (1974).

4. K. Fukunaga, *Introduction to Statistical Pattern Recognition*, Academic Press, New York, NY (1972).
5. P. Devijver and J. Kittler, *Pattern Recognition: a Statistical Approach*, Prentice-Hall, Englewood Cliffs, NJ (1982).
6. R.M. Bolle, A. Califano, and R. Kjeldsen, A complete and extendable approach to visual recognition, *IEEE Trans. on Pattern Analysis and Machine Intelligence*, Vol.14, pp.534-548 (1992).
7. R. Narasimhan, On the description, generation, and recognition of classes of pictures, in *Automatic Interpretation and Classification of Images*, A. Grasselli ed., Academic Press, New York, NY, pp.1-42 (1969).
8. E.J. Gibson, H. Osser, W. Schiff, and J. Smith, An analysis of cirtical features of letters, tested by a confusion matrix, in *A Basic Research Program on Reading*, Cooperative Research Project No. 630, U.S. Office of Education (1963).
9. D. Marr and E. Hildreth, Theory of edge detection, *Proc. of the Royal Society*, B, Vol.207, pp.187-217 (1980).
10. K. Koffka, *Principles of Gestalt Psychology*, Harcourt-Brace, New York, NY (1963).
11. G. Kanizsa, *Organization in Vision, Essays on Gestalt Perception*, Praeger Special Studies, Praeger, New York, NY (1979).
12. T. Pavlidis and Y.T. Liow, Integrating region growing and edge detection, *IEEE Trans. on Pattern Analysis and Machine Intelligence*, Vol.12, pp.225-233 (1990).
13. A.C. Shaw, A formal picture description scheme as a basic for picture processing systems, *Information and Control*, Vol.14, pp.9-52 (1962).
14. R.S. Ledley et al., FIDAC: film input to digital automatic computer and associated sintax-directed pattern recognition programming system, in *Optical and Electro-Optical Information Processing*, chapt.33, MIT Press, Cambridge, pp.591-614 (1965).
15. S.J. Dickinson, A. Rosenfeld, and A.P. Pentland, Primitive-based shape modeling and recognition, in *Visual Form: Analysis and Recognition*, Plenum Press, New York, NY, pp.213-229 (1992).
16. M.D. Levine, R. Bergevin, and Q.L. Nguyen, Shape description using geons as 3D primitives, in *Visual Form: Analysis and Recognition*, Plenum Press, New York, NY, pp.363-377 (1992).
17. P.J. Flynn and A.K. Jain, CAD-based Computer Vision: From CAD Models to Relational Graphs, *IEEE Trans. on Pattern Analysis and Machine Intelligence*, Vol.13, No.2, pp.114-132 (1991).
18. P.J. Besl and R.C. Jain, Three-dimensional object recognition, *ACM Comput. Surveys*, Vol.17, No.1, pp.75-145 (1985).
19. R.T. Chin and C.R. Dyer, Model-based recognition in robot vision, *ACM Comput. Surveys*, Vol.18, No.1, pp.67-108 (1986).
20. J. Brady, N. Nandhakumar, and J. Aggarwal, Recent progress in the recognition of objects from range data, *Proceedings 9th Int. Conf. Pattern Recognition*, pp.85-92 (1988).
21. C. Hansen and T.C. Henderson, CAGD-based Computer Vision, *IEEE Trans. on Pattern Analysis and Machine Intelligence*, Vol.11, No.11, pp.1181-1193 (1989).
22. I. Biederman, Human Image Understanding: Recent Research and a Theory, *Computer Vision, Graphics and Image Processing*, Vol.32, pp.29-73 (1985).
23. A. Chianese, L.P. Cordella, M. De Santo, and M. Vento, Classifying character shapes, in *Visual Form: Analysis and Recognition*, Plenum Press, New York, NY, pp.155-164 (1992).
24. A.L. Yarbus, *Eye Movements and Vision*, Plenum Press, New York, NY (1967).
25. D.J. Felleman and D.C. Van Essen, Distributed hierarchical processing in the primate cerebral cortex, *Cerebral Cortex*, Vol.1, pp.1-47 (1991).
26. L.G. Ungerleider and M. Mishkin, Two cortical visual systems, in *Analysis of Visual Behaviour*, MIT Press, Cambridge, pp.549-586 (1982).
27. M. Corbetta, F.M. Miezin, S. Dobmeyer, G.L. Shulman, and S.E. Petersen, Attentional modulation of neural processing of shape, colour and velocity in humans, *Science*, Vol.248, pp.1556-1559 (1990).
28. D.I. Perrett and M.W. Oram, Neurophysiology of shape processing, *Image and Vision Computing*, Vol.11, pp.317-333 (1993).
29. F. Attneave, Informational aspects of visual perception, *Psychological Review*, Vol.61, pp.183-193 (1954).
30. H. Blum, A transformation for extracting new descriptors of shape, in *Models for the Perception of Speech and Visual Form*, MIT Press, Cambridge, pp.362-380 (1967).
31. A. Chianese, L.P. Cordella, M. De Santo, and M. Vento, Decomposition of ribbon-like shapes, *Proc. 6th Scandinavian Conf. on Image Analysis*, Oulu, SF, pp.416-423 (1989).
32. S. Zhang, G.D. Sullivan, and K.D. Baker, The Automatic Construction of a View-Independent Relational Model for 3-D Object Recognition, *IEEE Trans. on Pattern Analysis and Machine Intelligence*, Vol.15, No.6, pp.531-544 (1993).
33. D. Marr and K.H. Nishihara, Representation and recognition of the spatial organization of three-dimensional shape, *Proc. R. Soc. Lond.*, B, Vol.200, pp.269-294 (1978).

34. M.R. Korn and C.R. Dyer, 3-D multiview object representations for model-based recognition, *Pattern Recognition*, Vol.20, No.1, pp.91-103 (1987).

35. L. De Floriani, Feature extraction from boundary models of three dimensional objects, *IEEE Trans. on Pattern Analysis and Machine Intelligence*, Vol.11, No.8, pp.785-798 (1989).

36. A.K.C. Wong, S.W. Lu, and M. Rioux, Recognition and shape synthesis of 3D objects based on attributed hypergraphs, *IEEE Trans. on Pattern Analysis and Machine Intelligence*, Vol.11, No.3, pp.279-290 (1990).

37. J.J. Koenderink and A.J. van Doorn, The singularities of the visual mapping, *Biol. Cybern.*, Vol.24, pp.51-59 (1976).

38. M. Pratt, Solid Modelling, *Survey and Current Research Issues* (1990).

THE OCULOMOTOR SYSTEM

Angelo Buizza

Dipartimento di Informatica e Sistemistica, Università di Pavia
Via Abbiategrasso 209, I - 27100 Pavia, Italy

1. INTRODUCTION

In many vertebrates, in particular man and primates, an important part, maybe most, of the information coming from the environment is collected and processed by vision. Then, these animals have developed both sophisticated visual systems and effective Eye Movement, or *OculoMotor,* Systems (OMS). Both in higher and lower animals, the main role of eye movement (EM) is to serve vision and ensure optimal conditions for it. Then EMs may be different in different animals, depending on the particular features and needs of the animal's visual system.

In *afoveate animals,* EM should mainly prevent the image of the environment from slipping on the retina, in order to obtain stable vision. Then OMS in these animals reduces to basic image-holding reflexes (vestibulo-ocular and optokinetic reflex) coupled with a fast position-resetting mechanism (afoveate saccade). Animals with *fovea* (or any specialized retinal area with high receptors density, such as cat's *macula)* need to stabilize retinal images and also to fix and pursue specific objects, that is, to bring the image of the selected object on the fovea and to hold it therein when the object moves. Then, *foveate animals* have got both image-holding reflexes and other kinds of eye movements aimed at orienting gaze, maintaining fixation, and pursuing moving objects. Finally, frontal-eyed animals (usually foveate) need vergence movements to look at different distances (e.g. when fixing an object getting nearer or farther) and to make the images of one object to fall onto corresponding areas of the right and left retina, in order to be fused.

Then, the oculomotor system is quite a complex system which philogenetically evolved from stabilizing reflexes (mainly the vestibulo-ocular and the optokinetic reflex), to either voluntary or involuntary gaze orienting and visual scanning movements (saccades), fixation control, pursuit, and vergence.

The link between eye movements and vision is not simply one of solution of mechanical and control problems. EMs give essential contribution to active vision. Like the movement of any other body segment, EM is the result of a complex process starting with information acquisition from the environment, passing through movement goal selection and motor programming, and eventually resulting in movement execution. Eye movements, by endlessly exploring the environment, allow the visual system to collect a continuous flow of data. These data are used by the central nervous system to build its internal representation of the world and to decide and programme appropriate motor (re)actions, including further eye movements aimed at gathering more information. Binocular vision provides depth information for these processes, although other mechanisms may take part in depth perception.

The fulfilment of the many diverse duties of the OMS may easily create conflict situations. For instance, foveal pursuit of a moving object is incompatible with scan movements which must stop during pursuit. A hierarchical control structure is then required, in which higher levels are in charge of conflict solution by selecting specific goals and deciding the consequent priorities. Conflicts may also arise from the interaction with head movements. Head is the platform supporting the oculomotor apparatus and, in principle, any head movement may disturb oculomotor control. In many situations, the presence of stabilising reflexes helps to reduce and possibly eliminate OMS sensitivity to head movements.

Eye movements may appeal to many researchers: neurophysiologists, neurologists, ophthalmologists, otolaryngologists, psychologists, psychiatrists, bioengineers, control engineers, people working on machine vision and robotics. This widespread interest has promoted many studies and favoured multidisciplinary approach, with consequent remarkable growth of knowledge. Nevertheless, understanding of many aspects of oculomotor behaviour, in particular hierarchical control and relationships with vision, is still unsatisfactory in many respects.

In the following sections a concise account of the overall structure of the oculomotor system and the main features of its components will be given. Some aspects more likely related with vision will be outlined.

2. EYE MOVEMENTS CLASSIFICATION

Eye movements may be either *vergence* (or disconjugate) or *version* (or conjugate) movements. The former allow fixation of objects moving away from *(divergence)* or toward *(convergence)* the observer. During pure vergence the eyes make equal and opposite rotations. Instead, during pure version the eyes rotate by the same angle in the same direction so as to change the orientation of the line of sight without changing the fixation distance. In natural conditions EM usually contains both vergence and version components, that can be easily separated in laboratory conditions.

Both vergence and version movements may be either slow or fast. This distinction does not simply rely on different eye speed during movement but also on different neural substrates and different control modes. Fast movements are commonly referred to as saccadic movements or *saccades,* are ballistic in nature, and behave as if they were controlled in discrete time. Slow movements are smoother and controlled in continuous time.

Like other kinds of movement, eye movements may be either *spontaneous, reflex,* or *voluntary.* Spontaneous movements are independent of both subject's will and external stimuli, reflex movements require physical external stimuli but are largely independent on subject's will, voluntary movements do require subject's will or strongly depend on it. Saccadic scan in the dark and EM during sleep are examples of spontaneous movements and so are most of the exploratory saccades in the light. A particular form of spontaneous movement is the so-called *physiologic nystagmus* which consists of continuous high frequency (30-80 Hz) small amplitude (less than one minute of arc) vibration of the eyeball around its position (a kind of tremor) with associated drift (amplitude of a few minutes of arc) and *microsaccades* (amplitude of 1 to 20 minutes of arc). It has been shown that stabilization of the seen object to the eyeball by means of contact lenses or other mechanical devices counteracting these eye micromovements results in image fade-out. Then the role of physiologic nystagmus seems to be continuous refreshing of the image on the retina in order to prevent fading due to the phasic behaviour of the retinal receptors.

Reflex movements are "automatic" responses to specific stimuli, independent of subject will, although in many situations voluntary control (e.g. enhancement or suppression, either total or partial) is possible. EMs produced by the vestibulo-ocular and the optokinetic reflexes are the most typical examples. The former is activated by stimulation of the vestibular receptors, the latter by visual stimuli.

Finally voluntary movements strictly depend on subject's will. Saccades may be executed voluntarily and independently of external stimuli in order to orientate the gaze to a new fixation point. In contrast smooth pursuit is the main example of voluntary movement requiring an external stimulus, i.e. a moving object to be pursued. Although it is admitted that smooth pursuit largely depends on will, it is not clear whether it should be considered a voluntary movement or rather a reflex particularly sensitive to voluntary control and/or requiring voluntary trigger.

3. ANATOMY OF THE OCULOMOTOR SYSTEM[1-4]

The *oculomotor plant* consists of the *eye-balls,* the *ocular muscles,* and the relevant *ligaments* and *connective tissue.* The eye rotates in the orbit as a sphere with three degrees of freedom and bounded movements. Translational movements are almost absent. Rotations are produced by coordinated contraction and relaxation of six ocular muscles surrounding the eyeball, arranged in three agonist-antagonist pairs. *Lateral* and *medial rectus* drive right-left rotation (i.e., rotation in the horizontal plane when subject's head is in its normal upright position), whereas up-down rotation and rotation around the line of sight (torsion) are jointly driven by the *vertical recti (superior rectus* and *inferior rectus)* and the *obliques (superior oblique* and *inferior oblique).*

Ocular muscles receive neural command from the *oculomotor nuclei* through the *oculo-motor nerves* (III, IV, and VI pair of cranial nerves). All neural commands from the different subsystems that control EMs eventually reach the *oculomotor nuclei.* Then at this level a neural control signal is constructed coding the eye movement to be done.

The eye movements elicited by visual stimuli are organized through a loop whose afferent pathway originates from *retinal receptors* and reaches higher centres *via* the *optic nerve* (II cranial nerve). The oculomotor nuclei are then reached by means of complex neural circuits involving the occipital and pre-occipital *visual areas* (areas 17, 18, 19, and 22), the *tectal* and *pretectal nuclei,* and the *superior colliculi. The frontal eye fields* (area 8) and the superior colliculi are the two major inputs to saccadic system.

The *vestibulo-ocular reflex arc* originates from the vestibular receptors *(semicircular canals* and *otolith maculae),* projects onto the *vestibular nuclei* (VN) through the *vestibular nerve* (VIII cranial nerve), and reaches the oculomotor nuclei both directly through the *medial longitudinal fasciculus* and indirectly through a complex pathway involving the *cerebellum,* the pontine *reticular formation,* and possible other structures. The vestibular nuclei also receive visual input, through the *optokinetic system,* and input from *neck receptors.* The optokinetic system involves the *nucleus of the optic tract* and the *nucleus reticularis tegmenti pontis.* Neck receptors measure the head movement with respect to trunk and drive a further reflex which is referred to as the *cervico-ocular reflex. VN* are considered as a centre for internal reconstruction of head movement in space.

The cerebellum is likely to have a regulatory function in movement coordination[5]. Then it would play a major role in coordinating the different kinds of eye movements with each other and with the movements of other body segments, in particular head and arms.

The overall structure of the oculomotor system may be synthesized as in the block diagram of figure 1. It is possible to identify in this scheme the typical components of a classical control system. The eyeball is the controlled plant with eye position or velocity as controlled variable (output). The ocular muscles are the actuator. The saccade, smooth pursuit, optokinetic, and vergence systems form the controller. The eye movements elicited by visual stimuli are controlled by a negative feedback loop with the visual reference movement as input variable. The closed loop structure results from measuring the reference movement by means of the retinal receptors, which move with the eye. Thus the retinal receptors cannot measure the reference absolute movement, but only its movement relative to the eye, that is the error signal that should be zeroed by the oculomotor control. Finally, the vestibulo-ocular and the cervico-ocular reflex can be seen as compensatory mechanisms aimed at counterbalancing the destabilizing effect of the head movements.

The description of OMS in terms of control theory[2] provided a sound interpretative framework and helped very much to understand the properties of its different parts. However, a comprehensive picture of the whole system including the interaction among different components is still lacking.

4. THE VESTIBULO-OCULAR REFLEXES

The vestibulo-ocular reflexes[2,4,6,7] represent the oldest oculomotor systems. They are aimed at compensating head movements by equal and opposite eye movements in order to keep the gaze orientation unchanged.

Head movements are sensed by the *vestibular receptors*. The *vestibule* represents the nonauditory part of the inner ear. It contains the *SemiCircular Canals* (SCC) and the *otoliths*. They basically are inertial accelerometers, the former are sensitive to head angular accelerations and the latter to accelerations with no rotary components (linear accelerations). In man and frontal eyed animals, canal-ocular reflexes compensatory for head rotations are much more important than otolith-ocular reflexes, which produce rudimentary compensatory eye movements only.

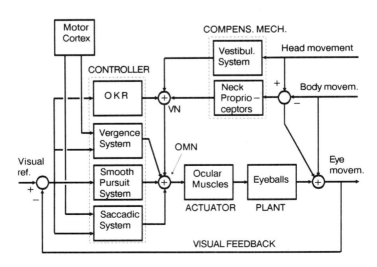

Figure 1. A functional scheme of the oculomotor control system.

On each side of the skull there is a group of three mutually orthogonal semicircular canals (see figure 2). When the head is in its normal upright position, the *horizontal canal* lies in an almost horizontal plane, the *anterior* and *the posterior canal* in vertical planes. The canals form a continuous hydrodynamic circuit filled with a fluid called *endolymph*. Each canal has a special compartment, the *ampulla,* containing a transverse crest *(crista ampullaris)* bearing the sensory epithelium. The sensory cells have a number of hair processes *(cilia)* embedded into a gelatinous mass, the *cupula,* that forms a water-tight barrier across the ampulla. The canal wall *(membranous duct)* is fixed to the skull. When the head rotates, inertia makes the endolymph to move with respect to the canal wall and to deflect the cupula and, consequently, the cilia. Through mechanisms which are still largely unknown, bending of the cilia produces electrical signals that modulate the discharge frequency of the primary vestibular neurons projecting onto the vestibular nuclei. At the level of the vestibular nuclei, a neural signal is made available which basically codes head rotation velocity. It is sent to the oculomotor nuclei both monosynaptically, through the *medial longitudinal fasciculus,* and polysynaptically, through the *pontine reticular formation.* Connections are organized so that head rotation in one direction produces a coordinated action of the ocular muscles resulting in eye rotation in the opposite direction. Due to its working principle, each canal is sensitive to rotations in its own plane and the geometric arrangement results in decomposition of head rotation in three orthogonal components.

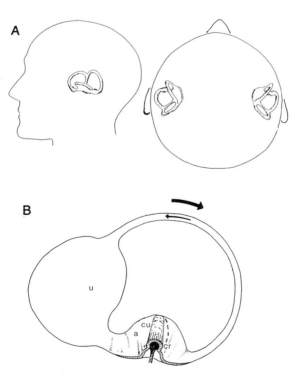

Figure 2. A - Location of the semicircular canals in the skull, lateral (on the left hand side) and top (on the right hand side) view. For the sake of clarity, canal dimensions have been magnified, since the curvature radius of each canal is of a few millimeters only. **B** - Anatomical scheme and working principle of the semicircular canal. **a** = ampulla, **cu** = cupula, **cr** = crista; **u** = utricule; the dashed curve shows the shape of the cupula when bent by the pressure produced by the endolymph flow (thin arrow).

The canal-ocular reflex can be tested by submitting subjects to controlled rotations[4,6]. Usually the horizontal reflex is tested, which requires rotations around a vertical axis only. To this purpose, the subject seats on a rotating chair, with his/her head restrained in order to prevent movements that could stimulate the neck receptors and the associated cervico-ocular reflex. Rotations are performed in the dark in order to avoid visuo-motor reactions. In these conditions a typical pattern of eye movement can be recorded which is referred to as *nystagmus* (strictly, *induced vestibular nystagmus*). It is made of alternating *slow and fast phases* (see figure 3). The former are *compensatory* in nature and represent the response of the canal-ocular reflex. The latter are *anticompensatory* and act as reset movements, although they probably are also manifestation of more sophisticated vestibulo-saccadic interaction[4,8]. By removing the nystagmus fast phases and fitting together the slow ones, the overall equivalent response of the canalocular reflex can be reconstructed (Slow Cumulative Eye Position, (SCEP), see figure 3B).

5. THE OPTOKINETIC REFLEX

Head movement compensation provided by the canal-ocular reflex may be uncomplete or inadequate for three main reasons: i) the inertial principle on which its function relies makes the canal scarcely sensitive to slow and/or constant velocity and/or low frequency head rotations; ii) the reflex gain (ratio of the eye to head rotation) may be less than one (0.4-0.7 in man, see also figure 3B); iii) the canal-ocular reflex is open-loop and cannot be as accurate as required for stabilization purposes. When subject moves in a still visual environment, uncomplete compensation would result in residual gaze shift and consequent image slip on the

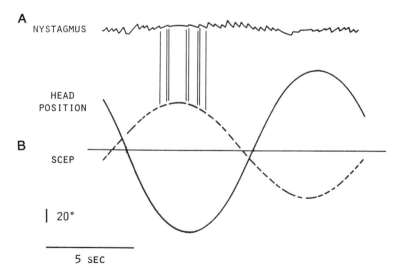

Figure 3. A - Example of horizontal nystagmus induced by harmonic subject oscillation (240° peak-to-peak at 0.07 Hz) around a vertical axis in the dark. **B** - By removing nystagmus fast phases and fitting together the slow ones, the equivalent eye movement (SCEP) produced by the canal ocular reflex can be reconstructed, which is almost phase opposite with respect to head rotation. Notice that in this case compensation is uncomplete because SCEP amplitude is less than head rotation amplitude.

retina. The slip on the retina of the image of a large portion of the visual field is the specific stimulus for a further stabilising reflex which is referred to as *optokinetic reflex*[2,4,6,7]. It produces eye rotation aimed at suppressing the image slip itself, so holding the gaze still. When stimulated by a continuous movement of the scene the optokinetic reflex too produces a nystagmic eye movement *(optokinetic nystagmus)*.

In physiologic conditions (subject's movement in a stationary visual environment) the eye movements induced by the vestibulo-ocular and by the optokinetic reflex are synergistic and cooperate in better gaze stabilization. The two reflexes interact at the level of the vestibular nuclei. They are so tightly coupled that the definition of *vestibulo-optokinetic system* has been proposed[4]. Actually, they have fully complementary roles. The vestibulo-ocular reflex is highpass and open-loop whereas the optokinetic one is low-pass and closed-loop, and provides a feedback signal also to its companion. Their cut-off frequencies are close to each other so that when they work together (as in physiologic conditions) head movement can be measured and compensated on a broad frequency band. The neural signal they produce at the level of the vestibular nuclei is likely to represent a central estimate of head movement in space[9]. This signal wouldn't be used for gaze stabilization only. It is also needed for reconstruction of the postural model, posture control, perceptual stability of the environment, self- and object-motion sensation, etc.

6. THE SACCADIC SYSTEM

Saccades[10] are fast, ballistic eye movements. They are stereotyped in nature and can rotate the eyeball with velocities as high as 600°/s in man. Many behaviourally distinct eye

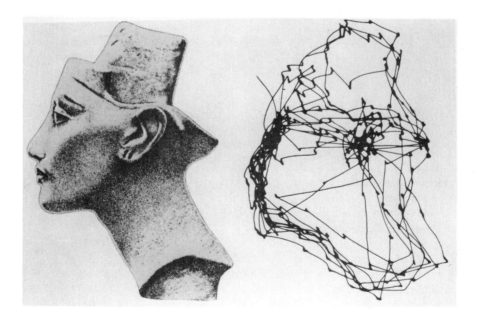

Figure 4. The right figure is the X-Y plot of a sequence of saccades made during visual inspection of the picture of the Egyptian queen Nephertitis' bust (Egyptian art, about 1360 B.C.) shown on the left (reproduced from Yarbus[11]).

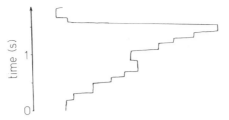

Figure 5. Sketch of a typical eye movement pattern during reading. The eyes scan the text line by means of successive saccadic jumps to the right, separated by fixation pauses. Backward saccades may sometimes occur as in the middle of this example. When the end of the line is reached a large, possibly double-step, saccade brings the eye to the beginning of the next line.

movements belong to the class of saccades, namely: visual scanning and search movements[11], fast tracking movements, eye movements produced during reading[12], spontaneous movements in the dark, and fast phases of nystagmus. Examples of scanning and reading movements are given in figures 4 and 5.

Saccades are used whenever fast re-orientation of gaze is needed. It has been suggested that the neural signal driving them implements a minimal-time, optimal control law[13].

Saccades may be produced over a large amplitude range, from a few minutes of arc (microsaccades) up to 80°-90°. In laboratory they are stimulated by means of controlled target jumps. In this case the saccadic response has a latency of about 200 ms with respect to stimulus onset. Then the saccade appears as a steplike movement with a highly repeatable sigmoid profile (see figure 6), very hard to modify after iniziation. Movements larger than about 20° are usually performed with two saccades (see figure 6), a primary one covering about 90% of the overall excursion, and a secondary one after about 130 ms, bringing the eye to the final position. Independently of the number of saccades used, the target can always be very accurately foveated.

As in the above example of double-step saccades, a new saccade cannot be produced before an interval of 80-200 ms from the previous one. Then saccade is commonly thought of as a movement controlled in discrete-time.

7. THE SMOOTH-PURSUIT SYSTEM

Continuous target movement cannot be accurately tracked by the saccadic system that could provide only a staircase approximation of the target displacement. To cope with this situation, foveate, frontal-eyed animals have developed a specialized system which is referred to as *smooth-pursuit system*[2,4].

Unlike saccades, that can be produced in the absence of moving targets (e.g. in the dark), smooth-pursuit movements require a moving stimulus. A visual target is the specific stimulus, but examples of pursuit of non-visual (e.g. tactile) stimuli have been described. This has lead to the hypothesis that pursuit movement is actually driven by an internal signal, representing the movement of an object of interest, and reconstructed from information provided by different sensory channels. Of course, visual information, when available, would play a major, dominant role in reconstruction.

Smooth-pursuit control is classically described as a velocity servomechanism. It has an initial response latency of 100-150 ms and can accurately match (gain close to one) regular and

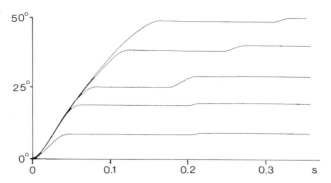

Figure 6. Time course of eye position during saccades of different amplitudes. Notice the typical sigmoid profile. Eye excursions larger than 20° are usually covered by two saccades (redrawn using data from Becker[21]).

continuous target movement, provided speed doesn't exceed 90°-100°/s and, in the case of alternating movement, frequency is limited to about 1 Hz. However, physical and geometrical stimulus features, stimulation conditions, background, target movement profile and predictability may influence system performance[14,15]. Due to speed and frequency limitations, tracking of fast targets often requires corrective saccades, whose number and relative contribution increase with target speed (see figure 7).

8. THE VERGENCE SYSTEM

In spite of their relevance in many fields (e.g. strabismus) the vergence movements[2,4,16] are still little known. They are found in frontal-eyed (usually foveate) animals and serve image fusion and stereopsis. They respond with a latency of about 160 ms. With respect to other foveation movements, they are less accurate and slower, *divergence* being slower than *convergence*.

Although vergence movements don't need visual stimulation to be voluntarily produced, specific stimuli are both retinal image disparity, that elicits *fusional vergence,* and retinal image blur, that elicits *accomodative vergence* in association with lens accommodation. Fusional vergence is closed-loop controlled as it results in reduction of the disparity that induced it. In contrast, accomodative vergence *per se* has no effect on its own cause, i.e. image blur, then it is basically open-loop controlled. The two mechanisms are easily separated in laboratory conditions, but in physiologic conditions they work together and this should result in almost closed-loop control.

In every-day life, vergence and version movements are usually produced simultaneously. An example of saccade-vergence coordination is given in figure 8.

9. COORDINATION AMONG DIFFERENT MOVEMENTS, EYE-HEAD COORDINATION AND HIERARCHICAL CONTROL

Head is the moving platform supporting the oculomotor apparatus. In turn, head is supported by a further moving structure, the body. This situation has two, seemingly opposite, major consequences. On the one hand, head and body movements may be detrimental to

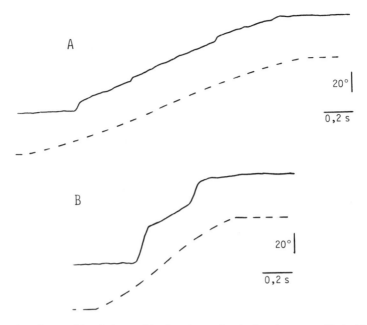

Figure 7. Examples of eye position during tracking (continuous lines) of moving targets (dashed lines). In **A**, the target moves from 45° left to 45° right with maximal speed of 60°/s and tracking is mainly performed by smooth pursuit movement with small corrective saccades. In **B**, the target moves as above but has maximal speed of 120°/s: most of the tracking is made of two large saccades.

oculomotor control. On the other hand, coordinated body, head, and eye movements may be used to extend the possibilities of gaze control.

During fixation of still objects, gaze stabilization against head movement is ensured, as already said, by the cooperative action of vestibulo-ocular and optokinetic (and possibly cervico-ocular) reflexes, which provoke eye rotations equal and opposite to head rotations. However the presence of these stabilizing reflexes may become undesirable in other conditions, in particular when gaze shift requires combined head *and* eye rotation. For instance, when pursuing by head and eye movement a target moving to the right in front of a non uniform background, the eyes would be driven to the left of the target by both vestibulo-ocular reflex, stimulated by head rotation, and optokinetic reflex, stimulated by the retinal slip of the background image. Actually it doesn't happen so and the eyes can quite easily keep target fixation if its motion isn't too fast. This suggests that *in these and similar conditions* reflex movements are suppressed, or at least reduced, to allow easy pursuit. How this is obtained it's still a matter for speculation. Possible mechanisms are cancellation by neural signals related to tracking signal[17,18], reflex gain control (in particular reduction)[17,18], and possibility of disconnecting the reflex[19].

Whatever the mechanism, the above example shows that conflicts may easily arise among the different oculomotor subsystems in the fulfilment of their tasks. Some conflicts are likely to be "automatically" solved thanks to the system organization, as in the case of gaze stabilization. In many other cases, conflict solution requires definition of goals and consequent priorities. This in turn implies that OMS control is organized in a hierarchical structure and must

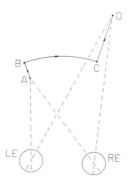

Figure 8. Typical combination of version and vergence during rapid gaze orientation. Due to shortest latency, the eye movement begins with vergence (A to B). Then a rapid versional saccade shifts the line of sight to the new direction (B to C). Final vergence brings the eyes on the new fixation point (C to D).

be coordinated with the control of other body segments, in particular the head. Hierarchical structure means a multi-level or multi-layer structure in which the higher levels/layers are given ability to make decision and can intervene and modify the behaviour of the lower ones in order to meet particular needs or to cope with particular situations. The organization of this structure is still unknown as well as its neural substrate. The cerebellum is likely to play an important role as a coordinator among different motor systems[5].

10. EYE MOVEMENTS AND VISION

It is commonly admitted that eye movements serve vision. As concisely stated by Carpenter[20], "The purpose of eye movements is after all not to move the eye, but to move the retinal image". In spite of such general agreement, relationships between eye movement and vision are still poorly understood. Many studies dealt with the influence of vision conditions (luminance, contrast, target dimension, target retinal location, background, etc.) on OMS performances. Even a brief discussion of the main results would be too long, and the reader is then referred to various reviews appeared in the literature (see for example[2, 4, 14, 20, 22]). In the present context, it seems more interesting to examine how eye movements influence vision. This topic has a long history characterized by many controversies. Nowadays it's generally accepted that eye movements are necessary to vision but say little about elaboration of visual information and, in particular, about cognitive processes. A few facts are listed below. For more thorough and comprehensive picture the reader is referred to the excellent book edited by Kowler[22] and to other reviews available in the literature (see for example Leigh and Zee[4], and Carpenter[20]).

Physiologic nystagmus prevents retinal image from fading. Its main role seems to counterbalance the effect of the phasic behaviour of the retinal receptors. It's not clear whether it also contributes to some aspects of image processing, for instance edge and contour detection.

During saccades, vision is at least partially suppressed. This is referred to as *saccadic suppression* and should be related with the *retinal blur* resulting from the high retinal slip velocity produced by saccade. For instance, suppression might be a means to prevent possible undesirable effects on perceptual stability of the environment. The hypothesis that it's a mere passive consequence of blur is contradicted by the fact that it begins before saccade onset, then it is likely to have a functional role.

It is admitted that *smooth pursuit* contributes to object motion perception. It has also been shown that changes of retinal slip velocity during *optokinetic nystagmus* are accompanied by modification of the perceived stimulus velocity[23]. Nevertheless retinal configuration change seems to provide much more effective information[24].

Binocular vision needs *vergence* movements and supports stereopsis, however vergence movements *in se* seem to give only minor, if any, contribution to depth perception[16].

It is known that saccades are used to scan figures and scenes, however recognition of complex figures is possible even if eye movement is prevented, for instance by short presentations, lasting less than the saccade latency[25]. Then saccades seem to be mainly used for detailed analysis of the visual scene, and it would undoubtedly be interesting to understand the relevant exploration strategy. However it's known that many factors (expectation, emotional involvement, context, task) may influence scan patterns and make their interpretation hard. The following conclusion by Viviani[25] seems appropriate to describe this situation: "empirical evidence from eye movement research, by itself, has been insufficient to specify models for the perceptual and cognitive processes involved in the visual exploration of the world".

As a conclusion of this brief review it's worth asking which suggestions can be drawn from eye movement research, useful in machine vision. More than a century ago, Lotze[26] realized that eye movements are needed to cope with the scarce resolving power of the peripheral retina. If the retina had uniform receptor distribution and uniform resolution, stabilizing reflexes against head movements and large saccades preventing object images from getting off the retina would be enough. That is, in animals provided with fovea, many features of the oculomotor control depend on the existence of that narrow, privileged visual area. Machines usually "see" by means of cameras with uniform resolution, then are similar to afoveate animals, and their "oculomotor" system can then reduce to basic, both reflex and voluntary, image-holding mechanisms. Whether or not tricks like micromovements to detect edges and vision suppression during saccades are useful in machine vision probably depends on the features and needs of the image processing. By the way, these and similar functions could be more easily realized *via* software, without requiring sophisticated mechanical and control solutions. Perhaps to get depth information from a measure of the convergence angle between two cameras seeing the same scene from different viewpoints is a more effective solution than to derive the same information from perspective or image disparity. Also in this case machine wouldn't closely mimic nature, where, as already said, vergence movements seem to give only minor, if any, contribution to depth perception.

In conclusion, basic differences between biological and artificial vision systems and resources make it difficult to find close parallelism between the respective "visuo-motor" behaviours, and the actual suggestion coming from nature is to develop oculomotor systems fully "adapted" to the visual system and meeting its needs as closely as possible.

REFERENCES

1. R.A. Moses ed., *Adler's physiology of the eye,* Mosby, Saint Louis, LO (1970).
2. D.A. Robinson, Control of eye movements, in *Handbook of Physiology,* Sect.1, Vol.II, Part 2, V.B. Brook ed., American Physiological Society, Bethesda, ML, pp.1275-1320 (1981).
3. J.A. Büttner-Ennever ed., *Neuroanatomy of the Oculomotor System,* Elsevier, Amsterdam, NL (1988).
4. R.J. Leigh and D.S. Zee, *The Neurology of Eye Movements,* 2nd ed., Davis, Philadelphia, PN (1991).
5. W.A. MacKay and J.T. Murphy, Cerebellar modulation of reflex gain, *Progr. Neurobiol.,* Vol.13, pp.361-417 (1979).
6. R.W. Baloh and V. Honrubia, *Clinical Neurophysiology of the Vestibular System,* Davis, Philadelphia, PN (1979).
7. V.J. Wilson and G. Melvill Jones, *Mammalian Vestibular Physiology,* Plenum Press, New York, NY (1979).

8. R. Schmid and D. Zambarbieri, Eye-head coordination during active and passive head rotations in the dark, in *The Head-Neck Sensory Motor System,* A. Berthoz, W. Graf, and P.P. Vidal eds., Oxford University Press, Oxford, UK, pp.434-438 (1992).

9. R. Schmid, A. Buizza, and D.Zambarbieri, Visual stabilization during head movement, in *Brain Mechanisms and Spatial vision,* D.J. Ingle, M. Jeannerod, and D.N. Lee eds., Martinus Nijhoff Publ., Dordrecht, NL, pp.211-227 (1985).

10. R.H. Wurtz and M.E. Goldberg eds., *The neurobiology of saccadic eye movements,* Elsevier, Amsterdam, NL (1989).

11. A.L. Yarbus, *Eye movements and vision,* Plenum Press, New York, NY (1967).

12. J.K. O'Regan, Eye movements and reading, in *Eye Movements and their Role in Visual and Cognitive Processes,* E. Kowler ed., Elsevier, Amsterdam, NL, pp.395-453 (1990).

13. M.R. Clark and L.W. Stark, Time optimal behavior of human saccadic eye movement, *IEEE Trans.,* Vol.AC-20, pp.345-348 (1975).

14. E. Kowler, The role of visual and cognitive processes in the control of eye movement, in *Eye Movements and theirRole in Visual and Cognitive Processes,* E. Kowlered., Elsevier, Amsterdam, NL, pp.1-70 (1990).

15. M. Pavel, Predictive control of eye movement, in *Eye Movements and their Role in Visual and Cognitive Processes,* E. Kowler ed., Elsevier, Amsterdam, NL, pp.71-114 (1990).

16. H. Collewijn and C.J. Erkelens, Binocular eye movements and the perception of depth, in *Eye Movements and their Role in Visual and Cognitive Processes,* E. Kowler ed., Elsevier, Amsterdam, NL, pp.213-261 (1990).

17. R.D. Tomlinson and D.A. Robinson, Is the vestibulo-ocular reflex cancelled by smooth pursuit?, in *Progress in Oculomotor Research,* A.F. Fuchs and W. Becker eds., Elsevier, Amsterdam, NL, pp.533-539 (1981).

18. D.A. Robinson, A model of cancellation of the vestibulo-ocular reflex, in *Functional Basis of OcularMotility Disorders,* G. Lennerstrand, D.S. Zee, and E.L. Keller eds., Pergamon Press, New York, NY, pp.5-13 (1982).

19. V.P. Laurutis and D.A. Robinson, The vestibulo-ocular reflex during human saccadic eye movements, *J. Physiol. (Lond.),* Vol.373, pp.209-233 (1986).

20. R.H.S. Carpenter, *Movements of the Eyes,* Pion, London, UK (1977).

21. W. Becker, Metrics, in *The neurobiology of saccadic eye movements,* R.H. Wurtz and M.E. Goldberg eds., Elsevier, Amsterdam, NL, pp.13-67 (1989).

22. E. Kowler ed., *Eye Movements and their Role in Visual and Cognitive Processes,* Elsevier, Amsterdam, NL (1990).

23. A. Buizza, A. Léger, J. Droulez, A. Berlhoz, and R. Schmid, Influence of otolithic stimulation by horizontal linear acceleration on optokinetic nystagmus and visual motion perception, *Exp. Brain Res.,* Vol.39, pp.165-176 (1980).

24. H. Wallach, The role of eye movements in the perception of motion and shape, in *Eye Movements and their Role in Visual and Cognitive Processes,* E. Kowler ed., Elsevier, Amsterdam, NL, pp.289-305 (1990).

25. P. Viviani, Eye movements in visual search: cognitive, perceptual and motor control aspects, in *Eye Movements and their Role in Visual and Cognitive Processes,* E. Kowler ed., Elsevier, Amsterdam, NL, pp.353-393 (1990).

26. Lotze (1884) cited in [22], Preface.

ACTIVE VISION

Massimo Savini

Dipartimento di Informatica e Sistemistica, Università di Pavia
Via Abbiategrasso 209, I - 27100 Pavia, Italy

ABSTRACT

In this talk I will try to show what Active Vision is, and what it is not. The dominant paradigm in past computer vision research will be illustrated, and the benefits emerging from the active approach will be underlined. Beyond that, I will describe the related approach of purposive and qualitative vision, and try to place the active approach in the perspective of an intelligent perception system.

1. INTRODUCTION

Research in machine vision started approximately three decades ago as part of the enthusiastic thrust toward artificial intelligence. That effort had been prompted by the fascinating idea of building machines that could show to possess "intelligent" human capabilities and behaviours.

At the beginning, since the late fifties, much of the effort had been directed toward making the machines to perform formal tasks, such as theorem proving and game playing.

When the power of the computers grew, and more knowledge could be put into the systems, it became possible to deal with another category of tasks: those requiring, to be performed by a human, some sort of expertise. Problems of this kind have been tackled in fields like engineering, medical diagnosis, scientific analysis, and finance. The amount of knowledge required to perform these tasks is respectable, but not too large; moreover it is specific knowledge on a well defined and restricted domain. This has allowed for a good success of the so called Expert Systems, or Knowledge Based Systems.

Both of these categories of tasks somehow match one aspect of the common-sense idea of "intelligence". That is, people skilled in them are generally regarded as being intelligent; mostly because they have learned to do things that the majority will never learn, or attempt to learn. But, if intelligence is related to "ease of learning", then which genius can compare his/her performances with those of any normal baby in his first two years of life? In fact the use of the word "intelligence" is misleading, and we would better talk about "mental faculties"; but for sake of simplicity we will continue to use the common denomination of Artificial Intelligence (AI) throughout this chapter.

It is not by chance that a third category of tasks has received a lot of attention since the beginning of AI history. Among these we find perception (vision and speech), natural language understanding, common-sense reasoning, and motion and action planning. All of these tasks are performed daily by any non-handicapped human being, who learned them in early infancy. Although these capabilities are so common, and not apparently connected with intelligence, according to the writer, they constitute the most difficult and interesting problems in the AI arena. They do not require specialistic knowledge, but sometime they require a much broader amount of generic or common-sense knowledge. And somebody says that "intelligent behaviour" does not require at all any explicit representation of knowledge.

It could seem at this point that all these discourses do not have much to share with the title of this chapter, namely "Active Vision". But, first of all, vision is probably the most complex perceptual problem we can deal with in AI, and its solution would shed light on many other problems, outside of perception. Second, it is important to see vision in the context of the overall behaviour of an intelligent system, in order to avoid the pitfall of considering this sense, mostly useful to humans and to many other animals, only as a separate functionality providing input to higher processes in an open loop fashion.

We will see in this chapter that active vision is actually an approach, a shift in perspective, and considers vision itself not as an isolated functionality, but as a set of tools that an intelligent agent uses at its best to reach a goal. So, active vision is not a part of vision; rather it is vision performed in an active way. And, once for all, it should not be confused with active sensing, in which the sensory system sends some sort of energy onto the observed scene in order to measure, through its use, some physical property. In active vision it is the observer that is active, while the sensors can be either active or passive.

2. COMPUTATIONAL VISION

Let's start with an apparently naive question: "What is vision for?". If asked without previous notice we would probably answer something like: "It is for seeing!". And many of the readers will not find it difficult to agree with it. But then the second answer comes: "What does it mean to see?". At this point we would probably start thinking and humming, in quest for a satisfactory and definitive answer that can stop the sequence of questions.

Let us see how one of the most influential works in computer vision research answered the same question. In his famous book "Vision" David Marr[1] faces it from the very first paragraph (the underlines only have been added to the original text):

> "What does it mean to see? The plain man's answer (and
> Aristotele's too) would be, to <u>know</u> <u>what</u> is <u>where</u> by <u>looking</u>.
> In other words, vision is the *process* of discovering from images
> what is present in the world, and where it is".

It is interesting to analyse this sentence for two reasons: because it explicitly points out what have been the important topics of computer vision research, and because it reveals, between the lines, the cultural influences that, beyond technological limitations, acted on the researchers.

Attention is explicitly focussed on WHAT is present in a scene and WHERE it is, and on the PROCESS for extracting this information from images. To know what is present in a scene means to recognize objects, and the most distinctive characteristic that allows to discriminate between different objects is shape. To define what is shape is difficult, but it is easy to accept its relationship with the topological and geometrical properties of the object. In the same way, to know where an object is can mean to know its geometrical relationship with the three-dimensional real world in which it is placed. Thus, as obvious consequence, the importance given to the problem of obtaining a good geometrical representation of the viewed scene. This approach has been called *reconstructionist* because it tries to reconstruct the three-dimensional structure of the scene from its two-dimensional projections on images. This reconstruction is part of the vision process as intended by Marr, and makes heavy use of representations.

As a matter of fact, in the Marr paradigm, representations play a very important role. This is due also to the rigorous computational approach followed by him: the solution to the vision problem is seen as decomposed in three levels.

First comes the *computational theory*, that must answer questions like: "What is the goal of the computation, why is it appropriate, and what is the logic of the strategy by which it can be carried out?".

Second comes the level of *representation* and *algorithm*, answering questions like: "How can this computation theory be implemented? In particular, what is the representation for the input and output, and what is the algorithm for the transformation?"

Third comes the level of *hardware implementation*, that must answer the question: "How can the representation and algorithm be realized physically?".

In this computational perspective the role of representations is correctly emphasized as that of formal schemes that make explicit some information about an entity. Moreover, formal schemes are the object that can be manipulated and transformed, one into the other, by computer programs. The visual process, or at least its prominent part, is seen as transforming an image into a different kind of representation. As image, or input representation, is generally taken a regularly spaced grid of pixels, each of which carrying a numerical measure of the irradiance, taken at the corresponding point on the sensor. In this way what is made explicit by the representation is the measure of light, while other information, such as reflectance of the surface, texture, shape, position and motion of the object are left in the background. The effort is then to devise a process, or a collection of processes, that extracts interesting information by transforming the input representation into a new one, that makes that information explicit.

But how should the output representation be organised? What kind of information, embedded in the input, should be made explicit? The answer depends on the task for which visual processing is performed. The suitability of a representation for a given task depends on wether it makes explicit and easy to use the information relevant to the task itself.

While recognizing that the input image contains, encoded into its luminance patterns, different kinds of information, Marr and the other researchers following the same approach gave absolute preference to shape and spatial relations. This choice has been due mainly to two reasons. First, the already mentioned usefulness of shape as distinct characteristic for object discrimination and recognition. Second, the support given to theories centered on shape description by experiments carried out on humans who had suffered parietal lesions in their brains[2]. These experiments had shown how the right and left parietal lobes have different roles in object recognition. While the left lobe is more concerned with language and symbolic description of object properties like its name and usage, the right lobe seems concerned with spatial reasoning and shape representation. Patients with lesions in their left parietal lobe, although not able to name a viewed object or tell its function, were able to show to have captured its shape. This evidence for separation between shape representation and logic-symbolic descriptions was a stimulus to attempt the development of object recognition techniques based solely on shape, and not carrying into the process all the amount of extra knowledge that many AI researchers thought would have been necessary.

If shape is the key point, then the final output representation is a three dimensional model of the observed scene. Model based recognition uses internal geometrical representations, often of the same kind of those used in CAD systems, to be matched against the resulting representation produced by the visual processing. This processing is seen as performed in several steps. First, a so called primal sketch is computed from the intensity image. It has things like edge segments, zero crossings, terminations and boundaries in the image as its primitives. Second, depth information is recovered and the $2\frac{1}{2}$-D sketch is constructed, with local surface orientation, distance from the viewer and discontinuities in both these properties as primitives. Third, the segmentation process and the matching of segments against the stored models lead toward recognition.

3. THE PASSIVE APPROACH

Computational vision has to face the difficulties posed by non trivial, real world images. One image provides a significant amount of data, that becomes enormous if we want to approach real time performance. Moreover, anything can happen in an image; not only noise, but the extreme variability of the real world constitute a very difficult problem at the computational theory and algorithmic levels. So, in search for a solution, a heavier and heavier burden is laid on the available computing resources. For these reasons the predominant approach has considered a static observer trying to recognise objects by analysing monocular images, stereo pairs or very short sequences. It was thought that observer motion would further complicate the already difficult job. This intuition would have been proved wrong.

4. THE ACTIVE APPROACH

The history of computer vision research provides strong evidence for the difficulty of developing artificial visual capabilities similar to those of human beings. The attempted strategy is to get suggestions and inspiration from the study of its biological counterpart and try to identify useful functions that could be computed through appropriate algorithms to be implemented in computer programs.

There are three points that represent possible motivations for the failure of this strategy, at least as it has been followed in the past.

Starting from the input to the computation, the huge amount of data/information contained in it overwhelmed the limited resources of the machines. Humans too have limited computational resources, but for the particular problem of the amount of input data, an effective solution adopted by nature seems to be a variable resolution, foveated retina. This solution combines the advantages of covering a wide field of view in the environment, while avoiding to flood the brain with an excessive amount of data, and maintaining a small high resolution area that enables a detailed analysis of restricted regions in the scene[3,4].

The other two points are concerned with functions and underlying hardware. Computer vision often tried to emulate biological visual capabilities by identifying the performed functions and explicitly deciding to implement them on a completely different hardware architecture. But in the biological domain it is difficult to separate functionality from morphology and physiology. Nevertheless, since we have chips and not cells in our computers, an attempt should be made in this direction. But at least we should consider also biological components other than the retina and the visual cortex.

Vision should not be considered as an isolated input functionality, but as part of an integrated system. We constantly move our eyes, head and body to act in the world; so maybe the observer motion is not such an obstacle on the road to vision; maybe it is more an advantage. For sure it is an effective way to collect an amazing amount of useful information while dominating the spatio-temporal huge dimension of the input data by sequentially directing the fovea only on small regions of interest in the scene.

And more about functions. If vision is not to be considered only as an open ended input channel to the conscious mind, then it is not a self contained problem anymore; and it does not represent the final purpose but a tool, or a set of tools, to be used by an intelligent agent to help it carrying out its own tasks. This means that not every visual information nor every part or object present in a scene should be respectively extracted or analysed, but only the single elements that are currently useful for the completion of the task at hand.

Active vision recognizes the essential role of motion in perception. It is interesting to remember that one of the first researchers to use the term "Active Perception" in this sense has been Ruzena Baicsy[5] who has been working for years on the use of robots tactile perception, in which the smooth shift of a finger on a physical surface allows to extract different kinds of information from the case of a stationary contact.

Going back to vision, a controlled movement made by the observer allows for the collection of an amount of information not only quantitatively richer, but also, and more importantly, qualitatively more suitable to the visual task at hand.

There are several ways in which an observer can take advantage of an active use of its sensors:
- a nearly insoluble, under-constrained computational problem can be easily solved by adding known constraints;
- lack of information or excessive uncertainty can be solved for by selective gathering of new data;
- visual processing resources can be concentrated on regions of interest in space and time by directing gaze;
- disturbing or distracting input signals can be under-emphasized by defocussing;
- a moving object of interest can be kept in sharp focus in a high resolution sensor area, while deemphasizing the background with blurring, by tracking it;
- a complete model/understanding of the environment can be incrementally built/achieved by sequential exploration;
- but also, the development of a complete internal model can be avoided by collecting, time to time, only information that is necessary at that moment.

In essence, an active observer uses its resources at its best to achieve a goal. And resources include sensors, computing machinery and actuators. Active vision achieves the best conditions for performing a given visual task by adjusting observer and sensors parameters.

5. ACTIVE RECONSTRUCTIONIST SOLUTIONS

In a seminal work first presented in 1987 Aloimonos and others[6,7] demonstrated that a controlled motion of the sensor(s) can considerably reduce the complexity of some classical vision problems of good importance for the reconstructionist approach. Some of these problems fall into the so called "Shape from X" group, in which local three dimensional shape of the surface of an object is recovered from two dimensional image cues. Among the cues that can instantiate the "X" are shading, texture and contour. For all these three cases the authors showed that adequately controlled observer motion results in a mathematical simplification of the computations.

Let us take for example shape from shading, that is, a process to infer local orientation of object surfaces from apparent luminance values[8,9]. It results that if we want to solve it with a single monocular view, we need to take additional assumptions about the kind of surface and illumination. These assumptions are, for example, the smoothness and uniformity in reflectance characteristics of the surface, and the uniformity of lighting. These often unrealistic assumptions limit the applicability of techniques and algorithms based on them, but they must be taken to regularize a generally under-constrained, ill-posed problem like passive shape from shading.

Things become easier with two images in stereo configuration, leaving aside the correspondence problem, and even better with three images. But the problem is still non linear. Moreover stereo images pose two different kinds of problems in short and long baseline configurations: a short baseline makes the correspondence problem easier and allows for a linearization of the equations to be solved by taking only up to the linear term of a Taylor series expansion, but the resulting accuracy is low; a long baseline reaches a good accuracy but suffers from the difficult correspondence problem, that can trap the search for solution in a local minimum.

Aloimonos and colleagues present an active technique that takes the advantages of both stereo configurations. This technique decouples the process of finding the unknowns (surface orientation and reflectance parameters, plus lighting distribution) in every single point in the image. The reference coordinate system adopted is fixed and is shown, along with sensor positioning, in figure 1. The camera moves by shifting the position of the optical center in the XY plane, that is parallel to the image plane. For the mathematical details the reader can see the works published by Aloimonos and others[6, 7].

The effect of camera motion is that more data are collected in a controlled way such that the equations for the different surface points become coupled along time (subsequent camera positions) and are no more coupled for adjacent points on the surface. This takes rid of the old unrealistic assumptions on maximal surface smoothness and uniform reflectance and lighting. In this way it is possible to separate shape information from reflectance and lighting variations, thus detecting and measuring all of them.

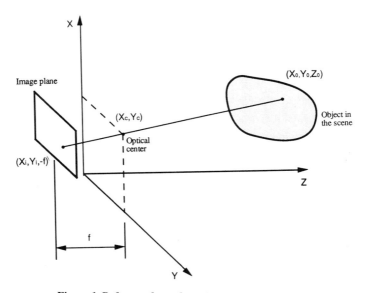

Figure 1. Reference frame for active shape from shading.

6. IMAGE UNDERSTANDING REVISITED

We have seen how an active approach can reduce the complexity or even solve otherwise unsolvable problems in intermediate level computer vision. We can appreciate more now the real meaning of the four words that I have underlined in the initially cited sentence: to know what is where by looking. Marr and colleagues did not realize the full extent of the importance of <u>looking</u>, or directing gaze. For sure Marr consciously limited his important contribution under the influence of the first of the four words: to <u>know</u>. This influence, stemming from an AI perspective that considers as central the logical-symbolic processing of the mind, often meaning its conscious manifestation, leads to consider vision as a functionality to be used mainly for the construction and maintenance of a conscious understanding of the visible world around us.

But image understanding is difficult, and an active approach might make it easier, through divide and conquer, by sequentializing the two problems of *what* and *where*, and not trying to solve the problem in parallel everywhere. We have already seen some of the advantages that can come to the object recognition problem from active vision; the *what* can be reduced to a deep analysis of a restricted region at a time. The *where* can be reduced to the problem of roughly detecting other regions of interest in the surrounding areas and sequentially follow a scan-path or some exploration strategy, in the same way as human beings do. Even if we need a precise measurement of spatial relations among different objects, we can take advantage of the capability of positioning gaze in an easy and measurable way.

There is another aspect of image understanding that deserves careful consideration. It has been shown (see the famous work by Yarbus[10]) that the visual exploration of a picture by a human observer follows different paths according to the particular task that has been assigned to him. Not only different global tasks may cause the exploration of different regions in the image, but some regions that are of particular significance for the given task are analysed more carefully; gaze is positioned on them with insistence, often coming back to these points of interest after a round trip to other regions. These experiments suggest two conclusions:
- what is important is to collect, in as short a time as possible, the information that allows to perform the task assigned to the system, leaving aside the rest;
- to answer some questions about the scene one needs to analyse few restricted areas for a long time, meaning perhaps that the extraction of certain kinds of information is difficult compared to the extraction of others.

One objection that could be moved to the drawing of these conclusions is that animal visual systems, and the human one is of course among these, have been shaped according to evolution and biological adaptation. In this way the visual apparatus is structured in such a way that some stimuli of more importance for survival, either of the individual or of the specie, get a faster response than others. Certainly this is true, but also an artificial system like an autonomous robot acting in an hostile environment faces the problem of "survival". And even if survival is not a problem, nevertheless it is important to identify what we really need to extract from an image and adhere to it in order to face the complexity problem.

This view is not completely new, and the idea that a complete internal representation of the world might not be the best approach to the vision problem can be traced for example in the works by Gibson[11,12,13]. According to him the world around us acts as a huge external repository of the information necessary to act, and we directly extract from time to time the elements that we need.

7. PURPOSIVE VISION

An important functionality that has been identified in computer vision research is the so called "Structure from Motion". It consists of inferring the 3D structure and motion parameters of a mobile object from an image sequence of it taken during a period of time.

Object Recognition and Image Understanding constitute only one of the two major visual problems of interest, the other one being Navigation. Research on both of these problems would receive great benefits from a working structure from motion module. Unfortunately we are still far away from having it.

On the other hand, as we have already mentioned, not all the information present in an image is always necessary to accomplish a task. There are many problems or sub-problems, that must be frequently solved by an intelligent, autonomous, perceiving agent, that need only some specific, task dependent information to be extracted from images. Some of these problems are: obstacle detection and avoidance, detection of independent motion, tracking of a moving object, interception, hand-eye coordination, etc. Trying to solve each of these problems separately can lead to much more economical solutions and to the realization of effective working modules to be activated on demand when need arises.

The alternative approach now sketched has been called "Purposive Vision", referring to the utilitarian exploitation of both system resources and surrounding world information.

We have already seen that an active approach can help in the recovery of the three-dimensional structure of the observed scene. And the results presented in the literature until now are only the first ones obtained by following this approach; research in active vision is flourishing all around in the world and will probably bring new insights and practical applications. Now, some questions arise. If there is convenience in selecting the information to be extracted from the scene, what kind of information is best suited to the task and to the resources at hand? How to combine perception with action, vision with behaviour? How to blend the requirements of different concurrent tasks? Should we develop anthropomorphic devices and systems, or should we take the principle and then go with it in an other direction? How much explicit representation do we need?

All these questions have today the attention of researchers in active vision.

8. QUALITATIVE VISION

Some problems in computer vision suffer from an intrinsic instability due to their noise sensitivity[14,15]. This consideration has led several researchers to consider the possibility of developing intrinsically robust methods. The active and purposive approach to vision has helped them to conquer freedom from a supposed necessity of precise measurements and to devise a new approach to their problems.

When we talk about qualitative vision we are of course out of the reconstructionist approach, since a qualitative appreciation of scene properties is much more suited to an approach that avoids the construction of a detailed 3D model of the world. The ideal framework for employing these techniques is a purposive vision system that works by activating specific, task dependent modules in each specific situation. These modules should implement visual behaviours suited to the gathering of required information, and nothing else.

It happens that needed information either is itself of a qualitative type or requires only approximate measurements, like for example in the problem: "Is there something moving in this room?" (problem of independent motion detection by a moving observer[16]).

9. GAZE CONTROL, AND MORE

One of the most important problems in active vision is gaze control, that is, the dynamic control of position and orientation of the camera(s).

At the base of the possibility of real camera positioning control is the evolution of hardware components like electric motors and controllers. Progress in these components has allowed the practical realization, at affordable costs, of experimental research set-ups. In this phase there is no agreed upon configuration, and probably there will never be only one agreed upon architectural solution, but proposals flourish that try to overcome all the limitations that can be encountered in the practical realization of an active vision system (see figure 2 for a sketch of three proposed active stereo configurations).

The required control techniques for these camera heads are often quite sophisticated, since the control loop includes some processing element, for visual computations, that introduces pure delays[17]. For this reason some prediction technique (e.g. Kalman filtering) is often necessary to manage the problem. In addition, it can happen that different control loops interact in a cooperative way to solve a particular problem[18,19].

Aside from the control problems that can arise, we should remember that in the human visual system there are two kinds of eyes movements: *smooth pursuit* movements and *saccades*. These two kinds of motion have very different characteristics, but may often interact. For example, in tracking a fast moving object the system should try to follow it with gaze, and normally this is done with smooth movements in closed loop control. In this control loop the error signal is encoded in the image and must be extracted by computation. When, because of delays produced by computations, gaze falls behind the tracked object of an amount that the system could not regain by smooth acceleration, a saccade can re-establish an acceptable pointing condition. In the same way a ballistic saccade could use, during its final portion, signals from another control to improve its precision in positioning gaze.

Beyond strict control problems, scheduling problems should be taken into consideration when developing an autonomous system. If we think of a robot, for example, it will have a mission assigned, but while pursuing this main accomplishment it will have to activate several other concurrent tasks. These tasks will be probably related to navigation, hand-eye coordination, or other sub-problems. Just think about what goes on when we are driving: we have a main task that is to reach a certain place, and to accomplish it we look at a map or look at direction indications along the road, but at the same time we keep our vehicle on the road, we stop at red semaphores, we keep our car speed under the speed limits, etc.; beyond all that we break abruptly if a pedestrian suddenly crosses the road or a mad driver crosses a read semaphore in front of us. All these controls make use of visual perception!

These processes can be seen as classified in three different categories according to their degree of visibility at the conscious level. Using a slightly different terminology, related to attention, the three classes of processes can be named *voluntary*, *task-driven* and *pre-attentive*.

The voluntary processes are started and driven by a conscious purpose. Looking at direction indications along the road or examining a map, fall into this class.

Processes that are at a lower level of consciousness are activated by the same fact that we are carrying out the task to which they are related. They often have been learned, maybe during early infancy, maybe later, but they have been mastered at such a level that we don't need to take care of them very much, they have become somewhat "automatic". Slowing down and stopping in front of a red semaphore or keeping the lane by an experienced driver could be inserted in this class.

Processes of the third class are unconscious, and normally continuously active. They are very much related with individual survival, and react to stimuli before any identification can

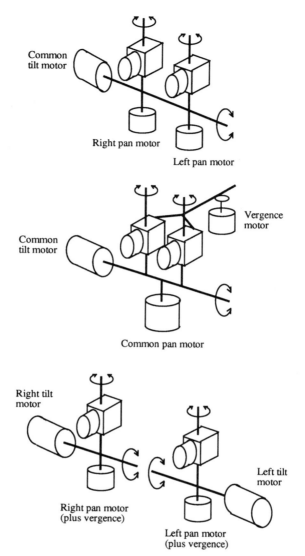

Figure 2. Some alternative configurations for binocular active vision systems.

take place. Some of the impulsive reactions by a driver, like braking if a pedestrian or some other object suddenly occupies the road in front of the vehicle, can be classified in this group. In the same way an object rapidly approaching from our side, and detected only by our peripheral vision, makes us turn or move our head or raise our arm in an attempt to identify a potential danger and to protect ourselves.

We should thus probably talk about "active processes" that interact in different cooperative ways to reach the global goal. Nearly all of these processes participate in some control loop in which both sensors and actuators, the observer and the world participate.

Some of these loops must work in hard real-time and must close at a very low level, without intervention of heavy computations. Ideally they should be implemented in hardware. Other loops close at higher levels and may comprise a significant amount of computation. These computations may frequently include functions typical of an intelligent system like planning, uncertainty management, hypothesis generation, different kinds of reasoning, and stored knowledge.

10. SOME CURRENT RESEARCHES

Several results have been presented on problems mainly related with the visual behaviours of an autonomous system engaged in a task, and with devices, systems and architectures that should facilitate research in active vision itself.

The research group in the Computer Vision Laboratory of the Center for Automation Research at the University of Maryland has proposed, among other results, a technique for robust detection of independent motion by a moving observer[16]. If an observer is moving in a fixed environment it can have the need to detect the presence of another moving object (think about for example to a mobile surveillance system). Due to egomotion it is not possible to simply detect non zero optical flow regions in the image, because optical flow will generally be non zero everywhere. At the same time the task requires a fast detection with no time for heavy computations. The technique is based on small movements of the camera that cause small shifts of the Focus Of Expansion (FOE) in the acquired images, and authors claim it can be implemented in real time since it uses only local information at every point, and that it has high insensitiveness to noise, since it makes use of only the sign of the normal flow (component of the optical flow parallel to the local intensity gradient). Other works carried out by the same group are concerned with the computation of relative depths from motion[20] and with the influence of tracking on the computation of 3D motion[21]. Their work seems directed to the exploration and development of active, purposive and qualitative solutions to problems related mainly to navigation.

Another research group at the Department of Computer Science of the University of Rochester has developed a system that uses a stereo pair with active control of vergence and gaze that performs real-time tracking of a moving object by smooth pursuit movements[18,19]. Another problem tackled by the same group is the use of Bayesian networks in the determination of the best sequence of gaze shifts for carrying out a given task[22].

There is an ESPRIT project on active vision (ESPRIT 5390: Real Time Gaze Control) that produced, among its various outcomes, a binocular vision head developed by some of the partners.

A group in the Division of Applied Sciences at Harvard University is also working on the control of vision heads but with a particular focus on attentive vision[23].

An impressive work on autonomous vehicle guidance has been carried out at the Bundeswehr University in Munich, Germany. A system has been implemented that is able to autonomously drive a van on normal roads, with normal traffic[24,25]. The system is capable of driving at speeds of up to 90 km/h, keeping lane, drive in convoy, stop in front of obstacles, etc. The authors declare that the van has accumulated, during the various experiments, around one thousand km of autonomously guided travel.

There are many other groups, working on active vision or related fields, that would deserve mention, but, for limits of space and time, and willing to avoid forgetting somebody, I prefer to stop here with this very restricted list and redirect interested people to the current literature and vision related conferences.

As a supplemental source of references I want to signal a special number of Image Understanding dedicated to active vision[26] and an interesting collection of research works in a book edited by Blake and Yuille[27].

11. CONCLUSIONS

Active vision is perhaps the most actual subject of research in the entire field of computer vision. This approach has already proved its usefulness in solving some hard problems in computer vision, but it is still in full development and more results are expected. It will not solve all the problems that still confront the researchers in this stimulating area, but it has already reached the result of revitalizing interest in the research community. Moreover, one of the aspects that the author sees with more favour is the change in perspective that has been produced: vision should not be considered as an isolated functionality; rather it should be viewed as a powerful tool, so strictly integrated with the capabilities of an intelligent system that any artificial separation between vision and intelligence should be banned. This interactions will probably bring new insights and advantages in the study of both disciplines.

REFERENCES

1. D. Marr, *VISION. A Computational Investigation into the Human Representation and Processing of Visual Information,* W. H. Freeman and Co. (1982).
2. E.K. Warrington and A.M. Taylor, The contribution of the right parietal lobe to object recognition, *Cortex,* Vol.9, pp.152-164.
3. E.L. Schwartz, Spatial mapping in the primate sensory projection: analytical structure and relevance to perception, *Biol. Cybernet.,* Vol.25, pp.181-194 (1977).
4. G. Sandini and V. Tagliasco, An anthropomorphic retina-like structure for scene analysis, *Comp. Graphics Image Proc.,* Vol.14, pp.365-372 (1980).
5. R. Baicsy, Active perception, *Proceedings of the IEEE,* Vol.76, pp.996-1005 (1988).
6. J. Aloimonos, I. Weiss, and A. Bandyopadhyay, Active vision, *Proc. 1st ICCV,* IEEE, London, UK (1987).
7. J. Aloimonos, I. Weiss, and A. Bandyopadhyay, Active vision, *Int. Journal of Computer Vision,* Vol.1, pp.333-356 (1988).
8. B.K.P. Horn, Understanding image intensities, *Artificial Intelligence,* Vol.8, pp.201-231 (1977).
9. B.K.P. Horn, *Robot Vision,* Mc Graw Hill (1986).
10. A.L. Yarbus, *Eye Movements and Vision,* Plenum Press (1967).
11. J.J. Gibson, *The Perception of the Visual World,* Houghton Mifflin, Boston, MA (1950).
12. J.J. Gibson, *The Senses Considered as Perceptual Systems,* Houghton Mifflin, Boston, MA (1966).
13. J.J. Gibson, *The Ecological Approach to Visual Perception,* Houghton Mifflin, Boston, MA (1979).
14. A. Verri and T. Poggio, Against quantitative optical flow, *Proc. 1st ICCV,* IEEE, London, UK (1987).
15. J. Aloimonos, Purposive and qualitative active vision, *Proc. 10th IAPR,* Atlantic City, NJ (1990).
16. R. Sharma and J. Aloimonos, Robust detection of independent motion: an active and purposive solution, *CAR-TR-534,* Univ. of Maryland (1991).
17. C.M. Brown, Gaze controls with interactions and delays, *IEEE Tr. on Systems, Man, and Cybernetics,* Vol.20, No.3, (1990).
18. D.J. Coombs and C.M. Brown, Cooperative gaze holding in binocular robot vision, *IEEE Control Systems,* pp.24-33 (1991).
19. D.J. Coombs, Real-time gaze holding in binocular robot vision, *TR 415,* Comp. Sc. Dept., Univ. of Rochester (1992).
20. L. Huang and J. Aloimonos, Relative depth from motion using normal flow: an active and purposive solution, *CAR-TR-535,* Univ. of Maryland (1991).
21. C. Fermuller and J. Aloimonos, Tracking facilitates 3-D motion estimation, *CAR-TR-618,* Univ. of Maryland (1992).
22. C.M. Brown, Issues in selective perception, *Proc. 11th IAPR ICPR,* Vol.1, pp.21-30 (1992).
23. J.J. Clark and N.J. Ferrier, Modal control of an attentive vision system, *Proc. 2nd ICCV,* IEEE, Tampa, FL, pp.514-523 (1988).
24. E.D. Dickmanns and T. Christians, Relative 3D-state estimation for autonomous visual guidance of road vehicles, *Intelligent Autonomous Systems,* Vol.2, pp.683-693 (1989).
25. E.D. Dickmanns, B. Mysliwetz, and T. Christians, An integrated spatio-temporal approach to automated visual guidance of autonomous vehicles, *IEEE Tr. on Systems, Man and Cybernetics,* Vol.20, pp.1273-1284 (1990).
26. J. Aloimonos, Special number on purposive, qualitative, active vision, *CVGIP: Image Understanding,* p.56 (1992).
27. A. Blake and A. Yuille, *Active Vision,* MIT Press, Cambridge, MA (1992).

PANEL SUMMARY
ALLOCATION OF ATTENTION IN VISION

Steven Tanimoto[1] *(chairman),* Angelo Buizza[2], Carlo Alberto Marzi[3],
Massimo Savini[4], and Sergio Vitulano[5]

[1] Department of Computer Science and Engineering, FR-35
   University of Washington, Seattle, WA - 98195, USA
[2,4] Dipartimento di Informatica e Sistemistica, Università di Pavia
   Via Abbiategrasso 209, I - 27100 Pavia, Italy
[3] Istituto di Fisiologia Umana, Università di Verona
   Strada Le Grazie, I- 37134 Verona, Italy
[5] Istituto di Medicina Interna, Università di Cagliari
   Via San Giorgio 12, I - 9124 Cagliari, Italy

ABSTRACT

The phenomenon of attention in human vision is a biological solution to the problems of complexity and overabundance of data. Attention is related to such concepts as spatial resolution, multi-resolution representations, and active vision. Attention is also meaningful in cognition and is associated with the idea of consciousness. This panel consisted of short presentations followed by a general discussion.

TOPIC INTRODUCTION (S. Tanimoto)

Hot Points

The panel addressed a variety of questions and ideas related to attention. As a basis for stimulating discussion, the following "hot points" were identified:
1. Definitions of attention in the context of the entire visual process:
    a) visual attention as fixation on a particular portion of the visual field;
    b) cognitive attention as holding an idea, issue, object, or image in short-term memory or as directly processing the attributes or representation of that idea, issue, etc.
2. Visual/non-visual phenomena that attract/guide attention:
    a) motion in the visual periphery;
    b) geometric constructs, contours of objects, unusual sounds.

3. Perceptual mechanisms that control attention.

4. Correlation between duration of attention and visual importance.

5. Relationship between attention and ambiguity.

6. Loss of attention in vision: masking, distraction, camouflage.

7. Computational models for attention in vision:
   a) multiresolution approaches;
   b) temporal models;
   c) region-of-interest operators;
   d) visual search techniques.

Not all of these hot points were addressed in the discussion, however. The topic of the panel was introduced by S. Tanimoto including a reading of the hot points. Dr. Tanimoto proceeded with a presentation of a definition of attention and two interpretations: one in terms of resolution pyramids, and another in terms of state-space search in artificial intelligence. Brief presentations were then made by C.A. Marzi and S. Vitulano. Comments were then made by M. Savini and A. Buizza.

Common Use of the Word "Attention"

Consider the sentence, "Please give your attention to Professor Cantoni." This could mean, "Please accept inputs from the chairman, and process them." While the implication here is that we are not currently accepting inputs from the chairman, this might only be the case because the chairman is being quiet and there are other events going on.

The sentence, however, must also imply the request, "Please avoid attending to other phenomena or sources of inputs... stop daydreaming." Paying attention to something seems to mean taking attention away from something else. And yet we humans have too many needs and other events to give attention to; we cannot afford to completely ignore every other possible input - a fire could break out in the building and we would have to pay attention to that even if we had decided to pay attention to the chairman. Thus the sentence might mean, "Please raise the priority of chairman-related signals higher than would be normal."

Attention can be thought of as a resource which must be shared between a number of competing needs or activities.

Why Attention?

The human visual system has a number of limitations. These include the following:
- size of the visual field - width, height, size of the fovea;
- resolution in the visual field - periphery, fovea;
- location(s) of viewpoints (at most one or two at a time);
- duty cycle on a daily basis is about 0.7; people sleep with their eyes closed;
- computational power-night driving in the city is confusing for the untrained. (Day driving on the Autostrada is confusing for the non-European).

Attention is a mechanism that serves to help overcome these limitations. It is a means to put the limited resources of the visual system into the right place (and the right orientation) at the right time, and to set the mind in the right context.

Attention as Part of Behavior

Attention can be thought of as a kind of trace through time of the locus of processing in the visual field or through a "conceptual field."

By a locus of attention we might mean any of the following:

- 2D: attention is a parameterized plane curve (the eye-trackings of Buswell[1] and Yarbus[2] give nice examples of this kind of attention in human vision);
- 3D: attention is a parameterized space curve (fixation point plus focal depth give three coordinates in space at a time);
- multiresolution: attention is a pyramidal function of time;
- conceptual: attention is a history of the contents of short-term memory;
- neural activation: attention is a history of a network's activation.

However, attention should not represent the history of the entire visual system; attention should not be everything. To make attention equal to all of vision would not shed any light on vision.

Therefore, we can take attention to be a kind of projection of the visual process onto a space of lesser dimension - the visual field or some low-dimensional conceptual field.

ATTENTION IN TERMS OF PYRAMIDS AND ARTIFICIAL INTELLIGENCE (S. Tanimoto)

I would like to present two fairly simple models or approaches to attention from a computational standpoint. The first approach combines the 2D locus-in-the-visual-field ideas with multiresolution aspects. The second treats attention as part of a search mechanism for finding paths to goal states in state spaces.

Attention in Pyramids

Let us examine the notion of attention in the context of one kind of multiresolution image structure: the pyramid. A pyramid is a set of images of decreasing resolution, e.g., 128 by 128, 64 by 64, 32 by 32, 16 by 16, 8 by 8, 4 by 4, 2 by 2, and finally 1 by 1. The images after the first, e.g., those of size less than 128 x 128, are generally reduced-resolution versions of the largest array. The hierarchical relationships among pyramid cells are illustrated in figure 1.

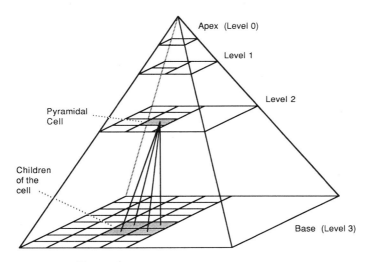

Figure 1. Structure of a pyramid with four levels.

Ignoring for the moment the actual brightness values in the cells of a pyramid, we may consider the cells as comprising a quadtree whose root is at the 1 x 1 pyramid level (the "apex") and where each cell in any reduced-resolution level has four "children" which represent it in a slightly higher resolution.

An attention scheme can be thought of as an assignment of times $t_o, t_1, \ldots, t_k$ to each of the cells of a pyramid. Thus, an attention scheme constitutes a temporal ordering the elements of a pyramid. However, we may allow the possibility of several cells having the same time or position in the ordering - this allows for possible parallelism.

Let us consider some of the possible attention schemes of this sort. They fall into two varieties: data-independent and data-dependent. The data-independent orderings are the following:

1. raster scan (base level only);
2. breadth-first, top-down;
3. quadtree depth-first;
4. iterative-deepening depth-first;
5. top-down, image-parallel;
6. pyramid-parallel.

Of these, the first four are serial and the last two are parallel. For example in the top-down, image-parallel scheme, the apex of the pyramid is attended to first, and then its four children are attended to together and at the same time, and then their 16 children are attended to, etc.

Two possible data-dependent orderings are the following. Note that not all cells are treated; attention is selective in these schemes.

1. Depth-first search with no backtracking in an average pyramid or maximum pyramid - this has been referred to as "bright-spot detection"[3].
2. Depth-first search for edges (as described by Tanimoto and Pavlidis[4]).

Variations on these schemes may be developed by combining these approaches with other selection mechanisms, such as region-of-interest operators or "a priori" interest criteria[5].

An Artificial Intelligence Approach to Attention

At the cognitive level, a useful model of attention comes from the field of artificial in-telligence. Let us assume that the system (human or machine) is attempting to solve some problem. This problem could be a traditional puzzle such as the "Fifteen" puzzle (having a 4 x 4 board and 15 tiles numbered 1 through 15, and one blank space, in which the object is to move the tiles into ascending raster-scan order, without lifting any tiles off the board); alternatively the problem could be to "recognize" an object in a visual scene. Whatever sort of problem we are dealing with, we assume that the problem is represented by identifying a starting state, a collection of operators for transforming states, and a predicate that can be applied to a state to determine if the state is a solution to the problem.

Within this state-space formulation for cognitive activity, a solution may be sought by exhaustively searching the state space or by searching according to a particular search algorithm. An exhaustive search method, such as depth-first search or breadth-first search, typically gets hopelessly bogged down in the enormous number of possible states that arise by applying the operators in various sequences to the starting state. The state space is usually far to large to explore blindly.

In order to overcome this challenge, artificial intelligence programs typically use a sort of attention mechanism to limit the portion of the state space that is actually searched[6]. The popular A* algorithm (pronounced "Eh-star") provides a standard framework in which a guiding principle (or "heuristic evaluation function") is applied to evaluate the relative merits of searching from one state which has been reached, as against from another state which has been reached.

This sort of intelligence mechanism may be able to serve not only at the cognitive (symbolic) level of machine vision, but also at the iconic, or image-processing level. In order for this to work, the image-processing computations must be embedded in a system of states and operators in such a way that a heuristic evaluation function can guide their application. A sketch of this part of such a system is given in the following:

1. the space of states S includes states that correspond to pictorial entities, such as an image in raster form, a list of boundary points, a polygon, or a symbolic representation of the shape of an object;
2. the set of operators includes image-analysis transformations T, each of which transforms some pictorial entity into another pictorial entity;
3. there is a means of computing the cost of an arc (s_1, s_2);
4. the start state is established to represent an input image to be analyzed;
5. a goal-state predicate is defined which returns *true* when a state represents a recognized object.

It is challenging to define heuristic evaluation functions that work effectively. One approach uses the notion of "planning" as described by Minsky in general terms and by Kelly[7] in a machine vision context. One builds a "plan" for a search problem by first reducing the dimensions and amount of detail of each of the pictorial entities, and then running the standard uniform-cost search algorithm in the reduced environment. In the reduced state-space, we end up with one or more paths that provide us with numerical estimates of the costs of original state-space paths to goal nodes. Then we may define a many-to-one correspondence between pictorial entities in the original space and pictorial entities in the planning space. This correspondence allows us to define values of a heuristic evaluation function for most of the nodes in the original space.

Thus, the process of "seeing" an object in image may be thought of as finding a path through a particular kind of graph. The path connects a node that corresponds to the starting image, to a node that represents the recognized object.

Each arc in the graph represents an image-analysis operation that transforms one pictorial object into another, possibly more symbolic, one. We assume there is an association between weights or costs and the arcs along the path, where a low cost means that the data is not greatly altered, while a high cost means that the data is being altered so drastically that there may be a misinterpretation of the result. A minimum-cost path from the start state to a goal state is therefore analogous to a maximum likelihood classification of the input image.

COVERT ALLOCATION OF VISUAL ATTENTION (C.A. Marzi)

It has become increasingly clear in the last two decades that allocation of visual and auditory attention to specified spatial locations is preceded and guided by covert attentional mechanisms. Therefore, formal models of human attention should include *implicit or covert* as well as explicit forms of attentional shifts.

In the visual modality, covert phenomena have been particularly well documented for spatial attention although they represent a fundamental feature of selective attention in general. Briefly, the speed of detection of an extra-foveally presented luminous signal varies as a consequence of the subject's allocation of covert attention to the site (or hemifield) of stimulus presentation (see Posner[8]).

At least three types of cues have been shown to be effective.

The most simple cueing paradigm is "Direct Voluntary Allocation of Visual Attention" (Tassinari et al.[9]) which involves (as is the case for all the paradigms mentioned below) the measuring of simple manual Reaction Time (RT) to very brief light flashes appearing at one of several visual field locations. Following direct verbal instructions, subjects are tested under three attentional conditions. In one they are supposed to attend to the location where the stimulus will be presented; in another condition, they are to attend to a different location and in a third condition, they attend to all possible locations (for example two extra foveal locations on the horizontal meridian on each side of the fixation point). The latter condition is a neutral condition which allows the calculation of attentional *benefits* and *costs*. The former are decreases in RT at attended locations (in reference to the neutral condition) while the latter are increases in RT in comparison to the neutral condition. With the above paradigm, we found[9] that benefits are strictly limited to the attended position while costs are apparent in the entire contralateral visual hemifield. This suggests a special role for the vertical meridian of the visual field in delimiting the area of costs when implicit attention is allocated to the left or the right hemifield.

Another cueing paradigm is represented by the "The Effect of Central Cues on simple RT." Typically, a symbolic cue, such as an arrow pointing to either the right or the left hemifield, is presented immediately above the fixation point about half a second prior to the presentation of a peripheral visual stimulus. On valid trials the arrow points to the opposite field. Subjects are informed that with a probability of 80 percent the arrow will signal the correct hemifield of stimulus occurrence and therefore their likely strategy is to attend covertly to the cued side. Accordingly, valid trials yield faster RTs than invalid and neutral trials, the latter condition occurring when the arrow is substituted by a symmetric central cue inviting the subject to pay attention to the whole field. In fact, it turns out that central cues are effective even when they are *non-informative*; that is, when there is only a 50 percent probability that the direction of the arrow indicates the (right or left) hemifield of stimulus occurrence. As is well known, Posner has proposed a so-called "torch-light model" of attentional focusing according to which the central cue induces an effortful, serial and time consuming shift of covert visual attention from the fixation point to the site of likely stimulus occurrence in one or the other hemifield. The successive step is an attentional *engagement* which allows an effective stimulus *detection*. Following stimulus detection attention must be *disengaged* for the subject to be ready to respond to other locations of stimulus presentation. Importantly, Posner and his colleagues have been able to show that each of these processing stages are subserved by specific neural centers. Other investigators, however, believe that focusing of spatial attention is not movement-like but represents a sort of "all-or-nothing" increase in detection speed in one hemifield at a time.

Finally, a third paradigm is represented by the use of "Peripheral Cues." As in the case for central cues, peripheral cues can be *informative or non-informative*. At variance with central cues, whether a peripheral cue informs on the likely site of stimulus presentation or not has substantially different consequences on the direction of subsequent covert attentional shifts.

Informative peripheral cues improve detection speed at the cued location with a mechanism that for cue-stimulus intervals of less than 150 msec is automatic or involuntary while for longer interval entails an effortful summoning of attention not unlike that of central cues. On the contrary, non-informative peripheral cues determine a slowing down of RT to successive stimuli presented to the same location or hemifield. Such ipsilateral inhibition is accompanied by a facilitation of stimuli appearing in the contralateral visual hemifield. The interpretation of this remarkably consistent phenomenon is still under dispute. Posner and Cohen[10] stress the importance of such a mechanism in reducing the probability of summoning attention to an already stimulated spot while facilitating responses to new locations. Tassinari et al.[11] instead suggest that ipsilateral inhibition represents a consequence of the suppression of natural motor responses to the cue and in particular of saccadic movements for stimulus foveation. Such inhibition would include also a slowing of the manual response used in a RT paradigm which represents one of the various possible motor responses to the cue to be inhibited.

On a more general ground, the effect of non-informative peripheral cues are in keeping with multi-channel theories of selective attention according to which the facilitation of a given channel is an obligatory accompaniment of the inhibition of other competing channels and vice versa.

ENTROPY AND ATTENTION (S. Vitulano)

The Gestalt Theory[12] asserts that in the perceptive processes, the human eye finds the contours of the objects in the picture, so the process tries to perceive each of the background and the objects as uniform. So, we can interpret the first perceptive acts as the result of two different processes: (1) the regions that are constant (texture, structures, etc) become homogeneous; (2) the regions of the picture that are separated have strong variations of gray level or strong variation of colors or strong variations of topology (right or acute angles, lines that break the homogeneity, etc.) or a variation of a certain order that characterizes the scene.

V. Braitenberg[13] asserts that in the perception process, the human eye is able to transmit to the brain only few tens of bits of the millions of bits that are contained in a TV picture. So, the perceptive process can be interpreted as an iterative and hierarchical process. If we see a sequence of TV pictures, we perceive only the variation that happens between a picture and the successive one, and we are not able to describe some regions of the picture, as for instance the background, a lady's dress or all the details contained in the picture.

If we consider a picture as a thermodynamic system we can associate with the image its *entropy*. So, the entropy of a picture containing only one gray level will be null; on the contrary, the entropy of an irregular puzzle will be higher. We affirm that during the first phase of the perceptive process the human eye considers as homogeneous all the regions of the picture whose entropy is null or nearly null, and it pays attention to those regions of the picture where strong variations of entropy are verified (contours of the objects, topologic variations, etc.). The shapes are generated by the variation of the entropy.

We have proposed[14] a visual perception model composed of two modules: (1) the first module identifies all the regions of the picture where the entropy is null or nearly null and it points out the regions where the entropy is higher; (2) the second module associates with each region identified by the first module a structure that describes the texture or the gray-level distribution of the region itself. The model has been fully tested on different kinds of pictures: ambiguous, biomedical, natural, producing significant and interesting results.

DIFFUSED VS FOCUSSED ATTENTION (A. Buizza)

The main point that I would like to make here is that it may be useful to consider different "kinds" or "levels" or "states" of attention, as in the following examples.

In everyday life, our eyes endlessly explore the visual environment in order to provide the Central Nervous System (CNS) with the information necessary to reconstruct our internal model of the world. This model, in turn, is used for many purposes, including programming appropriate motor reactions to external inputs. This process is mostly involuntary as it is independent of both the subject's will and consciousness; yet it may be considered as a form of attention that the CNS gives to what happens around it. This attention is "diffused" and "general," rather than directed to specific objects or concerned with particular goals. It would possibly be the base for our awareness of "being here" or "in a particular environment," and not "elsewhere." Depending on previous experience (including absence of experience) in the same or similar environment, awareness will result in particular feelings and attitudes to the environment that, in turn, will influence our behavior, including, for instance, our visual exploration strategies. During visual exploration something happens or a detail is caught that attracts our interest and therefore also our gaze. Then our eyes begin to explore selected, usually limited, parts of the visual environment. This is what is commonly referred to as (visual) attention. With respect to the above form of attention, it is purposive, goal-directed and possibly also conscious.

From a machine-vision point of view, both kinds of attention may be interesting, but they seem to apply to different situations.

The first kind (diffused attention) is more relevant in the exploration of unknown scenes as, for instance, when scanning an image in order to recognize it or to search for particular elements (that may or may not be present) in it. Here the main problem is to find out an optimal, or suboptimal, strategy for visual exploration and information acquisition. Previous experience might be unable to help, or provide only very general, and possibly scarcely effective, suggestions.

The second kind of attention (focused attention) seems more likely related with the exploration of scenes or environments about which information is already available (kind of environment, dimensions, objects that may be present, relative positions of the objects, etc.). In this case exploration and processing of the relevant visual information can (or rather must!) be guided by all the available information about the scene and its content and by the experience already acquired in previous experiences with similar scenes.

In order to understand the link between these two attentional levels, research needs to identify features that are able to trigger the focused attention. This seems quite a difficult problem to solve in general terms.

ATTENTION, FOVEATION, ALERTNESS AND COMPUTATIONAL RESOURCES (M. Savini)

When people speak about visual attention, they often refer to the direction of gaze. This comes from considerations about the visual systems of the primates, and of man in particular, in which a foveated retina is positioned in space to capture at high resolution highly informative and localized visual signals in the surrounding environment.

But we should argue that the process of directing gaze to areas of interest is better referred to as "foveation." For sure, this process is often guided, or accompanied, by an attentional mechanism, either voluntary or involuntary. But gaze control is not attention itself!

It is an experimental outcome that we can consciously and voluntarily shift our visual attention away from the foveal region, even if with some difficulties, probably due to not being used to it. Moreover, we can distribute our "attention" in the field of view to much wider areas than the foveal region. By the way, it seems that this is an important ability in different activities such as, for example, piloting planes and playing some sports, like basketball.

One important point is the limited amount of available computational power, both in our brain and in the computer that we actually have. On the biological side, this might be the reason for the usefulness of both a foveated retina and attentional mechanisms.

As second important point, we should not forget that all the information-gathering activities that we perform are to be seen in the context of the task(s) we are pursuing. Attention is given to information that is important, or assumed to be important, to the task at hand. This allows us to concentrate on some part of the environment and to discard the rest of the signals. What I would like to underline is that we often have more than one task going on concurrently; for example, at least we have the task of survival in the background. The attentional mechanism can then be seen as the tool for granting "computational resources" to the tasks, according to their (dynamical) priorities .

So, what is attention? The brain is a parallel machine and, if attention is to be defined as a grant for computational resources, we come to a point in which it can be distributed across several tasks. But this seems in contrast with the ordinary idea we have about it. Ordinary consciousness is sequential, and we are used to consciously and voluntarily shifting processing focus from one location to another in a serial fashion. But visual processes are mostly unconscious. So, maybe it is better to keep the term "attention" to indicate the "conscious" allocation of computational resources to a visual task, while we can indicate with "alertness" the allocation of computational resources to "unconscious" visual tasks.

In any case it seems to me that the concurrent aspect of the processes going on in a vision system, be it biological or artificial, and the sharing of resources among them, can be seen as a fixed point. This hypothesis can be corroborated by studies on the effects of concentration on "alertness" or "attention," but is a central point for the development of any advanced computer vision system.

CLOSING REMARKS (S. Tanimoto)

It is clear from the diversity of ideas presented in the panel that the subject of attention is a broad one. While we could not even address all the issues mentioned in the "hot points" list, we have brought out at least a few interesting perspectives on attention. Attention mechanisms can be designed for machine vision systems at either the image level or the symbolic (cognitive) level using such approaches as pyramidal organization of the image or heuristic evaluation functions in state-space search. In human vision, attention processes work at several levels as well, including at a conscious, voluntary level, and at a covert, subconscious level. Another view of attention is as a mechanism for determining regions of interest in an image; one basis for such interest is the information-theoretic entropy associated with each locality of an image. A predisposition to perceive or to react to a scene or situation in some particular way can be called a "diffused" sort of attention; in contrast, the selection of small portions of the visual field for analysis constitutes "focussed" attention, and both sorts of attention are important. One can also argue that the meaning of the term "attention" should be distinguished from gaze control (or "foveation"), and from "alertness," and that attention be thought of as the granting of computational resources.

Part of the difficulty of finding meaningful agreement on the nature and role of attention in vision is that it pervades the vision process in many ways and at many levels. Solving the attention problem (understanding thoroughly how it works) seems tantamount to solving the vision problem itself.

REFERENCES

1. G.T. Buswell, *How People Look at Pictures: A Study of the Psychology of Perception in Art,* University of Chicago Press, Chicago, IL (1935).
2. A.L. Yarbus, *Eyes Movement and Vision,* Plenum Press, New York, NY (1967).
3. R.P. Blanford and S.L. Tanimoto, Bright spot detection in pyramids, *Computer Vision, Graphics and Image Processing*, Vol.43, pp.133-149 (1988).
4. S.L. Tanimoto and T. Pavlidis, A hierarchical data structure for picture processing, *Computer Graphics and Image Processing*, Vol.4, pp.104-119 (1975).
5. A.K. Griffith, Edge detection in simple scenes using *a priori* information, *IEEE Trans. on Computers*, Vol.22, pp.371-381 (1973).
6. S.L. Tanimoto, *The Elements of Artificial Intelligence Using Common Lisp*, W.H. Freeman, New York, NY (1990).
7. M.D. Kelly, Visual identification of people by computer, Ph.D. Dissertation. Report No. CS168, Dept. of Computer Science, Stanford University (1970).
8. M.I. Posner, *Quarterly Journal of Experimental Psychology*, Vol.32, pp.3-25 (1980).
9. G. Tassinari, S. Aglioti, L. Chelazzi, C.A. Marzi, and G. Berlucchi, Distribution in the visual field of the costs of voluntary allocated attention and of the inhibitory after-effects of covert orienting, *Neuropsychologia*, Vol.25, pp.55-71 (1987).
10. M.I. Posner and Y. Cohen, Components of visual orienting, in *Attention and Performance X*, H. Bouma and D.G. Bowhuis eds., Erlbaum, Hillsdale, NJ, pp.137-157 (1984).

11. G. Tassinari, M. Biscaldi, C.A. Marzi, and G. Berlucchi, Ipsilateral inhibition and contralateral facilitation of simple reaction time to non-foveal visual targets from non-informative visual cues, *Acta Psychologica*, Vol.70, pp.267-291 (1989).

12. W. Köhler, *Gestalt Psychology*, Liveright, New York, NY (1947).

13. V. Braitenberg, *I Tessuti Intelligenti*, Boringhieri, Torino, I (1980).

14. C. Di Ruberto, M. Nappi, and S. Vitulano, Problem Solving in a Vision Model, *Proc. of the 2nd IASTED Intl. Conf. on Computer Appl. in Industry*, Alexandria, ET (1992).

NEW WINE IN OLD BARRELS: KNOWING HOW TO MOVE TELLS US HOW TO PERCEIVE MOVEMENT

Paolo Viviani

Department of Psychobiology
Faculty of Psychology and Educational Science, University of Geneva
Route de Drize 9, CH - 1227 Carouge, Switzerland

1. INTRODUCTION

Trained as an electrical engineer, I first approached the work of James Gibson in the late sixties, by reading his second major book "The senses considered as perceptual systems"[1]. Those were the years when the extravagant hopes raised by Information Theory as a possible clue for understanding sensory processing, and the equally extravagant hopes raised by Cybernetics as a possible clue for understanding action were being exposed for what they are, just extravagant and misleading. Yet, in many quarters the notion that the connection between perception and action had to be construed as some form of algorithmic symbolic transformation implemented by dedicated hardware was still quite popular.

I still remember the feeling of intellectual satisfaction that I experienced going through the chapters of that book. For the first time since I had begun my apprenticeship in psychology, I was being exposed to an articulated, convincing argument for the possibility to make sense of perception without invoking ad hoc mechanisms for integrating modality specific sensorial information. At about the same time, I got acquainted with the beautiful work of Johansson on event perception and with the experiments by Michotte on the perception of causality. What I found fascinating in all three cases was that relatively simple experiments could be devised to expose some far-reaching principles of perceptual organization. The few experiments that I will be illustrating in the following do not relate directly to the work of these great psychologists, and, in all likelihood, will not have the same impact. Yet, it should be clear, I hope, that their ideas and discoveries have played a major role in shaping the general conceptual framework within which my own experiments have been conceived.

The general theme of this contribution is the interaction between movement - and by this I mean active movement by the observer - and visual perception. Of course, this interaction has a rather trivial quantitative aspect: the simple fact that we are able to move around and to orient the eyes broadens enormously the range of events that we can apprehend. However, I will rather concentrate on some qualitative aspects of the interaction, and more specifically on the contributions that movement in its various forms can make to the structure of visual perception.

Since the term perceptuo-motor interaction is used to denote a wide range of phenomena, in referring to this contribution I will interchange the position of the two terms, and use instead the term *motor-perceptual* interactions.

2. FOUR LEVELS OF MOTOR-PERCEPTUAL INTERACTIONS

Ever since scientists and philosophers have begun to wonder how we come up with a usable representation of the visual world, it was realized that projecting a three-dimensional world onto a flat retina doesn't seem a priori a very smart solution. Yet, it is a solution that works surprisingly well. This is what one may call Berkeley's problem: vision affords information that it should not be able to afford:

"It is plain that distance is in it's own nature inperceivable, and yet it is perceived by sight"[2].

It so happens that Berkeley's solution to the specific problem of judging distance, namely the hypothesis that distance is estimated by integrating size judgments with the voluntary effort to accommodate and converge, is insufficient: it would have taken more than a hundred years to discover binocular disparity. What is important, however, is that this is one of the first explicit mentions of the fact that active movement can play a qualitative role in the construction of the visual world. In what follows, I will try to do three things. I will begin by suggesting a distinction between four general modes according to which motor-perceptual interactions can occur. Then, for each of these modes, I will try to argue that contemporary research is in more than one way the development of ideas that go back at least to the last century. Finally, I will describe some recent investigations that I've been involved with. The four types of interactions that I will be considering are the following:
- the ecological level;
- active exploration;
- the generation of expectations;
- implicit knowledge.

At the first level we find the interactions that are mediated by the laws of optics and by the physical properties of the objects. Of course, much of the interest for this level of interaction has been generated as a consequence of Gibson's analysis of information pick-up. *Sensation-based* theories of perception hold that perception depends wholly on sensations that are specific to receptors. Gibson's *information-based* theory of perception holds that perceptual experience is possible without underlying sensory qualities that are specific to receptors:

"According to the theory ... the neural inputs of a perceptual system *are already organized* and therefore do not have to have an organization imposed upon them - either by the formation of connections in the brain or by the spontaneous self-distribution of brain processes. ... Instead of postulating that the brain constructs information from the input of a sensory nerve, we can suppose that the center of the nervous system, including the brain, *resonates* to information"[3].

Gibson's approach applies to all possible ways of orienting attention, but visual perception is by far the domain that has been explored more thoroughly along the lines that he had suggested. Indeed considerable experimental evidence has been marshalled in the last twenty years in support of the view that the optic flow - in both its static and dynamic components - affords in many cases information that is sufficient to disambiguate the environmental contingencies and to guide behaviour accordingly. To mention just one particularly elegant demonstration of this possibility, I may recall the utilization that gannets make of the dilation rate of the visual field in order to optimize their fishing behaviour[4]. At the same time this is a very typical example in as much as the visual domain is also the domain in which the dynamic contribution to the global amount of potential information is largest. In my view, one of the most important contributions by Gibson has been precisely that of emphasizing the wealth of information afforded by the principled covariations that are introduced in the visual field by any relative motion between the observer and the environment.

I don't think I do any injustice to the importance of Gibson's insight by remembering that people who make a living out of studying the best way of arranging visual information have often been keenly aware of the intimate relation that holds between motion and light gradients even in the case of purely static displays. I cannot resist the temptation to quote one of these people who wasn't too bad at arranging visual information:

"Painting is philosophy, because philosophy deals with the increase and decrease through motion . Painting can be shown to be philosophy because it deals with the motion of bodies in the promptitude of their actions, and philosophy too deals with motion ... [However], while philosophy penetrates below the surface in order to arrive at the inherent properties, it does not carry the same conviction as the work of the painter who apprehends the foremost truth of these bodies, as the eye errs less". (Libro di Pittura di M. Leonardo da Vinci, pittore et scultore fiorentino. [Propositions 9-11]).

Maybe it is not too far-fetched to see in this "promptitude" that bodies have with respect to their actions a remote predecessor of what, following Gibson, we now call the affordance of a body.

Interactions of the first level arise inevitably, whenever either the environment or the observer (or both) are in motion, and quite independently of the nature of this motion. At the second level, instead, we find the interactions that arise as a consequence of our intentional, active use of movement as a way of complementing purely visual information. Even in this case the first analyses of the possible role of movement date from the eighteenth century. It is in the middle of that century in fact that Condillac[5], through his celebrated metaphor of the statue, poses the question of whether vision is sufficient to afford the consciousness of external objects. His answer is negative:

"It takes three conditions [for a man] to judge that there are external bodies: first, that his limb be determined to move; second, that his hand - the main organ of touch - move over his body and whatever is around him; lastly, that among the sensations that his hand affords, there be one that pertains to bodies in general"[5].

Two centuries later Gibson integrated a similar view in his global theory of information pick-up and, at the same time, attempted to provide experimental ground to Condillac's original intuition:

"Active touch is an exploratory rather then a merely receptive sense. ... Vision and touch have nothing in common only when they are conceived as channels for pure and meaningless sensory data. When they are conceived instead as channels for information pick-up, having active and exploratory sense organs, they have much in common"[6].

Let us now move on to the third level at which motor-perceptual interaction can occur, the level that I have dubbed "the generation of expectations". What characterizes this level is the fact that it is not movement per se that affects perception, but rather the motor plan that is set up for implementing that movement. The very idea of such a level is rooted in an intuition that, according to historians, must be credited to the tchek physiologist Purkinjie. In fact, most people now associate this so-called "outflow" or "efferent copy" hypothesis to the name of Helmoltz. The basic idea is that, whenever a motor command is issued to the muscles, a copy of the command is also forwarded to the cortical areas where the position of the body in space is represented. This copy is utilized by the central nervous system to generate a set of expectations concerning the postural consequences of executing the movement.

In other words, the brain is supposed to have an internal model of the various parts of the body. By feeding this model with a copy of the actual commands, the brain can run a fast simulation of the impending action, a simulation whose results are available even before the actual movement is executed. One advantage of this mechanism is supposed to be the possibility of executing equally fast correction to the motor plan if the original version does not match the expected result. However, already at the time of Helmholtz, it was also suggested that kinesthetic sensations associated with the will to act may have a role in the genesis of motor images. Here's how this possibility was evoked by the physiologist Stricker:

"When I imagine the flight of the clouds, I feel in the ocular muscles the same sensations as though I were actually following them; if I try to inhibit this muscular sensation, the image of the moving cloud stops immediately: the cloud appears still"[7].

In more recent times the notion that the mechanism of efferent copy may also play an important role in structuring perception has gained widespread acceptance. The late Hans-Luckas Teuber articulated this notion with the greatest clarity:

"The unity and stability of our perceptions might [...] be more understandable if we were to trace the information flow in the central nervous system not only in traditional fashion, from sensory to motor regions, but in the reverse direction as well, so that *intended and potential movements* are seen to enter into the structure of our perception"[8].

The reality of such a two-way interaction between motor and perceptual information has been demonstrated quite clearly in the case of saccadic eye movements. When we make a saccade, the retinal image undergoes an abrupt change in a matter of few milliseconds. Yet, phenomenally, we perceive the visual world as stable. What the experiments have shown is that, even before the saccade starts, the retinocentric frame of reference is translated to a new origin that coincides with the expected gaze position after the saccade is completed. This translation, which makes perceptual stability possible, is supposed to be mediated by the very saccadic commands sent to the extraocular muscles.

We come finally to the fourth and more abstract level of interaction, the one that I will discuss in more detail. So far, we have considered motor effects on perception that implied either an actual movement, voluntary or involuntary, or at least the planning stage that precedes an actual movement. The phenomena that are included in this fourth level are supposed to be rooted not in any real or intended action, but rather in the very nervous machinery that makes action possible. More specifically, it is supposed that this machinery, and whatever software is associated with it, embodies some implicit knowledge concerning rules and constraints that characterize the class of movements that we are capable of with certain body segments. Even in this case the hypothetical underpinnings of this class of perceptual phenomena were the object of intensive speculation long before any concrete experimental technique was available to test these ideas. Indeed, the matter was first taken up by physicists like Mach, and even by mathematicians as shown by this intriguing quote from Poincaré extraordinary book "Science and Hypothesis":

"[Motor space] may have as many dimensions as we have muscles. [Thus to localize an object in space] simply means that we represent to ourselves the movement that we have to make to reach this object"[9].

Few years before these words were written, in discussing the relation between the symmetry of the oculomotor system and the symmetry of visually perceived space, Mach hinted at the possibility that the decisive factor is not the actual motor experience, but the anatomo-functional properties of the system. In spite of the continuing lack of corroborating experimental results, the general idea that the structure of perception is influenced by the functional properties of the motor system remained alive during the period of glory of behaviourism, and in fact took up a distinctively abstract flavour as demonstrated for instance by the attempt that Piaget made to generate an ontology of objects based on the group properties of elementary displacements:

"It is clear, in effect, that the reversibility of the [displacement] group presupposes the notion of object and conversely. This is because, in order to recapture an object, one must have the possibility of a reversal (either by a displacement of the object or by a displacement of his own body): *the object is nothing but the invariant originating from the reversible composition of the group*"[10].

One had to wait until much more recent times, however, before a full-fledged motor-theory of perception was articulated. The theory, intimately associated with the name of Liberman, does not concern visual perception but rather the auditory perception of speech sounds. Given that speech sounds, like all other sounds, are continuously varying physical quantities, it is some kind of a mystery how the nervous system parses this continuous flow into categorically distinct phonemes. Liberman's solution to the puzzle is based on the fact that the listener shares with the speaker the same discrete repertoire of speech motor templates. Thus, in a manner that evokes Gibson's "resonance", the theory assumes a natural mapping between the physical stimuli and their potential causes. In Liberman's own words:

"The object of speech perception are the intended phonetic gestures of the speaker, represented in the brain as invariant motor commands that call for movements of the articulators through certain linguistically significant configurations. ... [Since] speech perception and speech production share the same set of invariants, they must be intimately linked. ... *The link is not a learned association ... [it] is innately specified*"[11].

The motor theory of speech perception hinges on the assumption that discrimination of speech sounds is indeed categorical. In fact, some of the sharpest criticisms levelled against the theory come from people who are skeptical about the reality of categorical discrimination. Clearly, it is impossible to envisage a motor theory of visual perception along the same lines followed in the case of speech. On the one side categorical discrimination in the visual domain is even more difficult to demonstrate than in the acoustic domain. On the other side, the connection between acting and perceiving action is certainly looser than the connection between speaking and listening. After all, the only reason for us to speak is to be understood by someone else. Ideally, the best way of validating the hypothesis of a direct influence of the motor machinery on visual perception would call for two ingredients. One should start by demonstrating some peculiar rule or constraint that characterizes unambiguously a certain class of movements. Then one should be able to show that certain visual phenomena bear the imprint of these rules and constraints.

A significant contribution along these lines has recently been made by Shiffrar and Freyd[12] who concentrated on the constraints that the geometry of the human body imposes on the relative displacements among limbs. Shiffrar and Freyd exploited very cleverly the well established phenomenon of apparent motion. It is known that whenever two images showing the same object in different positions are displayed in rapid sequence, one has the impression that the object moves from one position to the other. Moreover, it is also known that, whenever the displacement can occur along two different paths, one consistently sees the object following the shortest path. Consider now for instance the two images shown in figure 1.

There are obviously two ways for the left arm to move between the two positions depicted, and one rotation is much shorter that the other one. So, one would expect that if the two images were presented alternatively, with an appropriate tempo, one should perceive an apparent rotation of the arm through an acute angle. The trouble, however, is that in real life such a rotation would result into a fracture of the elbow. Somehow, the visual system seems to be aware of how painful this solution would be, and arrange things in such a way that what is perceived in most cases is the arm going through the larger angle. In a way, the visual system has interiorized that in real life certain motions are precluded by an intrinsic property of the skeletal system.

How can we improve on this elegant demonstration ? According to the two ingredients that I mentioned above, one should start looking for some property of biological movements that

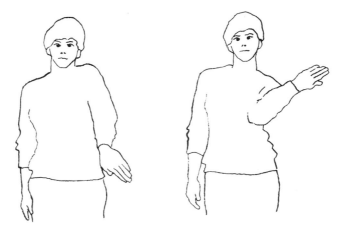

Figure 1. Schematic representation of the two pictures used to demonstrate the phenomenon of apparent motion for biological movements. In the actual experiments real photographs were used (modified from Shiffrar and Freyd[12]).

characterizes them in a strong way, some kind of biological signature. That properties of this kind must exist, follows almost inevitably from the results of the pioneering work of Johansson who showed that the visual system is sharply tuned to any form of biological movement. Even an extremely impoverished dynamic image of a moving body evokes a totally non-ambiguous perception of the underlying real motion. Moreover, as suggested by the work of Berthendal, Proffitt and Cutting this tuning does not seem to be the result of perceptual learning. In fact, children as young as 5 months of age already seem to have the same sharp tuning found in adults.

In the remaining part of this chapter I will describe the results of an effort made in my laboratory to expose this biological signature and to exploit it for demonstrating some perceptual effects that, we believe, are evidence of motor-perceptual interaction. The people who have been associated to this joint effort are Natale Stucchi at the University of Turin, Pierre Mounoud at the University of Geneva and Paola Campadelli at the University of Milan. Some of the earlier work on biological signatures was done in collaboration with Carlo Terzuolo at the University of Minnesota.

3. BIOLOGICAL SIGNATURE AND MOTOR-PERCEPTUAL INTERACTION

I will begin by recalling some simple facts about the physical motion of a point (see figure 2). In order to define the kinematics of the motion one needs to specify two things: the trajectory that is followed by the point, and the so-called law of motion along this trajectory. The trajectory may itself be represented analytically as either a direct relation between the x and y coordinates of its points or in a parametric form. The law of motion s(t) tells you how the length of trajectory travelled by the point increases as a function of time. In certain applications, including those that I will describe later, it is convenient to use two different but equivalent ways of describing these determinants of the motion. As for the trajectory, it can be shown that if you know, for each point, the radius of curvature R of the geometrical path followed by the point you have the same information that is provided either by the implicit or the parametric form of the trajectory. Likewise, if you know the time derivative of the law of motion V, to within a constant, you have

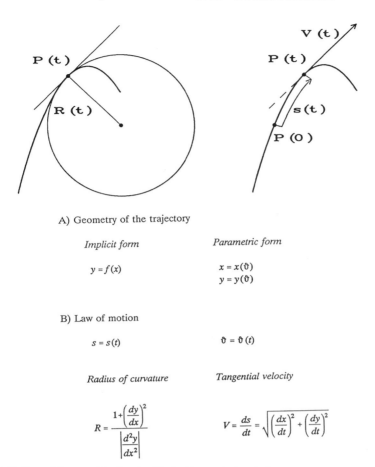

A) Geometry of the trajectory

Implicit form	*Parametric form*
$y = f(x)$	$x = x(\vartheta)$
	$y = y(\vartheta)$

B) Law of motion

$$s = s(t) \qquad\qquad \vartheta = \vartheta(t)$$

Radius of curvature   *Tangential velocity*

$$R = \frac{1+\left(\dfrac{dy}{dx}\right)^2}{\left|\dfrac{d^2y}{dx^2}\right|} \qquad\qquad V = \frac{ds}{dt} = \sqrt{\left(\frac{dx}{dt}\right)^2 + \left(\frac{dy}{dt}\right)^2}$$

Figure 2. Definition of the quantities involved in the kinematic description of a point motion in the plane.

the same information provided by the function s(t). This time derivative is called the tangential velocity of the motion. Now, the important point that I want to stress is that, in general, these two determinants of the motion, however we want to represent them, are independent of each other. Clearly, you can follow any given trajectory with an infinite number of different laws of motion. Conversely, any given law of motion can be superimposed to an infinite number of different trajectory. Thus, unless you know the causes of the motion, each one of these two determinants by itself tells you nothing about the other one.

A series of experiments have shown that voluntary, biological movements are special in this respect because in their case geometry and kinematics are not independent: whatever the trajectory, the associated velocity depends in a rather strict way on the radius of curvature of the path. To illustrate this important point with one example concerning drawing movements, let us consider the data in figure 3.

In figure 3A is shown a sample of a drawing movement tracing continuously a closed trajectory. In figure 3B is shown the result of plotting in a doubly logarithmic scale the instantaneous values of R and V. Clearly, a linear relationship interpolates quite accurately the

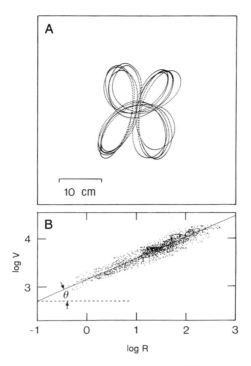

Figure 3. A demonstration of the two-thirds power law.

data points. This implies that V is a power function of R. In adults, the exponent of the power-law (slope of the linear regression) is almost exactly 0.33. In other words, the tangential velocity is proportional to the cubic root of the radius of curvature.

An equivalent way of describing this empirical relationship is to say that the angular velocity of the movement is a two-third power function of the curvature of the trajectory and, for this reason, the empirical relation between the kinematics and the geometry illustrated in figure 3 is known in the literature as the *two-thirds power-law*. The specific mutual dependency between geometry and kinematics that we observe in biological movements is not the only peculiar characteristics of these movements but it is certainly one of the most consistent and robust one. For this reason, it is a good candidate to be the biological signature that we want to exploit. Before going further, however, I would like to stress that the two-thirds power law has nothing whatsoever to do with biomechanics, that is with the fact that actual movements are generated by applying forces and torques to a physical system with mass, stiffness and viscosity. In fact, a recently published experiment by the group of Georgopoulos at the University of Minnesota[13] has shown that the structural relationship underlying the power law is already present at the level of the forces and torques. Moreover, an even more recent experiment by Schwartz in Tucson suggests that a similar relationship exists already at the level of the neural activity in the primary motor cortex. Thus, the power law captures a truly fundamental principle of organization of voluntary movements. Indeed, the principle is so fundamental that it cannot be overridden even when the movement is under strict visual control. This is demonstrated by the results of visuomanual tracking experiments[14, 15].

Consider for instance the situation in which one is asked to track with a stylus a visual point target that moves along an elliptical trajectory[15]. This is the two-dimensional generalization of the well-known visuo-manual pursuit-tracking task that has been studied extensively in the last forty years or so. In the task the ability to remain as close as possible to the target for as long as possible is known to depend on the velocity of the target, but, as long as the target motion is predictable, performances in the range of 10 to 40 cm/s are generally good. Now, let us contrast two situations. In the first one the velocity along the trajectory has the biological signature found in drawing movements. The tangential velocity oscillates between a minimum in the two poles of the ellipse where the curvature is minimum and a maximum in the flatter regions of the trajectory. Then, it can be shown that the x and y components of the motion are sine and cosine function and the motion itself follows the so-called Lissajous model.

In the second condition, that we call transformed, the tangential velocity oscillates between the same two extrema, but the location of the maxima and minima are interchanged so that the velocity is maximum at the poles and minimum in the flatter regions. The x and y components are no longer sine and cosine functions, but remain well-behaved functions of time. In fact, each one of them, considered as target in a conventional one-dimension tracking experiment would be pursued quite accurately. The Lissajous and transformed conditions, therefore, are not all that different. In fact, they only differ in one relational property. In the biological case, the velocity is proportional to the cubic root of the radius of curvature, whereas in the transformed case the relation is highly non linear. The way adult subjects perform under the two conditions is illustrated by the example shown in figure 4.

The left (N) and right (T) panels describe the pursuit of a true Lissajous target and of a trasformed target, respectively. In both cases, the lower panels compare the pursuit trajectory with that of the target. The geometric accuracy of the performance is quite good in the N condition, whereas large distortions are present in the T condition. The difference is even more dramatic when one considers the relative timing between target and pursuit (upper panels). For all intent and purposes, the target that does not have the biological signature is not tracked at all in the usual sense of this word. If one looks at the pursuit movement, it is obvious that the reason for this failure is that subjects cannot depart from their usual behaviour, even when they are visually guided in a very constrained manner. The signature is not an optional feature.

Let us now move on to describe the second step of our strategy, that is the way in which this biological signature can be exploited to demonstrate that motor processes interfere with the visual perception of motion. I will begin by summarizing the results of two experiments, both of which demonstrate the interference in the form of a visual illusion. Consider first the situation in which an observer is shown a light-point stimulus that traces continuously an elliptic trajectory. The persistence of the screen is such that only a very short stretch of the path remains visible at any one time. There are two possibilities: either the major axis of the ellipse is vertical or it is horizontal, and the subject's task is to decide between the two possibilities in a forced-choice paradigm. When the eccentricity of the trace is fairly large, say 0.7 or 0.8, the task is trivial and the subject never goes wrong. However, after each correct response, the eccentricity of the stimuli is progressively reduced making the task increasingly difficult. In the limit, the eccentricity goes to zero, the stimulus becomes a circular motion, and there is no correct answer any more. Thus, if subjects had no bias in the perception of circularity, at this limit point, discriminability should be at chance level. Conversely, if answers become random when eccentricity is not yet zero, one can conclude that there is a bias in the perception of circularity because either the vertical or the horizontal extent is subjectively stretched.

Subjective estimation of circularity was measured under three experimental conditions concerning the tangential velocity of the light spot[16]. Here it will be sufficient to illustrate the results for only two of them. In the first condition, the tangential velocity was always constant,

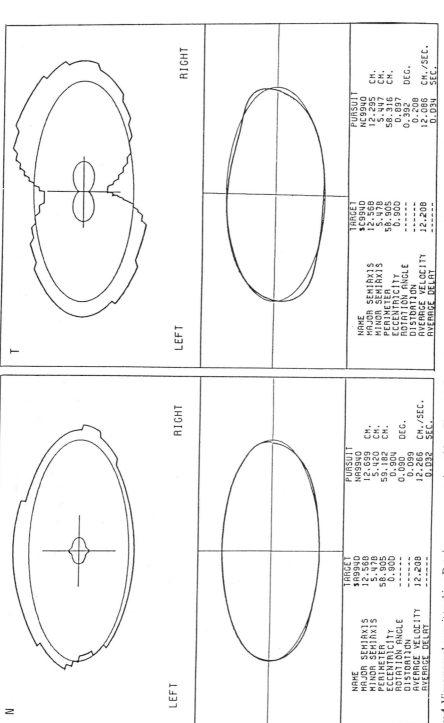

Figure 4. Visuomanual pursuit tracking. Data in one representative subject. The two left panels are relative to the pursuit of an elliptical motion that follows the so-called Lissajous model (x and y components are sine and cosine functions of time). The two right panels are relative to a target that violates the two-thirds power law. The diagrams at the bottom compare the trajectories of the target (thin lines) with that of the pursuit (thick lines). The upper diagrams represent in elliptic coordinates the delay between the target and the pursuit (radial distance between the traces and the elliptic reference curve). When the trace is inside the reference curve, the pursuit is actually leading the target. The diagrams at the center of the reference ellipse represent in polar coordinates the variations of the angular velocity of the target (modified from Viviani and Mounoud[15]).

whatever the eccentricity and the orientation of the trace. Thus, when the trajectory was elliptical, independently of the orientation of the major axis the stimulus violated the constraint described by the two-thirds power law: the motion couldn't possibly be a biological one. However, as the traiectory approached a circle, the discrepancy was reduced because circles have a constant radius of curvature and therefore are indeed traced at constant velocity. Experiments showed that in this condition some subjects have a slight tendency to contract the vertical direction, while others show the opposite bias so that, on average, no consistent distortion emerges. Consider now the second condition in which, whatever the eccentricity and the orientation of the trace, the tangential velocity was always modulated as prescribed by the biological signature in the case of an ellipse with eccentricity 0.9 and with a vertical major axis. In this case there is no condition in which the stimulus agrees with the two-thirds power law. Actually, the smaller the eccentricity, the larger is the discrepancy with respect to the biological signature. Under this condition it is observed that the incompatibility results into a large bias. The vertical direction undergoes a highly significant contraction so that what is subjectively perceived as a circle is in fact a rather elongated horizontal ellipse.

I would like to argue that the most natural interpretation of this bias is in term of motor-perceptual interaction. The reasoning is the following. If we make the assumption that biological movement provides a priori the prototypical model for interpreting visual motion then we must admit that the second condition introduces a conflict that become worse and worse as one approaches circularity: the velocity just doesn't square with the geometry. Distorting geometry (and possibly, velocity as well) can then be construed as an effort by the perceptual system to reduce this conflict. In other words, I would like to suggest that the subjective distortion of geometry is a manifestation of the fact that the visual system takes into account some implicit knowledge about motor performances.

It should be stressed that the illusion that I have illustrated in the case of visual perception also occurs in kinesthetic perception. An experiment by Fagg, Tillery and Terzuolo reported at the 1992 Neuroscience Meeting, has demonstrated a similar velocity-induced bias in a condition where the hand of the subject is passively driven by an artificial arm.

I move now to describe a second experiment in which we have interchanged the role of the controlled and measured variables[17]. In this case, in fact, the geometry of the stimuli is invariable and the subject is asked to make judgments about the velocity. Consider the situation in which a light-point stimulus is shown which follows a given trajectory which can be either regular, or pseudo-random (see figure 5).

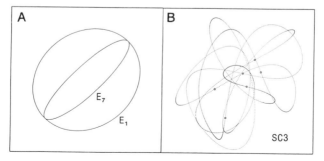

Figure 5. Examples of the trajectories of the dynamic stimuli used to demonstrate the presence of visual illusions (reproduced from Viviani and Stucchi[16]).

In both cases, the task of the subject is to indicate whether he perceives the velocity of the spot as constant. If that is not the case, he can change the distribution of the velocity along the trajectory by acting on a control. The subject is told that moving the control in one direction reduces the velocity in the points of highest curvature and increases it proportionally in the flatter regions of the trajectory. Moving the control in the other direction has the opposite effect. He is also told that the average velocity remains the same and that the game is a fair one, meaning that there is one position that truly produces a constant velocity. What the subject does not know is that in all cases the velocity is modulated by the curvature according to a power law that has the same general form that the relation observed in biological movements. He also ignores that the action of the control is to change the exponent of the power law. Thus, there are two significant positions of the control: the one that corresponds to an exponent equal to zero, and, therefore, to a true constant velocity, and the one that corresponds to an exponent equal to one

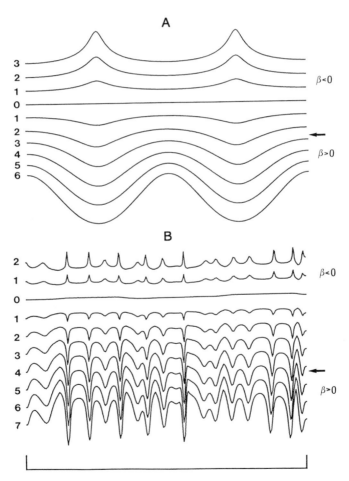

Figure 6. Variation of the velocity along an elliptic (upper panel) and pseudorandom (lower panel) trajectory as a function of the exponent the power law relating velocity and radius of curvature. When the exponent is zero, velocity is constant. When the exponent is 0.333, the velocity varies as it would if the movement were generated biologically (as in free-hand drawing) (reproduced from Viviani and Stucchi[17]).

third which produces a biological movement. Figure 6 illustrates how the velocity changes along the trajectory as a function of the value of the exponent. The upper diagram is relative to an elliptic trajectory and the lower one to a pseudo-random movement. Notice that in both cases the biological velocity (indicated by an arrow) is far from constant. In fact the oscillations between maxima and minima can be as large as 200%. The task was perceived as quite straightforward and easy. In fact there was little variability in the results both within and between subjects. Figure 7 illustrates the average results for 13 subjects in the case of pseudo-random trajectories.

Panel A shows, for five different trajectories, the cumulative probability distributions of the exponent that subjects indicated as the one which corresponds to a constant velocity. Panel B plots again the same results after applying the z-transform to the probability values and evaluating the mode of the distribution (crosses). Panel C resumes the results by plotting averages and standard deviations of the exponent for each trajectory. Each data point in this panel corresponds to a different pseudo-random movement. It can be seen that the estimated exponents, instead of being close to zero, as expected if velocity were perceived correctly, are in fact close to the biological value 0.333. Because of what I mentioned before, this demonstrates a huge perceptual illusion. In fact, if the light point with an exponent 0.333 is moved along a straight line, the accelerations and decelerations are immediately perceived. Moreover, subjects were incredulous when we demonstrated a true constant velocity along a curved trajectory. The results of this experiment confirm those of the previous study and suggest the same interpretation. On the one side we have a further confirmation of the fact that the visual system is sharply tuned to the biological model. On the other side, for some reason, the implicit knowledge that we have about this model produces an illusory distortion in the perception of velocity.

So far, the stimuli that we have been considering were rather special instances of real bodily movements. In fact, they were all instances of drawing end-point movements. To conclude, I would like to present some data concerning more realistic stimuli. Following the technique pioneered by Johansson, we presented dynamic stimuli that describe complex movements of the all body. The technique consists of recording the position of 13 targets placed at the major articulations of the body: ankles, knees, shoulders and so on. Figure 8 illustrates this technique in the case of a person that throws a ball as in the game of bowling.

It is well known that even such an impoverished version of the real image, when presented dynamically on a computer screen, elicits a vivid perceptual representation of the action. The task in this experiment was for the subject to indicate whether the perceived movement appeared as natural. If that was not the case, they could again use a control to change the distribution of velocities along the trajectories. As in the previous experiment, the control modified the exponent of the underlying power law. Here, however, the 13 points cannot be controlled independently for otherwise the geometrical aspect of the movement would be modified. Thus, we modulated only the exponent of the movement of a single point, namely the hand that throws the ball. The kinematics of all other points was affected only indirectly through the constraint that the shape of the movement remains the same. Figure 9 shows the result of this experiment for one typical subject.

The staircase plots on the left illustrate the series of adjustments the subject made starting from several initial values of the exponent, before arriving to a decision. On the right we have the histogram of the final values of the exponent. As before, we see that the average exponent is rather close to the value corresponding to the biological signature. In other words, even in the case of complex naturalistic stimuli, movement perception is finely tuned to a parameter that characterizes movement production.

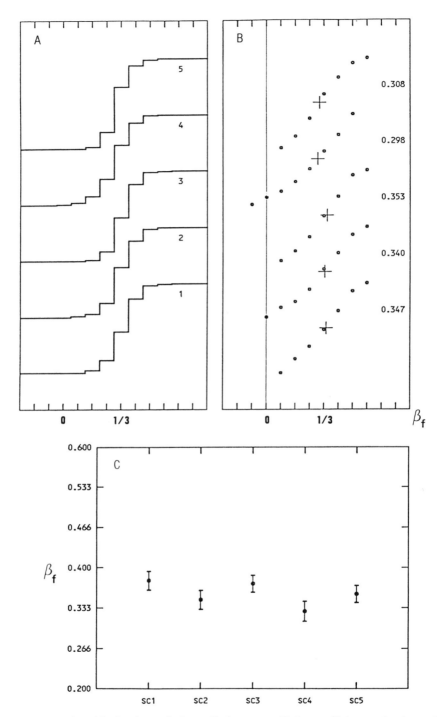

Figure 7. Demonstration of the fact that a velocity profile that agrees with the two-thirds power law is perceived as constant. In fact, as shown in figure 6, velocities vary by as much as 200% (reproduced from Viviani and Stucchi[17]).

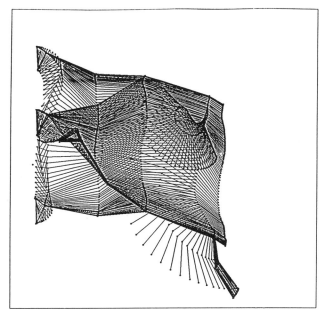

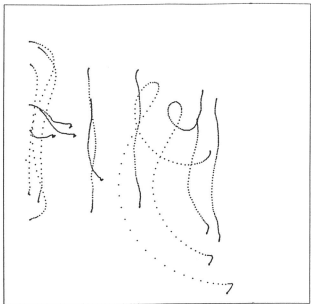

Figure 8. Schematic representation of a complex bodily movement obtained by recording with the ELITE system the coordinates of 13 points placed at the major articulations of the body. These displays presented dynamically on a computer screen are perceived as a veridical representation of a natural event only if velocities and radii of curvature covary as indicated by the two-thirds power law.

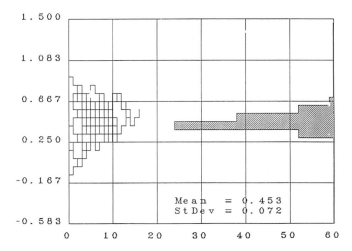

Figure 9. Results of experiment illustrated in figure 8.

4. CONCLUSIONS

Let me summarize briefly the message that I have tried to convey in this chapter and come to a conclusion. Quite sometime ago, Mach wrote that the dispute concerning the hypothesis that perception is based on action was over, and that the existence of a motor-perceptual connection had to be considered as a well-established scientific fact. Little he could suspect, at the time he was expressing this view, that the dispute was in fact going to remain open for the next hundred years. I have outlined four possible levels at which the connection that Mach alluded to can in fact be established. In spite of the fact that some consensus has been reached on certain points, the very reality of many of the processes and phenomena that characterize each level is still the subject of much controversy. Mach also believed that the question of how the basis for motor-perceptual interactions is established would be a matter of little extra time. Also on this count however, he might have been overly optimistic. The few results that I have summarized represent only a contribution to clarifying the nature of these interactions and a great deal of work remains to be done before the question is put to rest. I do believe, however, that the available evidence is already sufficient to accept as a valid working hypothesis the notion that, for both ontogenetic and philogenetic reasons, the structure of perception is intimately related to that of the motor control system.

REFERENCES

1. J.J. Gibson, *The Senses Considered as Perceptual Systems*, Houghton Mifflin, Boston, MA (1966).
2. G. Berkeley, *An essay towards a new theory of vision*, Dublin, IR (1709), reprinted J.M. Dent and Sons, London, UK (1969).
3. J.J. Gibson, *The ecological approach to visual perception*, Houghton Mifflin, Boston, MA (1979).
4. D.N. Lee and P.E. Reddish, Plummeting gannets: a paradigm of ecological optics, *Nature*, Vol.293, pp.293-294 (1981).
5. E.B. Condillac, *Traité des sensations*, Paris, F (1754), reprinted Delagrave, Paris, F (1905), English translation *Philosophical writings of Etienne Bonnot*, F. Philip and H. Lane eds. and trans., Abbé de Condillac, Erlbaum, Hillsdale, NJ (1982).

6. J.J. Gibson, Observations on active touch, *Psychological Review*, Vol.69, pp.477-491 (1962).
7. J. Soury, *Les fonctions du cerveau*, V.ve Babet Libraire Editeur, Paris, F (1892).
8. H.L. Teuber, *James R. Killian, Jr. Award Lecture* (1977).
9. H. Poincaré, *La science et l'hypothèse*, Flammarion, Paris, F (1905), English translation *Science and hypothesis*, Dover, New York, NY (1952).
10. J. Piaget, *La psychologie de l'intelligence*, A. Colin, Paris, F (1947), English translation *The psychology of intelligence*, Rutledge and Kegan, London, UK (1950).
11. A.M. Liberman and I.G. Mattingly, The motor theory of speech perception revisited, *Perception*, Vol.21, pp.1-36 (1985).
12. M. Shiffrar and J.J. Freyd, Apparent motion of the human body, *Psychological Science*, Vol.1, pp.257-264 (1990).
13. J.T. Massev, J.T. Lurito, G. Pellizzer, and A.P. Georgopoulos, Three-dimensional drawings in isometric conditions: relation between geometry and kinematics, *Experimental Brain Research*, Vol.88, pp.685-690 (1992).
14. P. Viviani, P. Campadelli, and P. Mounoud, Visuo-manual pursuit tracking of human two-dimensional movements, *Journal of Experimental Psychology, Human Perception and Performance*, Vol.13, pp.62-78 (1987).
15. P. Viviani and P. Mounoud, Perceptuomotor compatibility in pursuit tracking of two-dimensional movements, *Journal of Motor Behavior*, Vol.22, pp.407-443 (1990).
16. P. Viviani and N. Stucchi, The effect of movement velocity on form perception: geometric illusions in dynamic displays, *Perception and Psychophysics*, Vol.46, pp.266-274 (1989).
17. P. Viviani and N. Stucchi, Biological Movements look constant: evidence of motor-perceptual interactions, *Journal of Experimental Psychology, Human Perception and Performance*, Vol.18, pp.603-623 (1992).

MOTION ANALYSIS

Alberto Del Bimbo and Simone Santini

Dipartimento di Sistemi e Informatica, Università di Firenze
Via S. Marta 3, I - 50139 Firenze, Italy

ABSTRACT

In this paper, we review motion analysis from monocular views under perspective transform, with reference to significant experiences. We consider two schemes for motion estimation: the Correspondence scheme and the Gradient scheme, discuss their pros and cons and their suitability for implementation in the algorithmic and connectionist computing paradigm.

1. INTRODUCTION

Motion estimation is a key research area in image analysis. The relevance of the subject is related to the fact that many applications requiring automatic interpretation of image contents, actually deal with dynamic scenes and sequences of images. Incorporation of motion analysis in a vision system makes it able to track objects, estimate their speed and direction, evaluate potential collisions and support navigation in a way similar to how humans do. It is therefore no wonder that results in neurophysiology and psychology have influenced research on motion analysis.

Motion analysis is done by living organisms in a number of ways, depending on the complexity of the organism (and, thus, on the computing resources available) and on the role of motion analysis in the vision process. On one side we have organisms with a very simple visual system, such as the frog. A frog sees only moving objects. It could starve to death surrounded by food if the food stands still[1]. On the other side there are complex systems, such as the human's, in which motion processing is just a part of a more general-purpose system.

This difference is reflected in architectural and organizational differences between the two systems. In the frog, most of the visual processing is done by the eye. Lateral suppression in the retina processes the image and sends to the brain only information about objects of well defined size (the size of a bug...) moving at a well defined speed (the speed of a bug...). Motion detection in mammals is done mainly in the parietal areas V1 and MT: quite a high level in the visual hierarchy. In mammals, the visual system is *active*: fixation plays a fundamental role in it. Fixation also generates movements in the retinal image that are not due to movement of objects in the scene. For instance, during a saccade, the eye sees all the scene move, yet we don't perceive any motion: some mechanism must turn off the motion perception during saccades.

This distinction between simple, narrow purpose systems and complex general purpose systems is found in artificial vision as well. In general, we can place the distinction between *low level* and *high level* subsystems in correspondence with the focus of attention.

Current approaches to motion analysis may be classified according to the technique used, the number of frames that are needed to perform motion estimation, the use of *2D data* or *3D data* at each time instant, the law of projection, and the computational paradigm used (see table 1).

Table 1. Approaches to motion detection

Technique	Dimension of the scene	Number of frames	Projection law	Computational paradigm
Correspondence	*2D*	*2*	*Orthographic*	*Algorithmic*
Gradient	*3D*	*> 2*	*Perspective*	*Connectionist*

Use of 2D data at each time instant assumes that only one image is available at each time step *(motion from monocular views);* motion is estimated using a variable number of images taken at different time steps. Use of 3D data is related to the development of stereo systems and range finders: at each time instant, two or more images are taken by distinct sensors, providing 3D information about the scene in the form of a 3D frame *(motion from binocular views or stereo motion);* motion is derived based on a variable number of 3D frames taken at different time instants.

Techniques for the determination of motion from sequences of images fall into two schemes. In the *Correspondence scheme,* few features of the moving object are tracked in subsequent frames and 3D motion parameters are derived from relations between couples of points under projection laws. In the *Gradient scheme,* motion is measured based on changes of brightness between subsequent frames at each pixel in the image. Estimation of motion from monocular views both with the Correspondence and the Gradient schemes gives ambiguous solutions for motion. In fact, the absolute value of distance between the moving object(s) and the camera is determined only up to a scaling factor. Stereo pairs may be used to determine this scaling factor and provide absolute values for motion parameters.

The computational approach can be algorithmic or connectionist. Connectionism tries to replicate in artificial vision some of the characteristics determined by the architecture of the biological vision systems, such as the subsymbolic nature of the computation[2], the distributed character of the representation[3], etc.

Reviews on motion analysis techniques have been provided by several authors. Among these, Aggarwal and Nandhakumar[4] have extensively discussed methodologies for motion estimation based on feature correspondence, optical flow and stereo images. In this review, the main problems with the different approaches are pointed out with brief references to related researches. A complete survey on motion analysis is reported by Huang and Tsai[5]. Thevenaz[6] compares Gradient-based techniques with the Correspondence techniques, deriving some interesting conclusions partially reported in this paper. His review contains many details on the computational techniques used for the different approaches. Cappellini, Del Bimbo and Mecocci[7] gave an overview of noticeable motion estimation techniques, including also motion understanding. The reader should refer to these reviews for a more extensive bibliography on the subject.

In this review, we will address only the subject of motion analysis from monocular views and perspective projection. For motion analysis based on stereo pairs the reader should refer to Aggarwal and Nandhakumar; Huang and Tsai; Cappellini, Del Bimbo and Mecocci where references are made to primary literature. Traditionally, motion analysis is strictly connected to object structure estimation *(structure from motion)*, but this subject is not addressed in this review. The reader should refer to Aggarwal and Nandhakumar for reference to significant approaches.

Organization of this paper distinguishes the two basic techniques for motion estimation from monocular views. For each technique, significant algorithms are discussed in detail. Common assumptions and drawbacks are explicitly pointed out so as to highlight limitations and potential use of these techniques in real contexts. The Correspondence scheme and the Gradient scheme are discussed in sections 2 and 3, respectively. The use of the connectionist approach for object tracking and optical flow computation is discussed in section 4. Focus of Expansion and Time to Collision, which are important parameters derived from motion analysis, are discussed separately in section 5. A comparison between the Correspondence and Gradient techniques with reference to several performance figures is finally reported in section 6.

2. THE CORRESPONDENCE SCHEME

Techniques based on the Correspondence scheme, typically solve the 3D motion estimation problem in two distinct steps. The first step is to estimate the shifts of several feature points in subsequent images. The second step is to determine motion parameters based on these shifts. This second step originates a set of nonlinear equations with the motion parameters as unknowns. Methods for solving such equations are usually iterative.

2.1. Common assumptions

No method developed so far can cope with the infinite variety of the visual world in its wholeness. In order to develop a manageable system, the designer must usually restrict its scope — the types of images the system can cope with — selecting a suitable set of constraints or assumptions. Here follow the most common assumptions for systems based on points correspondence:
- most of the methods in the literature are restricted to rigid motion;
- it is usually assumed that the scene contains only one object or equivalently that the sensor is moving but the environment is stationary;
- most of the approaches make an inefficient use of the available information. In most systems, only two frames at a time are considered;
- it is usually assumed that motion through the relevant portion of the sequence is constant or changing slowly;
- in order to find point correspondences, images must contain points that are distinctive, such as sharp corners, or local maxima of gray level variations;
- typically, matching is successful between two frames only if the amount of rotation is relatively small so that perspective distortion is small.

2.2. Evaluation of features correspondence

The determination of correspondences between feature points in consecutive frames is in general a very complex task. The presence of a variable background, occlusions between moving objects, changes in objects size and orientation due to motion, all contribute to the complexity of the task.

The methods developed by several authors to find correspondences can be grouped into two categories: those based on *iconic models* and those based on *structural models*[8].

The methods in the first class, extract templates of features in the first frame, and then search the same templates in the following frames. Different iconic representations are used, such as a rectangular support, a binary mask, a binarized edge map. Matching is done with traditional similarity measures.

The methods in the second class are based on a generalization of the Hough transform to arbitrary shapes evaluated at a set of points that characterize the object structure.

2.3. Significant experiences

After correspondences of feature points of a moving object in consecutive frames have been determined, motion parameters can be explicitly evaluated from the perspective projection of the displacements.

Aggarwal and Duda[9] discuss a method for determining linear and angular velocities of rigid polygons from a sequence of images, with applications to cloud motion estimation from satellite imagery. The model consists of a collection of rigid planar figures moving in several parallel planes. The observed image at any instant is the union of the polygons in the various planes, with possible additive random noise smoothing vertices. The basic idea is to match polygon vertices in subsequent frames. Computational procedures are outlined for the estimation of linear and angular velocities.

Roach and Aggarwal[10] deal with the determination of 3D motion parameters from a sequence of images of moving objects.

With reference to figure 1, the relations between the 3D coordinates (x, y, z) and the coordinates (x', y') of the corresponding point on the image plane, with respect to the coordinate system (X_0, Y_0, Z_0), can be expressed as the equations of a straight line on which, given (x', y'), we can find the corresponding (x, y, z).

Considering the two coordinates of the point on the image plane, and a free variable z', we

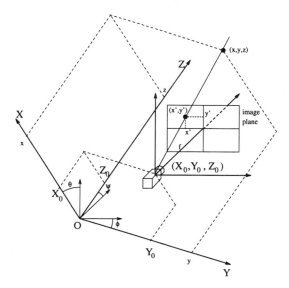

Figure 1. The system geometry.

have, for a focal length f:

$$x = X_0 + \frac{f}{f - z'}(a_{11}x' + a_{21}y' + a_{31}f)$$

$$y = Y_0 + \frac{f}{f - z'}(a_{12}x' + a_{22}y' + a_{32}f) \qquad (1)$$

$$z = Z_0 + \frac{f}{f - z'}(a_{13}x' + a_{23}y' + a_{33}f)$$

where:

$$a_{11} = \cos\phi\cos\psi$$
$$a_{12} = \sin\theta\sin\phi\cos\psi + \cos\theta\sin\psi$$
$$a_{13} = -\cos\theta\sin\phi\cos\psi + \sin\theta\sin\psi$$
$$a_{21} = -\cos\phi\sin\psi$$
$$a_{22} = -\sin\theta\sin\phi\sin\psi + \cos\theta\cos\psi$$
$$a_{23} = \cos\theta\sin\phi\sin\psi + \sin\theta\cos\psi$$
$$a_{31} = \sin\phi$$
$$a_{32} = -\sin\theta\cos\phi$$
$$a_{33} = \cos\theta\cos\phi$$

θ, ϕ and ψ being rotation angles as in figure 1. A consequence of the presence of the free parameter z' in equation (1) is that, given only the point (x', y') on the image plane, it is not possible to recover the full three dimensional position of the object point.

Given a point (x, y, z) in space, the camera parameters (X_0, Y_0, Z_0), and focal length f, the projection (x', y') of the point onto the image plane is:

$$x' = f\frac{a_{11}(x - X_0) + a_{12}(y - Y_0) + a_{13}(z - Z_0)}{a_{31}(x - X_0) + a_{32}(y - Y_0) + a_{33}(z - Z_0)}$$

$$y' = f\frac{a_{21}(x - X_0) + a_{22}(y - Y_0) + a_{23}(z - Z_0)}{a_{31}(x - X_0) + a_{32}(y - Y_0) + a_{33}(z - Z_0)} \qquad (2)$$

Roach and Aggarwal[10] make the hypothesis that the object is *not* moving and that the displacements in the scene are caused by the motion of the camera. Under this hypothesis, they find that two different views of 5 non coplanar points are necessary and sufficient to solve the problem. In this way, they obtain a system of 20 nonlinear equations that relate the image space (x,y,z) coordinates on the object with the camera position. Variables in this problem are the 15 (x,y,z) coordinates of the 5 feature points and 12 additional variables corresponding to the global coordinates and orientation angles $(X_0, Y_0, Z_0, \theta, \phi, \psi)$ of two subsequent camera positions. Numerical techniques are described with methods to deal with noise in the data. Setting 7 variables to known values (assuming, e.g., that the initial camera position is at the origin of the coordinate system and that the z component of any of the 5 points is equal to a given constant), the problem size can be reduced to 18 equations in 18 unknowns. A drawback of this approach is that the authors solve the motion parameters through a single system of equations and therefore have a large solution space. Reduction of the solution space has been discussed by Nagel[11].

Huang and his research group have published several papers on motion analysis with the Correspondence scheme[12,13,14]. The basic geometry of the problem, as considered by these authors, is sketched in figure 2 where object space coordinates are expressed with lowercase letters and image space coordinates with uppercase letters. The two views are taken at time instants t_1 and t_2. Coordinates at t_2 are primed. Rotation is assumed to be around an axis passing through the origin of the coordinate system. From kinematics, object coordinates of a generic point P at different time instants t_1 and t_2 are related by:

$$\begin{bmatrix} x' \\ y' \\ z' \end{bmatrix} = \mathbf{R} \begin{bmatrix} x \\ y \\ z \end{bmatrix} + \mathbf{T} = \begin{bmatrix} r_{11} & r_{12} & r_{13} \\ r_{21} & r_{22} & r_{23} \\ r_{31} & r_{32} & r_{33} \end{bmatrix} \begin{bmatrix} x \\ y \\ z \end{bmatrix} + \begin{bmatrix} \Delta x \\ \Delta y \\ \Delta z \end{bmatrix} \tag{3}$$

where $\Delta x = x' - x$ and $\Delta y = y' - y$, $\mathbf{R}$ is a rotation matrix, and $\mathbf{T}$ represents a translation.

Considering the unit vector $\mathbf{n} = (n_1, n_2, n_3)$ (for which $n_1^2 + n_2^2 + n_3^2 = 1$ holds) along the axis of rotation and the angle of rotation θ from t_1 to t_2, the elements of $\mathbf{R}$ can be expressed in

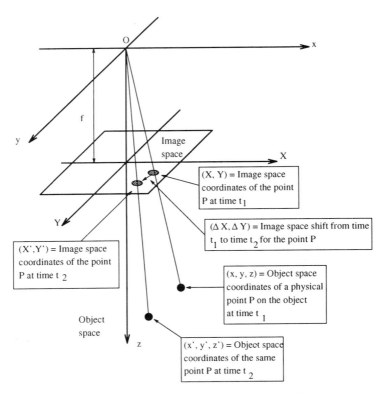

Figure 2. Basic geometry of the Correspondence problem.

terms of n_1, n_2, n_3 and θ, and therefore motion parameters that must be determined are n_1, n_2, n_3, θ, Δx, Δy, and Δz. However, from the two perspective views it is impossible to determine the magnitude of the translation: if the object were twice as far away from the image plane, twice as big and translating at twice the speed, we would get exactly the same two images. Translation can be determined to only within a scale factor. The value of this scale factor can be found if we know the magnitude of $\mathbf{T}$ or the absolute range (z coordinate) of any observed object point.

In the case of *pure translation*, each point correspondence gives rise to one linear homogeneous equation in 3 unknowns (Δx, Δy, Δz). Two point correspondences are enough to determine these unknowns within a scale factor. Actually, due to the presence of noise, more than 2 point correspondences are needed and the resulting system of equations is overconstrained and can be solved for instance by using a least square solution.

In the case of *pure rotation*, each correspondence gives rise to 2 relations with the 9 rotation parameters. Four point correspondences are needed to solve a set of 8 linear equations for the unknown r_{ij}. However it has been shown that only 2 point correspondences suffice. Since z does not appear in these equations, ranges of points cannot be determined for pure rotations around an axis through the origin.

For *general three-dimensional motions*, each correspondence gives rise to one nonlinear equation in the 6 unknowns n_1, n_2, θ, Δx, Δy, and Δz. Since translation can be obtained only to within a scale factor, it can be considered as nonlinear into 5 unknowns. Five point correspondences suffice for the determination of motion parameters and well known iterative techniques can be used to find solutions.

Use of several frames for motion estimation has been addressed by few authors:

Broida and Chellappa[15] consider a body under rotation and translation, with two correspondence points per frame, subjected to noise. Object dynamics is modelled as a function of time and model parameters are estimated by the use of an extended Kalman filter. Experiments are derived for a 2D moving object. The method is confirmed to work well when images are extremely noisy, but for a significant increase in accuracy, the point correspondence must be available for 20-30 frames.

Shariat and Price[16] discuss motion estimation from several frames. All motion in the scene is not assumed to be the same and keeps constant during the relevant part of the sequence (5 frames).

2.4. Drawbacks

Drawbacks of the Correspondence scheme for 3D motion estimation derive from the presence of nonlinear equations. Solving these equations poses both computational and theoretical problems.

Computational problems arise because of the numerical techniques employed in solving the nonlinear equations. Since, usually, solutions are iterative, they strongly depend on initial guesses: depending on this, the iteration may not converge or converge to an acceptable solution.

Theoretical problems arise because of the difficulty of proving properties such as the existence and uniqueness of the solution, i.e. if, given the required number of feature points, can there be more than one set of motion parameters consistent with the equations. Uniqueness of motion parameters obtained by solving the system of nonlinear equations has been addressed by many authors. Tsai and Huang[13] relate the image space coordinates of a feature point on a planar patch in two successive frames through the 8 *pure parameters* (which are proved to be unique). Once pure parameters are obtained, motion parameters can be computed by solving a sixth-degree equation in one variable; only 2 real solutions have been found through numerical examples. Theorems are also proved, which relate the multiplicity of singular values and uniqueness of motion parameters for a planar rigid patch. If the multiplicity of singular values is 2, then motion parameters are unique; if singular values are all distinct, then solutions exist; if singular values are all equal, then rotation parameters are unique and translation parameters are identically zero.

With point correspondences, small noise in the input causes large errors in the motion parameters. Effects of noise in the extraction of correspondence between points are discussed by Fang and Huang[12]. In the literature, researchers have recommended the use of more than the minimum required number of feature points. Alternatively, a greater number of frames can be used with 2 or 3 correspondence points per frame. This is especially needed for images with poor resolution but available at a high rate. To achieve a good noise immunity, approaches with a higher number of frames have been proposed by Broida and Chellapa[15], Shariat and Price[16], Lin and Huang[17]. Extension to more than two frames is complex for 3D motions.

In most of the techniques proposed, unnatural translation and rotation parameters are derived, depending on the assumptions made for the reference system. The reference system is assumed to be fixed with center of rotation at the origin in Tsai and Huang[13], Roach and Aggarwal[10], etc.; it is fixed — attached to the object at a point with motion considered as a rotation around an axis passing through that point, followed by translation — in Nagel[11]. These assumptions are sometimes violated even by simple moving objects. Consider figure 3, where a wheel rolling without slipping on the x axis is shown. If we assume that the center of rotation is fixed in the origin of the coordinate system, we require that all the motions be made up of a rotation around the origin composed with an arbitrary translation. In this case, every frame in figure 3 gives rise to different motion parameters. In particular, the angular speed of the wheel is seen as changing with every frame. Further analysis would be required to determine the actual motion parameters.

On the other side, assuming a coordinate system attached to the object (say at point P_I in figure 3), equals to considering the motion of the wheel as a rotation around an axis passing through P_I followed by a translation. In this case, we would obtain the description of the motion as that of a wheel rotating at a constant speed around a center placed at the periphery of the weel, while the center translates along a cycloidal path.

Solutions to this problem have been proposed by Shariat and Price[16], where the center of rotation is taken as an unknown in the system of equations.

Finally, a high image sampling rate is commonly required. In fact, especially for rotation, small motions are required between consecutive frames (rotation angle less than 5 degrees in Fang and Huang[12]).

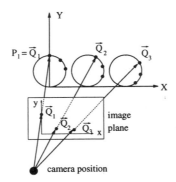

Figure 3. Analysis of the motion of a wheel rolling on the horixontal axis.

2.5. Other features used for the Correspondence scheme

Features other than points may be used to determine correspondences in subsequent frames. The use of lines as observables has been attempted by many authors. Four lines in three views are used by Mitchie, Seida and Aggarwal[18]. The orientation of lines is first recovered, then the motion rotational component and finally the translational component. The use of line correspondence has the advantage over the use of point correspondence that extraction of lines in images is less sensitive to noise than extraction of points. Also, it is easier to match line segments between images than it is to match points.

Other typical features used for motion estimation in the Correspondence scheme are corners and closed contours.

3. THE GRADIENT SCHEME

The relative motion between a camera and the environment gives rise to shifts of gray value structures in the image plane. Therefore, conceptually, motion could be recovered from images, by estimating these shifts and hence by interpreting the result of the estimation.

Optical flow is defined as the observable positional shifts of gray level structures evaluated at each time instant. According to this definition, optical flow does not model the true *2D motion field* on the image plane $\mathbf{v}(p) = (\dot{p}_1, \dot{p}_2)$ originated from the *3D velocity field* $\mathbf{V}(P) = (\dot{P}_1, \dot{P}_2, \dot{P}_3)$ at point P in 3D space through perspective transformation (see figure 4), but it models the distribution of *apparent velocities* $\mathbf{u}(p) = (u_1, u_2)$ of motion brightness patterns $E(x,y,t)$, at the point p in the image. Optical flow is close to the 2D motion field, but differs from this of a finite quantity which only under certain conditions may be estimated. Optical flow seems to be one of the fundamental low-level mechanisms in biological vision. In lower animals (most noticeably some insects such as the bee), it is probably the only visual mechanism. Its relevance to any general purpose perception system can hardly be overestimated. The work of Schunck on the subject[19] was admittedly influenced by the work of Braddick on human motion perception[20].

In his experiments, Braddick considered apparent motion using pairs of images with random black and white pixels. A central rectangle with no texture superimposed is shifted in two subsequent frames from its initial position. Pairs of images of different pixel sizes and

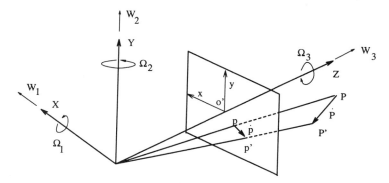

Figure 4. The relation between the true velocity field and motion field. Optical flow is an approximation of the velocity field.

displacement distances were shown to several people. They were requested to identify whether the rectangle was horizontal or vertical and to score the velocity and the clarity of motion perception. Results of the test showed that rate estimation performance and clarity decrease with the size of displacement irregardless of the pixel size. Small displacements of 5' of arc motion were perceived in the right way while motion beyond 20' of arc was not perceived at all and the bounding contour of the rectangle was not clearly recognized. These results demonstrated that human vision does not use a mechanism matching pixels of one type (black or white) to close pixels of the same type but rather a sort of short range process based on changes of image brightness. If a matching algorithm were used, then correct perception should have occurred at displacements up to 1' of arc.

Under the assumption that brightness in a particular point of the image pattern is constant over time, optical flow may be modelled by the following equation — referred to as *Optical Flow Constraint* (OFC) — which relates partial derivatives of image brightness with respect to x, y and t (time) axes and flow velocity components:

$$(\nabla E)\, \mathbf{u} + E_t = 0 \tag{4}$$

Researches by Schunck[21, 19] proved that this relation holds provided that the perceived change in image irradiance at each point in the image plane is *entirely due to the motion of image pattern* (and not to any reflectance effect) and that the image is *smooth* except at a finite number of discontinuities. These requirements are satisfied if the moving object is *diffuse* (i.e. with Lambertian surfaces) *with distant light sources* (i.e. under uniform illumination), and admittedly are too constraining for the derivation of precise object structure or motion from optical flow in most practical situations.

Nevertheless, optical flow provides important information about the spatial arrangement of the observed objects and the rate of change of this arrangement, as well as for segmenting images into regions corresponding to different objects and qualitative motion analysis. 3D motion may be determined from optical flow by exploiting correspondences of flow vectors which belong to the same moving object.

This equation alone, however, does not provide enough constraints to determine the vector $\mathbf{u}$. In fact, it contains the complete $\mathbf{u}$ vector explicitly and represent a single relation for the two unknown components u_1 and u_2 of $\mathbf{u}$ at each point in the image. This fact forces to introduce additional constraints to fully evaluate optical flow vector components. OFC may be also regarded as the equation of a line in the u, v coordinate space (referred to as *Constraint Line*) whose points all represent velocities that satisfy the constraint equation. Considering the Constraint Line, it is evident that motion component in the normal direction (i.e. parallel to the gradient of irradiance) does not depend on the polarity of the gradient and can be directly derived from the OFC as:

$$u_\perp = -\frac{E_t}{\|\nabla E\|} \frac{\nabla E}{\|\nabla E\|} \tag{5}$$

The magnitude of the velocity component which is tangent to the edge cannot be recovered directly from the image sequence and some more constraints are needed.

Assumption of different constraints, algorithms, and criteria for closeness in the velocity field, give rise to different types of optical flow, each with its own properties and behaviour.

A more general motion constraint equation, including a term representing the divergence of the optical flow field vector, and dubbed Extended Optical Flow Constraint (EOFC), was proposed recently by Nesi, Del Bimbo and Sanz[22], following researches by Schunck[23] and Nagel[24]:

$$\nabla(E \cdot \mathbf{u}) + E_t = 0 \tag{6}$$

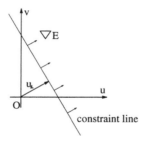

Figure 5. Constraint line and normal optical flow component.

Solutions obtained with EOFC have been proved to produce better results than OFC in the presence of rotational motion with non-zero components around the x- and y-axes. It has been shown that OFC-based solutions are superior in the central part of the image, while EOFC is better in the outer part. Comparisons of optical flow estimations with OFC and EOFC constraints, using the same computational techniques are presented togheter with a performance comparison to other solutions in Del Bimbo, Nesi and Sanz[25].

3.1. Common assumptions

- Since optical flow is defined on the basis of observable image brightness changes, it can be reliably computed only if a texture pattern exists on moving objects, so as to have non uniform images: this automatically gives rise to some intensity gradient. The detection of a sheet of white paper sliding on top of another sheet of white paper can be considered as an intractable case.
- The presence of discontinuities in the local velocity, originated by discontinuities of image brightness, affects seriously estimation of optical flow. In real images, such discontinuities result from the presence of noise, high velocity of detected objects and occlusions between moving objects. Source images are commonly required to have smooth velocity flows.
- Another problem of motion analysis from image brightness changes arises when motion is observed within some small window in the image. The only movement which can be recovered through this small aperture is a movement which is perpendicular to the edge, or equivalently parallel to the gradient (see figure 6). An infinity of different optical flow fields could fit the data obtained through the finite aperture window.
- Image sampling has effects on the accuracy of the estimation. Specifically, the spatial sampling distance must be smaller than the scale of image texture. The temporal sampling period must in turn be much shorter than the scale of time over which changes in the velocity field occur, so as to allow measurement of spatial displacements with no ambiguities.

3.2. Techniques for Optical Flow estimation

As previously noticed, many techniques and algorithms exist for the computation of optical flow, each giving rise to a different optical flow and yielding different approximations to the true 2D velocity field. Broadly speaking, three distinct approaches exist, referred to as regularization-, multiconstraint- and multipoint-based.

3.2.1. Regularization techniques. In this approach, solutions are obtained by defining as an additional constraint to the OFC (or EOFC), a functional including a smoothness constraint

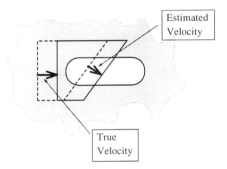

Figure 6. Illustration of the aperture problem.

and searching for a minimizing solution. In this approach, solutions are obtained iteratively. Regularization methods yield flow fields which are dense; the drawback with this approach is that optical flow field propagates also inside the shapes of moving objects and is not limited to object contours. In the presence of occlusions, discontinuities of the velocity field are found at objects boundaries, and these propagate inside object shapes blurring object contours. Depth of propagation typically depends on the value of the regularization factor and on the number of iterations.

Horn and Schunck defined a functional of the type[21]:

$$F = \int \int \left[\left(E_x u + E_y v + E_t \right)^2 + \alpha \left(u_{1_x}^2 + u_{1_y}^2 + u_{2_x}^2 + u_{2_y}^2 \right) \right] dxdy \tag{7}$$

In this functional, the first term gives a measure of the approximation of OFC, the second term is a measure of departure from the smoothness assumption and α is a weighting factor controlling the influence of this term.

Other approaches in the regularization method are found in Nagel[26], and Nagel and Enkelmann[27]. Other regularization-based solutions rely on optical flow evaluation at contours only, to avoid propagation effects inside object boundaries[28, 29]: in this case identification of individual moving objects is more complex.

3.2.2. Multiconstraint techniques. Multiconstraint techniques define additional constraints to OFC (or EOFC) at each pixel of the image. In this approach, an overdetermined system of equations is defined for each pixel and solved in the unknown components of optical flow with traditional numerical techniques. Definition of these additional constraints may be based on the consideration that stationariety condition assumed for image brightness also holds for different motion-invariant functions such as contrast, entropy etc[30]. Alternatively, additional constraints may be defined considering second order partial derivatives of image brightness such as in Haralick and Lee[31], and Tretiak and Pastor[32].

Haralick and Lee derive for each pixel a system of equations of the form[31]:

$$\begin{cases} E_x u_1 + E_y u_2 + E_t = 0 \\ E_{xx} u_1 + E_{xy} u_2 + E_{xt} = 0 \\ E_{xy} u_1 + E_{yy} u_2 + E_{yt} = 0 \\ E_{tx} u_1 + E_{ty} u_2 + E_{tt} = 0 \end{cases} \tag{8}$$

which is solved using the pseudo-inverse technique.

Tretiak and Pastor use for each pixel a system of equations of the form[32]:

$$\begin{cases} E_{xx}u_1 + E_{xy}u_2 + E_{xt} = 0 \\ E_{xy}u_1 + E_{yy}u_2 + E_{yt} = 0 \end{cases} \tag{9}$$

which is solved through the inversion of the Hessian, provided that this is not singular.

The same set of equations was also derived by Verri, Girosi and Torre[33] considering a linear approximation of the velocity field. Verri, Girosi and Torre showed that optical flow as computed from the system of equations (9) differs from the true 2D motion field according to

$$\mathbf{u} = \mathbf{v} + \mathbf{H}^{-1}\left(\mathbf{J}_v^T \nabla E - \nabla \frac{dE}{dt} \right) \tag{10}$$

where $\mathbf{J}_v^T$ is the transpose of the Jacobian of $\mathbf{v}$:

$$\mathbf{J}_v = \begin{bmatrix} \dfrac{\partial \dot{p}_1}{\partial x} & \dfrac{\partial \dot{p}_1}{\partial y} \\[2mm] \dfrac{\partial \dot{p}_2}{\partial x} & \dfrac{\partial \dot{p}_2}{\partial y} \end{bmatrix} \tag{11}$$

and $\mathbf{H}$ is the Hessian of $E(x,y,t)$:

$$\mathbf{H} = \begin{bmatrix} \dfrac{\partial^2 E}{\partial x^2} & \dfrac{\partial^2 E}{\partial x \partial y} \\[2mm] \dfrac{\partial^2 E}{\partial x \partial y} & \dfrac{\partial^2 E}{\partial y^2} \end{bmatrix} \tag{12}$$

Provided that dE/dt and $\mathbf{J}_v$ are bounded (that is true except in those points near object boundaries or motion discontinuities), closeness of optical flow to motion field is evaluated by simply evaluating the inverse of the Hessian $\mathbf{H}$, and optical flow is close to motion field when entries of the inverse of $\mathbf{H}$ are small. This happens whenever the two real eigenvalues λ_1 and λ_2 of $\mathbf{H}$ are large.

Techniques using equations (9) guarantee the numerical stability of the optical flow computation if the inversion of $\mathbf{H}$ is numerically stable. This is true when the determinant of $\mathbf{H}$ is large and $\lambda_{min}/\lambda_{max}$ is close to 1. Therefore a large determinant of $\mathbf{H}$ and $\lambda_{min}/\lambda_{max}$ close to 1 ensure both numerical stability in the computation of $\mathbf{u}$ and similarity between optical flow and motion field[34].

3.2.3. Multipoint techniques. Multipoint techniques are based on the consideration that, under the assumption that OFC (or EOFC) has a uniform solution in a spatial neighbourhood of the optical flow estimation point (i.e. that constraints evaluated for each pixel in a neighbourhood, all define the same velocity) and that the spatial gradient of brightness varies over the neighbourhood (i.e. the constraint lines computed at the neighbourhood are not mutually parallel), the solution can be computed in the velocity space as the unique finite point of intersection of all the constraint lines of the neighbourhood[19].

Multipoint techniques are viable, provided that the flow field is smooth. A number of identical constraint equations equal to the neighbourhood size is defined for each pixel, thus yielding an overdetermined system in the two unknowns u, v of the optical flow vector. If the size of the neighbourhood over which the constraint line clustering is performed is not large enough, a badly conditioned set of intersections of constraint lines is obtained. Increasing the neighbourhood size of a factor k may not be desirable since it increases correspondingly the size of computational effort by a factor k^2. In this case, downsampling the arrays of $\mathbf{u}_\perp$ and α values of constraint lines is expected to provide the right solution. Downsampling by a factor of k, increases the effective neighbourhood area by k^2 without increasing the amount of computation. Alternatively, overlapping sparse neighbourhoods may be a solution, since it leaves the number of operations unchanged, while increasing the effective area of the neighbourhood.

Examples of multipoint techniques are reported, among the others, by Schunck[19]. In practice, the constraint lines fail to intersect at a single point in the velocity space; the assumption on uniform solution is not satisfied in the presence of rotational motion and occlusions between objects, and estimation errors occur in the presence of noise, mainly due to the use of differential operators. These potential error situations are faced by selecting as a solution the velocity vector which minimizes the error on the cluster according to a suitable error measure. Effects of noise can be reduced by the use of appropriate filterings, applied either to the raw sequence data or to the optical flow field.

3.3. Problems with Optical Flow estimation

Using local temporal and spatial gradient of an image for recovering motion only requires that image sequences are analytical. But, since derivatives are numerically unstable, optical flow equations typically produce many vectors with an inconsistent direction or magnitude. Some form of filtering is commonly needed to overcome this problem.

Filtering can be performed on the original image, by using spatial and temporal smoothing. Spatial filtering is obtained by convolving the original image with a circular Gaussian filter

$$\exp\left(-\frac{x^2 + y^2}{2\sigma^2}\right) \tag{13}$$

Similarly, temporal filtering is obtained by convolving the original sequence of images at each pixel with the Gaussian filter

$$\exp\left(-\frac{t^2}{2\sigma^2}\right) \tag{14}$$

The extent of spatial and temporal filtering is dependent on the variance σ of the two Gaussian functions which is expressed in number of pixels and frames respectively and can be controlled.

If only preliminary information on the kind of motion is required, erroneous vectors are removed by smoothing the computed optical flow by linear filtering (i.e. making a convolution of the two components of optical flow with a Gaussian filter). However, in this way, motion discontinuities are blurred and an apparent movement is assigned to them. This appears to be almost a paradox: intuitively, motion estimates should be obtained easily at flow discontinuities. If for certain points, conditions are satisfied for which optical flow estimation may be assumed to represent exactly the velocity field, a sparse optical flow may be used, evaluated at those points, avoiding problems in the presence of motion discontinuities. From this sparse optical flow, a dense optical flow can be obtained through a fill-in procedure and linear interpolation. Nonlinear filtering preserves motion discontinuities much better that linear filtering.

Several types of optical flow can can be obtained from the same sequence: a *raw optical flow*, a *smooth optical flow*, a *sparse optical flow*, a *filled optical flow*, and a *smooth-filled optical flow* if a further smoothing is applied on the filled optical flow. The choice among the different types depends on the specific task. To detect motion discontinuities, a dense optical flow is necessary; if one wishes to recover motion parameters, a sparse optical flow suffices. A smooth optical flow may be useful for a qualitative analysis of the apparent motion. In many applications, such as in robotics and industrial machines, it may be useful to have less accurate optical flow, coupled with the possibility of real time estimation. Optical flows may be computed on a reduced grid or on a single scan line.

As previously noticed, velocity of moving objects has effects on the computed optical flow. In the presence of large velocities, it is useful to increase the amplitude of spatial filtering, by increasing variance σ. Nevertheless, too large spatial filtering has an excessive blurring effect on the image, and worse motion estimates are obtained as a result. For normal indoor scenes a good tradeoff is found for values of $\sigma = 3$ in De Micheli, Torre and Uras[35].

3.4. Derivation of 3D motion and motion parameters from Optical Flow fields

Optical flow fields only provide estimation of brightness changes in successive image frames. 3D motion parameters and object structures can be derived from these estimations.

Adiv[36, 37] discusses methods to derive 3D motion and structure from optical flow. Two steps are required. In the first step the flow field is partitioned into connected segments such that each segment is consistent with a rigid motion of a planar surface. This constraint is a necessary condition to make a segment likely to be associated with one rigid object and vectors in each set to satisfy (approx) the same transformation which describes the rigid motion of a planar surface. In the second step, these segments are grouped under the hypothesis that they are induced by a single, rigid, moving object.

De Micheli, Torre, and Uras recover 3D motion parameters, such as Time to Collision and angular velocity with the analysis of dependency of the accuracy on the optical flow estimation[35]. In the case of pure axial rotation (i.e. the rotation axis is orthogonal to the 3D surface), the angular velocity $\Omega = (\Omega_1, \Omega_2, \Omega_3)$ can be obtained from $\Omega^2 = \det \mathbf{J}_v(P_R)$, where P_R is the immoble point of the pure rotation and $\mathbf{J}_v$ is the Jacobian of the velocity $\mathbf{v}$ (see equation (11)), by using a least squares approximation of the flow. In the case of general motion, no unique solution to recovering the 3D motion parameters exists.

4. CONNECTIONISM

Connectionist (or Neural Network) research tries to make systems inspired to the same computational paradigms as biological systems. This includes the distributed computation and representation of knowledge, the analogical and imprecise computation, and the cooperative computation of a great number of relatively slow and simple computational units.

The connectionist approach postulates that perception and, in general, all the non-reasoning activities of the human brain are better described through a *nonsymbolic* paradigm[2, 38]. Symbolic (or quasi-symbolic) reasoning is just a particular type of activity that arises under certain circumstances to solve a well defined class of problems. Based on this — somewhat holistic — view, connectionism assumes that certain characteristics of peripheric processes such as vision can be more effectively replicated by systems in which the same features — such as the distributed representation, and the imprecise computation — are obtained.

Many efforts in connectionist vision have been dedicated to perceptual organization, attention direction and pattern recognition. Only a few works have been published dealing specifically with motion tracking and estimation.

4.1. Fixation and tracking

High level motion processing is mainly studied in conjunction with fixation and saccades. Fixation has been investigated — especially in connections with robot vision — as a way to reduce the computational burden of recognition algorithms. It allows to *focus* a region of the image where something interesting is happening and limit the high level processing to that region.

Fixation of a moving object requires tracking. Tracking an object results in a wider range of situations. While a low level process distinguishes only between areas in which there is retinal motion and areas in which the image stands still, when the camera is considered as an active part of the vision system, we must also consider the motion generated by movements of the camera[39].

These ideas have been applied by Viola, Lisgerger and Sejnowski[40], where a system for tracking of an object on a simulated retina is presented. The system is inspired by the organization of the visual system in primates and, though simpler, exhibits many of the characteristics of biological systems.

A self organizing system based on local interaction has been proposed by Marshall[41,42]. In this case, a grid of clusters adapts by self organization so that, when the unit i,j is activated by motion, at time t, it makes a projection trying to determine where the motion will be at time $t + \Delta t$ and, after a delay Δt, sends a pre-activation signal to all the clusters that could be reached by the moving feature at $t + \Delta t$. Only one of these pre-alerted features will receive activation at $t + \Delta t$ also from the local feature detector, and will be therefore activated.

4.2. Computation of the Optical Flow

The work of most visual systems rely on the presence of a *lower layer* of units that are sensitive to local motion. These units must be distributed around the image, and perform a very simple (and possibly unprecise) computation at every point in which motion must be estimated. A second layer of units enforces constraint so as to obtain, from the local measurement, a consistent estimation of the optical flow.

This lower layer is not well suited for complex digital hardware. Because of this, many researchers have studied retinal sensors alternative to the current TV cameras. Most of these sensor arrays are analogic.

In Delbruck and Mead[43], every sensor gives an output whenever it sees a change in intensity that is, if we consider the usual optical flow constraint equation

$$\nabla E + E_t = 0 \qquad (15)$$

Delbruck's retina computes the term E_t.

The system in Tanner and Mead[44] is an analog sensor that does the complete optical flow estimation, assuming that only one object is moving in the scene. A sensor like this can be used in the foveal area of a retina to estimate the motion of the object currently under fixation and pilot the fixation direction mechanism to track the object.

Lee, Sheu, Fang, and Chellappa[45] propose a VLSI optical flow estimation circuit. This is an array of processors, each processor being sensible to a particular velocity range. This organization resembles quite closely that of the hypercolumns, that play a basic role in the human visual system[3].

Sereno[46] showed that a solution to the aperture problem could be learned by self organization by a network connected randomly to a grid of units sensitive to motions locally along a given direction.

Colombo, Del Bimbo, and Santini[47] evaluated optical flow from an array of computational units based on relaxation due to local interaction that is, from units operating on a purely local basis. The system they present does no learning, but the connections between neighbour units are predetermined.

4.3. Features and limitations

From the architectural point of view, connectionist systems, and especially self-organizing systems, are extremely appealing. They have the potentiality to replicate the robustness characteristics of the living visual systems, since many of their characteristics —fuzzyness, cooperative computation, etc. — are present in connectionist models as well.

There are, however, many problems to be solved in order to achieve a practical applicability of these systems.

The system in Marshall[41], for instance, is particularly interesting for its local and self-organizing nature, but presents several problems for a practical implementation. First of all, every cluster should contain as many detectors as needed to detect local motion in all possible directions. This makes a cluster sensitive to all local motions at a particular retinal position, giving it a role similar to that of hypercolumn in the mammal parietal cortex, but requires a considerable number of units. The number of units could be reduced by using partially overlapping, loosely tuned sensors, as happens in the cortical maps in the visual cortex[3], but the problem remains serious.

This is an example of a more general problem encountered when trying to build circuits similar to those found in the brain. As a general rule, in the brain, units are cheap, and a lot of them are available even to perform simple functions. The most common structure in the visual system is the retinotopic map, in which the presence of a given feature in a given position of the visual field is represented by the activation of a few units on a map. This is a very inefficient way to represent information. If we have, for instance, a visual field discretized in 256 x 256 units, and we want to represent the position, say, of a lightspot, we can do it on a digital computer with 16 bits (8 bits for the x and 8 bits for the y), while in the brain it would be done by activating a single unit (or a small group of units) on a map of $256 \times 256 = 65,636$ units, that is, 8,192 times the number of bits (assuming a unit contains just a bit) used in the computer representation.

A similar problem is encountered when applying the paradigm of imprecise computation. Because of the imprecise nature of the computatiop, algorithms based on precise mathematical properties cannot be relied upon. For instance, in Mac Lennan[48], it is argued that a convenient means for the decomposition of image sequences is the expansion in Gabor kernels, although these are not an orthogonal basis of the space of image sequences. Orthogonality, in fact, is one of those sharp mathematical properties that an imprecise computing system cannot rely upon.

5. NOTICEABLE MOTION PARAMETERS

Motion parameters that are of interest in many applications are Focus of Expasion and Time to Collision.

5.1. Focus of Expansion

Focus of Expansion (FOE) is a commonly used motion parameter which is defined in the case of pure translation. Determining Focus of Expansion, yields the direction of the translation and many other inferences may be delived about 3D object structure and motion. Under perspective geometry, consideling a moving camera with a fixed background, if this camera moves forward along a straight line in space (pure translational motion), image points seem to diverge from this singular point which is the FOE (see figure 7). Conversely, if the camera moves backwards, lines seem to converge towards it which in this case is referred to as Focus of Contraction (FOC). According to the perspective transforms, the FOE is located, on the image plane at:

$$P_{FOE} = \left(f \frac{W_1 / Z}{W_3 / Z}, f \frac{W_2 / Z}{W_3 / Z} \right) \tag{16}$$

where W_1, W_2, W_3 are the components of the velocity of translation of the camera.

Methods for the derivation of FOE differ in the density of the optical flow used: FOE from dense optical flow uses the fact that all the flow vectors converge at the FOE and requires previous calculation of optical flow. FOE from sparse optical flow uses few points that are tracked in the subsequent frames and computation does not rely on a previous derivation of optical flow.

Burger and Bhanu use dense optical flow and define a *fuzzy FOE* that has an associated area around it[49]. Fuzzy FOE concept is introduced because locating the FOE precisely is difficult when displacement vectors are corrupted by noise and errors. Performance is improved if a region is considered, where FOE may be located, instead of determining a single point. The shape of this region explicitly gives an indication of the quality of the estimation. They demonstrate how 3D motion and structure can be inferred from fuzzy FOE for general motions. They consider a single moving camera *(egomotion)* and analyze effects of 3D camera rotation and translation with reference to FOE. In the case of pure translation the straight line passing through the center of the lens and the FOE is also parallel to the 3D motion vectors. The actual translation vector is determined by the vector from lens center to FOE up to a scale factor.

Jain provides a method for direct computation of FOE from a sequence of images using few noticeable points in the image[50]. Assuming that a sequence of frames has been preprocessed and contains only prominent feature points, and considering corresponding point pairs $(P_1, P_1'), (P_2, P_2')$ in consecutive frames, (see figure 7), the extended lines $P_1 - P_1'$ and $P_2 - P_2'$ will pass through the FOE. The following triangle inequalities hold for two points O'' and O''' (L stands for the Euclidean distance):

$$L(O'', P_1') - L(O'', P_1) \le L(P_1, P_1')$$
$$L(O''', P_2') - L(O''', P_2) \le L(P_2, P_2') \tag{17}$$

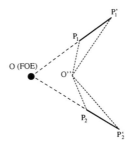

Figure 7. Focus of Expansion (FOE).

where the equalities hold only if O'' lies on the line $P_1 - P_1'$ and O''' on the line $P_2 - P_2'$. A point O is the FOE if it lies at the intersection of the two lines, that is, if substituting O' for O'' and O''' both equalities hold. Summing these two disequalities and substituting O'' and O''' with O' we get:

$$L(O',P_1') - L(O',P_1) + L(O',P_2') - L(O',P_2) \leq L(P_1,P_1') + L(P_2,P_2') \qquad (18)$$

which can be used for the determination of FOE. Only FOE will satisfy equality of the two sides, and it can be obtained by searching the point which maximizes the left side of the above disequality. Gradient techniques are proposed for deriving such maximum. Simulations presented show that FOE can be derived also by using only 2 or 3 strong feature points in the case of approaching or receding observer. In the case of motion parallel to the image plane, all disparity vectors will be parallel and therefore FOE is located at infinity. In this case there is no maximum at a finite point and the nature of motion can be derived only studying function (18).

FOE can be used for the interpretation of motion trajectories[51]. Considering an extended sequence of frames, which originates a sequence of trajectories between two frames, the locus of FOEs, evaluated on such sequence is closely related to the motion of the object under observation (it can be interpreted as the perspective projection of the space curve represented by equations: $X = \dot{P}_1, Y = \dot{P}_2, Z = \dot{P}_3$) and can be used to make qualitative assertions on the type of motion such as whether the object is approaching or receding, or the object is changing direction of motion.

5.2. Time to Collision

Time to Collision (TTC) is defined as the time between the instant in which the frame is grabbed and the possible collision of the camera with a moving object in the scene. TTC is a useful parameter in several applications, especially in robot navigation and can also be used to reconstruct a 3D scene.

TTC values depend only on the translational motion component along the Z-axis and the distance ΔZ between the observer and a point p of the moving object and is expressed as:

$$TTC = \frac{\Delta Z}{W_3} \qquad (19)$$

where W_3 is the translational component of the camera motion along Z.

A relationship exists between the divergence of motion and TTC expressed as:

$$\text{div } \mathbf{v} = -\frac{2}{TTC} + \frac{3x\Omega_2}{f} + \frac{3y\Omega_1}{f} \qquad (20)$$

which reduces to

$$\text{div } \mathbf{v} = -\frac{2}{TTC} \qquad (21)$$

in the presence of pure translational motion. This can also be used for any kind of motion, provided that it is evaluated close to the center of the reference system.

6. COMPARISON

As noticed by the reader, formulations of motion with reference to the image are considerably different in the two schemes of Correspondence and Gradient computation: in the Correspondence approach, motion is seen as a set of displacements, opposite to the velocity-based formulation that characterizes the Gradient method. Differences between the two approaches can be synthetized considering their performance with respect to some indices, such as spatial and temporal density, range of motions, behaviour in the presence of the aperture problem and computational robustness and complexity. Analogies between motion estimation techniques and human vision, and suitability to connectionist implementation are also shortly discussed.

- Considering spatial density, in the Gradient scheme, optical flow is defined almost every-where, thus originating a very dense flow field. The flow field can be made sparser by subsampling. In the Correspondence scheme the flow field is defined only at few significant points: the flow field is sparse and cannot be made denser due to inherent complexities related to the increased number of feature points.
- Temporal density is also different in the two schemes. In the Gradient scheme derivatives must be evaluated over a short time interval, thus resulting into a very fine temporal estimation. Temporal precision in the Correspondence scheme is constrained by the require-ment that the spatial displacement has to be finer than the temporal displacement. This may originate coarser temporal grain.
- As to ranges of motion that are detected in the two schemes, in the Gradient scheme, the upper and lower velocities directly depend on the image characteristics. If this has high contrast, then it is easy to detect slow movements, while high velocities are badly estimated. The contrary holds for smooth images. In the Correspondence scheme, detection of slow motion is constrained by the accuracy of the feature point estimation. This accuracy can be at most equal to the spatial quantization step.
- In the presence of the aperture problem, the Gradient scheme computes the component of velocity in the direction of the local gradient only and is therefore consistent with human perception. Since in the Correspondence scheme a limited number of features is selected in each frame, if their number in the window is too small, then it may happen that they are insufficient to reliably detect motion.
- Correspondence is the main limiting factor for the reliable computation of motion in the Correspondence scheme. In the Gradient scheme reliability of estimation is tightly bounded to the use of derivatives; it can be improved with smoothing techniques. Differently from the Correspondence scheme, the Gradient scheme has theoretical limitations that make results of computation only an approximate solution of motion estimation.
- From the computational point of view, complexity is almost the same in both the approaches. While difficulties are persisting in the Correspondence scheme, mainly concentrated in the task of finding correspondences between feature points, the information contained in discrete displacement fields is of more practical use than dense flow fields. In the Gradient scheme, the complexity of computing the local component of motion along gradient direction is low, but that of computing the other component of motion is higher.
- An analogy can be established between techniques for motion estimation and human vision processes. The Correspondence scheme can be associated with some long range process, where sampling can be relatively loose (2 frames per second, approx). The Gradient scheme can be associated with a short range process with a higher sampling rate. In humans, the range of displacement over which coherent motion is perceived in the short range motion process is limited by the size of the largest channel. In machine vision, the maximum displacement for which an image flow algorithm can be applied depends on the size of the smoothing filter

applied to the image or to the size of aperture function in the absence of smoothing.

- Connectionist systems seem to have a good potential to solve noise robustness, at least if we make a parallel with other fields —such as robotics, or speech recognition — in which the introduction of connectionist systems represented a progress over the state of the art. At present, there are several connectionist systems appeared in the literature dealing with gradient based methods. Gradient based methods are preminent in connectionist research since their character-istics—computation based on local interaction, parallelism—fit well the connectionist paradigm. Correspondence based methods are more algorithmic in nature, and don't seem to fit well with the distributed nature of connectionism. In spite of that, connectionist solutions have been effectively used for feature tracking.

REFERENCES

1. J.Y. Lettvin, H.R. Maturana, W.S. McCulloch, and W.H. Pitts, What the frog's eye tells the frog's brain, *Proceedings of the Institute of Radio Engineers*, Vol.47, pp.1950-1961 (1959).
2. B. MacLennan, Flexible computing in the 21st century, *Vector Register*, Vol.4, No.3 (1991).
3. E.I. Knudsen, S. du Lac, and S.D. Esterly, Computational maps in the brain, *Annual Review of Neuroscience*, Vol.10, pp.41-65 (1987).
4. J.K. Aggarwal and N. Nandhakumar, On the computation of motion from sequences of images - a review, *Proceedings of the IEEE*, Vol.76, No.8, pp.917-935 (1988).
5. T.S. Huang and R.Y. Tsai, *Image sequence analysis: motion estimation,* Springer-Verlag (1981).
6. P. Thevenaz, Motion analysis, in *Pattern Recognition and Image Processing in Physics,* R.A. Vaugham ed., Adam Hilger, pp.129-166 (1990).
7. V. Cappellini, A. Del Bimbo, and A. Mecocci, Motion analysis and representation in computer vision, *Journal of Circuits, Systems and Computers,* Vol.3, No.3, pp.1-35 (1993).
8. J.K. Aggarwal, L.S. Davis, and W.N. Martin, Correspondence processes in dynamical scene analysis, *Proceedings of the IEEE,* Vol.69, No.5 (1981).
9. J.K. Aggarwal and R. Duda, Computer analysis of moving polygonal images, *IEEE Transactions on Computers*, Vol.24, No.10, pp.966-976 (1975).
10. J.W. Roach and J.K. Aggarwal, Determining the movement of objects from a sequence of images, *IEEE Transactions on Pattern Analysis and Machine Intelligence*, Vol.2, No.6, pp.554-562 (1980).
11. H.H. Nagel, Representation of moving rigid objects based on visual observations, *Computer*, pp.29-39 (1991).
12. J.Q. Fang and T.S. Huang, Some experiments on estimating the 3D motion parameters of a rigid body from two consecutive image frames, *IEEE Transactions on Pattern Analysis and Machine Intelligence*, Vol.6, No.5, pp.545-554 (1984).
13. R.Y. Tsai and T.S. Huang, Estimating three dimensional motion parameters of a rigid planar patch, *IEEE Transactions on Acoustics, Speech and Signal Processing*, Vol.29, No.6, pp.1147-1152 (1981).
14. R.Y. Tsai, T.S. Huang, and W.L. Zhu, Estimating three dimensional motion parameters of a rigid planar patch II: Singular value decomposition, *IEEE Transactions on Acoustics, Speech and Signal Processing,* Vol.30, No.4, pp.525-534 (1982).
15. T.J. Broida and R. Chellapa, Estimation of object motion parameters from noisy images, *IEEE Transactions on Pattern Analysis and Machine Intelligence,* Vol.8, No.1, pp.90-99 (1986).
16. H. Shariat and K.E. Price, Motion estimation with more than two frames, *IEEE Transactions on Pattern Analysis and Machine Intelligence,* Vol.12, No.5, pp.417-434 (1990).
17. Y. Lin and T.S. Huang, Vehicle-type motion estimation from multi-frame images, *IEEE Transactions on Pattern Analysis and Machine Intelligence*, Vol.15, No.8 (1993).
18. A. Mitchie, S. Seida, and J.K. Aggarwal, Interpretation of structure and motion from line correspondences, in *Proceedings IEEE 8th International Conference on Pattern Recognition,* Paris, F, Vol.2, pp.1110-1112 (1986).
19. B. Schunck, Image flow segmentation and estimation by constraint line clustering, *IEEE Transactions on Pattern Analysis and Machine Intelligence*, Vol.11, No.10, pp.1010-1027 (1989).
20. O. Braddick, A short range process in apparent motion, *Vision research*, Vol.14 (1974).
21. B.K. Horn and B.G. Schunck, Determining optical flow, *Artificial Intelligence*, Vol.17, pp.185-204 (1981).
22. P. Nesi, A. Del Bimbo, and J.L.C. Sanz, Multiconstraint-based optical flow estimation and segmentation, in *Proceedings of the CAMP workshop,* Paris, F, pp.419-426 (1991).

23. B.G. Schunck, The motion constraint equation for optical flow, in *Proceedings of the 7th IEEE International conference on Pattern Recognition*, pp.20-22 (1984).

24. H.H. Nagel, On a constraint for the estimation of displacement rates in image seqeunces, *IEEE Transactions on Pattern Analysis and Machine Intelligence*, Vol.11, No.1, pp.490-498 (1989).

25. A. Del Bimbo, P. Nesi, and J. L. C. Sanz, Optical flow computation using extended constraints, *Technical Report DSI-5-93,* Dipartimento di Sistemi e Informatica, Università di Firenze, submitted to *IEEE Transactions on Image Processing* (1993).

26. H.H. Nagel, Displacement vectors derived from second order intensity variations in image sequences, *Computer Vision, Graphics and Image Processing*, Vol.21, pp.85-117 (1983).

27. H.H. Nagel and W. Enkelmann, An investigation of smoothness constraints for the estimation of displacement vecto fields from image sequences, *IEEE Transactions on Pattern Analysis and Machine Intelligence*, Vol.8, No.9, pp.565-593 (1986).

28. L.S. Davis and T.R. Kushner, Road boundary detection for autonomous vehicle navigation, *Technical Report CS-TR-1538,* Center for automation research, Univ. of Maryland (1985).

29. E.C. Hildreth, *Computing the velocity field along contours in motion: representation and perception*, Elsevier Science (1986).

30. A. Mitchie, Y.F. Wang, and J.K. Aggarwal, Experiments in computing optical flow with gradient-based multiconstraint methods, *Pattern Recognition*, Vol.16, No.6, pp.173-179 (1983).

31. R.M. Haralick and J.S. Lee, The facet approach to optical flow, in *Proceedings of Image Understanding Workshop*, Arlington ed. (1993).

32. O. Tretiak and L. Pastor, Velocity estimation from image sequence witgh second order differential operators, in *Proceedings of the 7th IEEE Int. Conference on Pattern Recognition,* pp.16-19 (1984).

33. A. Verri, F. Girosi, and V. Torre, Differential techniques for optical flow, *Journal of the Optical Society of America*, Vol.7, No.5, pp.912-922 (1990).

34. H.H. Nagel, Extending the oriented smoothness constraint into the temporal domain and the estimation of derivatives of optical flow, in *Proceedings of the 1st European Conference on Computer Vision*, Nice, F (1990).

35. E. De Micheli, V. Torre, and S. Uras, The accuracy of the computation of optical flow and of the recovery of motion parameters, *IEEE Transactions on Pattern Analysis and Machine Intelligence*, Vol.15, No.5, pp.434-447 (1993).

36. G. Adiv, Determining three dimensional motion and structure from optical flowgenerated by several moving objects, *IEEE Transactions on Pattern Analysis and Machine Intelligence*, Vol.7, No.5, pp.384-401 (1985).

37. G. Adiv, Inherent ambiguities in recovering 3d motion and structure from a noisy field, *IEEE Transactions on Pattern Analysis and Machine Intelligence*, Vol.11, No.5, pp.477-489 (1989).

38. B. MacLennan, Characteristics of connectionist knowledge representation, *Technical Report CS-91-147,* Computer Science Department, University of Tennessee, Knoxville, TN (1991).

39. J.J. Gibson, *The Perception of the Visual World*, Houghton Mifflin, Boston, MA (1950).

40. P.A. Viola, S.G. Lisberger, and T.J. Sejnowski, Recurrent eye tracking network using a distributed representation of image motion, in *Advances in Neural Information Processing Systems 4*, J.E. Moody, S.J. Hanson, and R.P. Lippmann eds., Morgan Kaufman, San Mateo, CA (1992).

41. J.A. Marshall, Self-organizing neural network for perception of visual motion, *Neural Networks*, Vol.3, pp.45-74 (1990).

42. J.A. Marshall, Challenges of vision theory: self-organization of neural mechanisms for stable steering of object-grouping data in visual motion perception, in *Stochastics and Neural Methods in Signal Processing, Image Processing and Computer Vision - Proceedings of the SPIE*, Su-Shing Chen ed., pp.200-215 (1991).

43. T. Delbrück and C.A. Mead, An electronic photoreceptor sensitive to small changes in intensity, in *Advances in Neural Information Processing Systems 1*, D.D. Touretzky ed., Morgan Kaufmann, San Mateo, CA, pp.720-727 (1989).

44. J. Tanner and C. Mead, Optical motion sensor, in *Analog VLSI and Neural Systems*, C. Mead ed., Addison-Wesley, pp.229-255 (1989).

45. J.C. Leeo, B.J. Sheu, W.C. Fang, and R. Chellappa, VLSI neuroprocessors for video motion detection, *IEEE Transactions on Neural Networks*, Vol.4, No.2 (1993).

46. M.I. Sereno, Learning the solution to the aperture problem for pattern motion with a hebb rule, in *Advances in Neural Information Processing Systems*, D.S. Touretzky ed., Morgan Kaufmann, San Mateo, CA, pp.468-476 (1989).

47. C. Colombo, A. Del Bimbo, and S. Santini, A massively parallel architecture for the determination of optical flow, in *Proceedings of the 11th International Conference of Pattern Recognition,* Le Hague, NL, pp.209-213 (1992).

48. Bruce MacLennan, Gabor representations of spatiotemporal visual images, *Technical Report CS-91-144*, Computer Science Department, University of Tennessee (1991).

49. W. Burger and B. Bhanu, Estimating 3D egomotion from perspective image sequences, *IEEE Transactions on Pattern Analysis and Machine Intelligence*, Vol.12, No.11, pp.1040-1058 (1990).

50. R. Jain, Direct computation of the focus of expansion, *IEEE Transactions on Pattern Analysis and Machine Intelligence*, Vol.5, No.1, pp.58-64 (1983).

51. K. Rangarajan and M. Shah, Interpretation of motion trajectories using focus of expansion, *IEEE Transactions on Pattern Analysis and Machine Intelligence*, Vol.14, No.12, pp.1205-1209 (1992).

DYNAMIC PERCEPTION OF THE ENVIRONMENT

Walter Gerbino *(chairman)*

Dipartimento di Psicologia, Università di Trieste
Via dell'Università 7, I - 34123 Trieste, Italy

TOPIC INTRODUCTION

Dynamic concepts enter the explanation of perception in several ways. And probably, some degree of ambiguity exists whenever one encounters statements in which perception is claimed to be "dynamic" (in opposition to something else). This panel is an attempt of elucidating the different meanings relevant to this area.

Let us list five different contexts in which dynamic concepts are used in the study of perceptual phenomena:
1) motion as a primitive;
2) kinematic organization;
3) primacy of dynamic information and recovery;
4) kinematic specification of dynamics;
5) dynamic vs. topographical representation.

This ordering reflects a gradual shift from the simple notion of dynamic as opposed to static (i.e., pertaining to motion) to the notion of dynamic as anything which has to do with the concepts of force and energy. Most ambiguities derive from the difficulty of extracting from the context which notion is invoked in a specific instance.

MOTION AS A PRIMITIVE

One can say that perception is dynamic, in its basic nature, because motion is a primitive property of visual experience. This issue was an essential part of the anti-reductionistic position of Gestalt theory, and more recently it reappeared in the criticism of snapshot vision by Gibson[1].

The paper which is commonly considered as the starting point of the Gestalt revolution, Wertheimer's study of perceived motion[2], applies an original methodology - the phenomeno-logical demonstration - to a classical problem: whether motion can be derived from the concatenation of static, disparate views. According to a reductionist position, the perception of a continuous change of position over time is an illusion, generated by ratiomorphic mechanism, which postulates motion as the best solution to the problem posed by the disparity between the actual sensation and the memory of a previous sensation. This type of explanation belongs to the category of the "taking-into-account" hypotheses, as they are now called by Rock[3, 4]. Wertheimer moved a direct attack against the problem-solving explanation, showing that the rationale of such an approach was phenomenologically wrong. Using stroboscopic motion, he was able to show that inter-stimulus intervals longer than optimal abolish the identity of the object (observers perceive a succession of two distinct objects in two distinct spatial positions), but do not abolish the perception of motion. The perception of motion without a moving object, called the "phi-phenomenon", contradicts the ratiomorphic explanation, which postulates that motion is generated to preserve identity. If identity is not preserved, why motion should be perceived?

Contemporary psychophysics has generated another, powerful demonstration of the untenability of a reductionist account of motion: random-dot cinematograms[5]. Each static frame is perceived as a random arrangement of dots. The alternation of two correlated frames supports the perception of a moving figure corresponding to the cluster of displaced dots. In this case, it is shown that the very existence of an object in each static frame is not a necessary condition for the perception of motion. Therefore, motion cannot be considered as the result of a match between the current object in the current position, and the memory of a similar object in a similar position.

Other types of evidence (for instance, on visual search) point to the same conclusion. Visual perception is intrinsically dynamic, as opposed to static, in the sense that motion is one of its primitive properties. Despite the fact that such a belief is widely accepted, the most common representation of the input is still in terms of a succession of static images. Such a "cinematographic" model of vision[6] can be deeply misleading, because it tends to suggest that perception of geometrical properties is given in the stimulus, whereas motion involves processing derived from the matching of successive images. The input is an entity distributed in the spatio-temporal domain. Models of processing which capture information contained in the optical input must take into account the possibility that change over time is registered as directly as change over space.

KINEMATIC ORGANIZATION

The environment is perceived as a complex dynamic entity in the sense that perception of velocities and trajectories of moving objects is hierarchically organized. The primary evidence for such an organization at a kinematic level derives from the work of Duncker[7], Rubin[8], Wallach[9], Johansson[10, 11], Metzger[12].

In the previous section we analysed if being stationary vs. moving can be considered as a primary property of the visual scene. Here we analyse two related questions: when will objects be perceived as stationary or moving? What determines the perceived trajectory of a moving object?

The general rules are the following. Motion is always perceived relative to the nearest frame of reference. Independent of retinal displacement, the framing object (in the simplest case, the ground) tends to appear stationary. These principles of organization can be treated as syntactic rules for parsing the visual scene, capable of explaining the perception of various dynamic configurations, at different levels of complexity.

The simplest relevant phenomenon is known as "induced motion", studied by Duncker[7], which in nature occurs when the moon is seen as shifting across the clouds, despite its retinal immobility (if the observer fixates it) and the absence of displacement relative to the far buildings. The small included object is perceived as carrying most of its immediate relative motion; i.e., the motion with respect to the near background. The far background is almost ineffective until a hierarchically intermediate object is present.

More complex phenomena can be observed with rotary motion[13]. The paradigmatic effect is the camouflage of the cycloidal path of a point on the circumference of a rolling wheel. The real path of such a point remained a scientific mystery until the time of Galileo, probably because the observer cannot visually relate its motion from the motion of the center of the wheel. And relative to the center, the point on the circumference simply describes a circular path. Therefore, rolling is perceived as a composition of rotation and translation. The cycloidal path becomes visible when the wheel is visually removed and the only visible objects are the luminous point and the terrain. Then, the real motion becomes a funny jump, with maximum speed at the top and minimum speed near the terrain.

According to Johansson[11], perceptual mechanisms accomplish a sort of vector analysis, which in many cases successfully decomposes retinal vectors into their original components. It is important to note that the inclusion principle, previously mentioned, subserves an ecological goal. In our environment, inclusion is a good indicator of belongingness. Think of the relationships between humans, vehicles and the ground: this ordering normally corresponds to inclusion (a man in a bus moving on a road) and to energy distribution (man saves energy by moving by bus). If the man is stationary on the moving bus, we usually perceive him as such, despite his retinal motion. This type of perception can be qualified as "veridical", not in a purely kinematic sense, but in a kinetic sense. One must take into account action between bodies, not simply the changing geometry of the situation. At a mere kinematic level, different combinations of trajectories of objects are equivalent, provided that the relationships between the carrier and the carried object (between the frame and the object belonging to it) are changed accordingly.

One question remains open. How much of vector decomposition is actually accomplished, during normal observation, by eye movements? In fact, one can regard vector analysis as a model of a perceptual process or as an abstract operation which can be only partly performed at a perceptual level. Most of the job would be ordinarily done by eye movements, which have the function of nullifying irrelevant component of motion (irrelevant because not related to the motion of the object/frame subsystem under observation at a given moment). Such a position has been held by Wallach[14].

Whatever the nature of the mechanisms which support hierarchical organization of motion in humans, it remains clear that perception can be qualified as dynamic in the sense that motion organization reflects a structure of relationships between objects in term of "actions", not simply of displacements.

PRIMACY OF DYNAMIC INFORMATION AND RECOVERY

The critique of snapshot vision by Gibson[1] was motivated by the assumption that dynamic information (i.e., optic flow) can support direct perception of the environment. Problems which are ill-posed in a static frame, like the recovery of form, can be solved when transformations are taken as the proper input to the visual system (position of Verri in this workshop). This assumption is connected to the issue of ecological validity of visual research.

The phenomenon which epitomizes this position is the kinetic depth effect[15]. The shadow of a 3D structure looks two-dimensional when it is stationary; whereas it looks three-dimensional when moving. According to ecological and computational approaches, such a radical perceptual change made possible by motion follows from the availability of invariants which specify 3D spatial structure over time[16].

The superiority of dynamic over static information (relative to the role of recovering the structure of the distal world) is often related to the superiority of active vs. passive vision, although it is clear that active vision implies the obtaining of dynamic information, but not vice versa. In fact, also the passive observer can benefit from dynamic information, as it happens when relative motion between objects is available. However, it is true that the observer is normally active, and therefore can obtain rich information about the structure of the environment by introducing changes in the optical flow which are contingent to the motion.

KINEMATIC SPECIFICATION OF DYNAMICS

The possibility of finding kinematic constrains for the perception of causal events was explored in the classical work by Michotte[17]. He argued that spatio-temporal patterns are spontaneously seen as mechanical actions, following general principles of perceptual organization.

A contemporary development of such a notion is represented by Runeson's principle of "Kinematic Specification of Dynamics (KSD)"[18]. According to this principle, there is more to perceive in dynamic events than mere collisions. For instance, the kinematic pattern can specify the ratio of the mass of the two inanimate objects, or more complex feature like intentions of animate objects.

A part from the logical problems connected with the idea that perception of causal relationships can be "direct", it is important to consider the possibility that higher-order invariants, available in rich and articulated patterns of stimulation, support the perception of properties very different from those of the optical medium. In this sense, a great deal of vision is a matter of "kinematic specification of...", as beautifully shown by Paolo Viviani position during our workshop. The most stimulating field, in this respect, is the perception of expressions conveyed by complex patterns like faces and human bodies.

DYNAMIC VS. TOPOGRAPHICAL REPRESENTATION

So far, we have been concerned with the nature of information available in the optical input, and with the properties of vision. However, dynamic concepts were also proposed as an integral part of the representation that underlies perception. A dynamic model of visual processing plays a central role in Gestalt theory. Such a model lost popularity in the scientific community, mainly because of its linkage with an unsatisfactory physiological doctrine.

However, the Gestalt criticism of vision as topographical representation of the stimulus remains intact. As such it has inspired generations of researchers interested in the rejection of simplistic hypotheses based on retinal topography. But Wertheimer[2], Köhler and Koffka[20] had in mind a very ambitious hypothesis: the idea that interactions in a physical medium (the brain) could explain the discrepancies from retinal topography and distal geometry as well. Visual organization was conceived as the effect of forces originated by the stimulus distribution and forces depending on intrinsic tendencies.

A case in point is the shrinkage of 3D space observed in the *Ganzfeld*[19, 20]. When stimulation is completely homogeneous, space tends to appear as small as possible. Inhomogeneity is needed to expand it, by counteracting inner forces which tend to minimize energy expenditure. This hypothesis does not apply only to the rather artificial case of the *Ganzfeld*; it applies equally well to more articulated conditions, like observation of the sky in the natural environment. The apparent shape of the sky reflects the expanding action of distance information, which has the effect of counteracting the endogenous tendency to locate the sky as close as possible to the observer.

Phenomena as different as geometrical illusions occurring in the pictorial world, perception of open space, evolution of visual memories can be equally accommodated within a dynamic framework to perception. In all cases the distinctive feature of the Gestalt proposal is the assumption that mental processes are based on the interaction of forces, and possess an autonomous direction. When external forces derived from the stimulation are reduced, the operation of internal forces becomes salient, and structural principles like the tendency to simplicity or to maximum homogeneity become evident.

Finally, it is worth mentioning that a different meaning of dynamic representation is currently available (for a recent review see Freyd[21]). It refers to the influence of an underlying representation of events in terms of actions on the perception of static scenes. This notion is implicit in the previous section on kinematic organization and related to the KSD approach, but plays a specific role in the area of visual cognition. Whatever the format of the representation (in terms of forces as supposed by Gestalt theory or in terms of symbols as supposed by formally-oriented cognitive theories), the key issue here is the influence of knowledge about natural events on perception of the actual geometry of a static scene. Freyd and Finke[22] proposed a construct, "representational momentum", to describe the basic dynamic schema activated in the perception of objects. Still instances or depictions of objects captured "during natural motion" are represented in visual memory as actions; this determines characteristic memory distortions consistent with the direction of natural events.

REFERENCES

1. J.J. Gibson, *The Ecological Approach to Visual Perception*, Houghton Mifflin, Boston, MA (1979).
2. M. Wertheimer, Experimentelle Studien über das Sehen von Bewegung, *Zeitschrift für Psychologie*, Vol.61, pp.161-265 (1912), English selection in *Classics in Psychology*, T. Shipley ed., Philosophical Library, New York, NY, pp.1032-1089 (1961).
3. I. Rock, In defense of unconscious inference, in *Stability and Constancy in Visual Perception*, W. Epstein ed., Wiley, New York, NY (1977).
4. I. Rock, *The Logic of Perception*, MIT Press, A Bradford Book, Cambridge, MA (1983).
5. B. Julesz, *Foundations of Cyclopean Perception*, University of Chicago Press, Chicago, IL (1971).
6. W. Gerbino, *La Percezione*, Il Mulino, Bologna, I (1983).
7. K. Duncker, Über induzierte Bewegung (Ein Beitrag zur Theorie optisch wahrgenommener Bewegung), *Psychologische Forschung*, Vol.12, pp.180-259 (1929), English selection in *A Source Book of Gestalt Psychology*, W.D. Ellis ed., Routledge e Kegan, London, UK, pp.161-172 (1938).
8. E. Rubin, Visuell wahrgenommene wirkliche Bewegungen, *Zeitschrift für Psychologie*, Vol.103, pp.384-392 (1927).
9. H. Wallach, Über visuell wahrgenommene Bewegungsrichtung, *Psychologische Forschung*, Vol.20, pp.22-380 (1935).
10. G. Johansson, *Configurations in Event Perception*, Almqvist and Wiksell, Uppsala, SW (1950).
11. G. Johansson, Visual motion perception, *Scientific American*, Vol.232, No.6, pp.76-88 (1975).
12. W. Metzger, *Gesetze des Sehens*, Kramer, Frankfurt, D (1975).
13. D.R. Proffitt and J.E. Cutting, An invariant for wheel-generated motions and the logic of its determination, *Perception*, Vol.9, pp.435-449 (1980).
14. H. Wallach, Eye movement and motion perception, in *Tutorials on Motion Perception*, A.H. Wertheim, W.A. Wagenaar, and H.W. Leibowitz eds., Plenum Press, New York, NY (1982).
15. H. Wallach and D.N. O'Connell, The kinetic depth effect, *Journal of Experimental Psychology*, Vol.45, pp.205-217, reprinted in *On Perception*, H. Wallach, Quadrangle, New York, NY (1976).
16. H. Wechsler, *Computational Vision*, Academic Press, New York, NY (1990).
17. A. Michotte, *Causalité, Permanence et Réalité Phénoménales*, Publications Universitaires, Louvain, B (1946), English translation in *The Perception of Causality*, Methuen, London, UK (1963).
18. S. Runeson and G. Frykholm, Kinematic specification of dynamics as an informational basis for person and action perception: expectation, gender recognition, and deceptive intention, *Journal of Experimental Psychology: General*, Vol.112, pp.585-615 (1983).
19. W. Metzger, Optische Untersuchungen am Ganzfeld: II. Zur Phänomenologie des homogenen Ganzfelds, *Psychologische Forschung*, Vol.13, pp.6-29 (1930).
20. K. Koffka, *Principles of Gestalt Psychology*, Harcourt Brace, New York, NY (1935).

21. J.J. Freyd, Five hunches about perceptual processes and dynamic representations, in *Attention and Performance XIV*, D.E. Meyer and S. Kornblum eds., MIT Press, A Bradford Book, Cambridge, MA (1992).

22. J.J. Freyd and R.A. Finke, Representational momentum, *Journal of Experimental Psychology: Learning, Memory, and Cognition*, Vol.10, pp.126-132 (1984).

VISUAL COGNITION AND COGNITIVE MODELING

Lorenzo Magnani, Sabino Civita, and Guido Previde Massara

Dipartimento di Filosofia, Università di Pavia
P.zza Botta 6, I - 27100 Pavia, Italy

ABSTRACT

The general objective is to consider how the use of visual mental imagery in thinking may be relevant to hypothesis generation. There has been little research into the possibility of visual and imagery representations of hypotheses, despite abundant reports (e.g. Einstein and Faraday) that imaging is crucial to scientific discovery. Some hypotheses naturally take a pictorial form: the hypothesis that the earth has a molten core might be better represented by a picture that shows solid material surrounding the core. We will discuss the particular computational imagery representation scheme proposed by Glasgow and Papadias. They have suggested an interesting technique for combining image-like representations and processing with linguistic information. The system they describe can make inferences using such cognitive representations as visual mental imagery ones. It seems to have some of the characteristics of Johnson-Laird's mental representations - achieving conclusions without laborious chains of inferences. We plan to explore whether this kind of hybrid imagery/linguistic representation can be improved and used to model image-based hypothesis generation, that is to delineate the first features of what we call *visual abduction*.

1. INTRODUCTION

The general objective is to consider how the use of visual mental imagery in thinking may be relevant to hypothesis generation and scientific discovery. To this end we treat imagery as a problem-solving paradigm in Artificial Intelligence (AI) and illustrate some related cognitive models. In this research area the term "image" refers to an internal representation used by humans to retrieve information from memory. Many psychological and physiological studies have been carried out to describe the multiple functions of mental imagery processes: there exists a visual memory[1] that is superior in recall; humans typically use mental imagery for spatial reasoning[2]; images can be rebuilt in creative ways[3]; they preserve the spatial relationships, relative sizes, and relative distances of real physical objects[4]; for a more complete list, see Tye[5].
Kosslyn introduces visual cognition as follows: "Many people report that they often think

by visualizing objects and events [...] we will explore the nature of visual cognition, which is the use of visual mental imagery in thinking. Visual mental imagery is accompanied by the experience of seeing, even though the object or event is not actually being viewed. To get an idea of what we mean by visual mental imagery, try to answer the following questions: [...] How many windows are there in your living room? If an uppercase version of the letter *n* were rotated 90° clockwise, would it be another letter?"[6].

We can build visual images on the basis of visual memories but we can also use the recalled visual image to form a new image we have never actually seen. Certainly, imagery is used in everyday life, as illustrated by the previous simple answers, nevertheless imagery has to be considered as a major medium of thought, as a mechanism of thinking relevant to hypothesis generation. Some hypotheses naturally take a pictorial form: the hypothesis that the earth has a molten core might be better represented by a picture that shows solid material surrounding the core. There has been little research on the possibility of visual imagery representations of hypotheses, despite abundant reports (e.g. Einstein and Faraday) that imaging is crucial to scientific discovery. It is well-known that the German chemist Kekulé used spontaneous imagery to discover the structure of benzene; Watson and Crick have reported the use of mental imagery in the interpretation of diffraction data and in the determination of the structure of the DNA molecule[7-10].

Thus, after illustrating the computational imagery representation scheme proposed by Glasgow and Papadias[11], together with certain cognitive results, we will explore whether a kind of hybrid imagery/linguistic representation can be used to model image-based hypothesis generation, that is to delineate the first features of what we call *visual abduction*.

2. KNOWLEDGE REPRESENTATION SCHEME

The central theme of the recent imagery debate in the cognitive science has concerned the problem of representation.

How can we represent images? Are mental images represented depictively in a picture or like sentences of descriptions in a syntactic language?

According to Kosslyn's *depictionist* or *pictorialist* view[12], mental images are *quasi-pictures* represented in a specific medium called the visual buffer in the mind. Kosslyn builds the analogy that visual information stored in memory can be displayed as an image on a CRT screen. Kosslyn's model of mental imagery proposes three classes of processes that manage images in the visual buffer: the *generation* process forms an image exploiting visual information stored in long-term memory, the *transformation* process (for example, rotation, translation, reduction in size, etc.) modifies the depictive image or views it from different perspectives, the *inspection* process explores patterns of cells to retrieve information such as shape and spatial configuration. According to Pylyshyn's *descriptionist* view[13] mental imagery can be explained by the tacit knowledge used by humans when they simulate events rather than by a pictorialist view related to the presence of a distinctive mental image processor. As shown by certain experimental results, Hinton proposes that imagery involves viewer-centered information appended to object-centered structural descriptions of images[14]. Finke[15] proposes five unifying principles of mental imagery to summarize some of the properties of imagery that a computational model should represent. Finally, Tye has recently suggested a theory, related to some of the aspects of Marr and Nishihara's theory of visual perception[16], in which mental images are represented as interpreted, symbol-filled arrays[5].

In Glasgow and Papadias' computational model[11] the imagery processes are simplified by a depictive knowledge representation scheme that involves inferencing techniques related to generation, inspection and transformation of the representation. According to Kosslyn's cognitive model, the knowledge representation scheme of mental imagery (see figure 1) is composed of two different levels of reasoning, *visual* and *spatial*, the former concerned with what an image looks like, the latter depending on where an object is located relative to other objects. The different representations of these ways of reasoning exist at the level of working-memory and are generated from a *descriptive* representation of an image stored in long-term memory in a hierarchical organization. Information is accessed from long-term memory by means of standard retrieval, procedural attachment and inheritance techniques.

The spatial representation explicitly outlines the image constituents in a multi-dimensional, symbolic array[17], that maintains spatial and topological properties and avoids distracting details, in accordance with the features of the image domain and the questions involved; moreover, the arrays are *nested*, to express multiple levels of the structural hierarchy of images. Of course the array can be interpreted in different ways depending on the application (see figure 2).

The spatial representation is processed using primitive functions that *transform* and *inspect* the symbolic array. The "uninterpreted" visual representation depicts the space occupied by an image as an *occupancy array*. The related primitive functions concern operations for manipulating and retrieving geometric information, such as *volume, shape* and *relative distance* (rotate, translate and zoom, which change the orientation, location or size of a visual image representation) (see figure 1).

According to Kosslyn's theory of imagery, the separate long-term memory contains visual and spatial information as descriptions[4]. These descriptions are considered in Glasgow and Papadias' computational model as a hierarchical network model for semantic memory: the implementation is done using a frame system where each frame contains notable information concerning a particular image or a class of images. A Kind Of (AKO), used for property inheritance, and PARTS, to indicate the structural decomposition of complex images, are the two kinds of image hierarchies in this scheme[18].

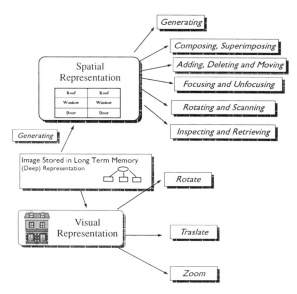

Figure 1. The multi-level knowledge representation scheme.

					Sweden		
	Scotland						
Wales	England			Denmark			
			Holland	Germany	Germany		
			Belgium				
			France	Italy	Yugoslavia	Yugoslavia	
Portugal	Spain			Italy		Greece	

Embedded symbolic array representation

Figure 2. Embedded symbolic array representation (reproduced from Glasgow et al.[11], copyright, Cognitive Science Society Incorporated, used by permission).

For example, we can generate as a symbolic array the following descriptive representation of *spatial* relations[18]:

> *the spoon is to the left of the knife;*
> *the plate is to the right of the knife;*
> *the fork is in front of the spoon;*
> *the cup is in front of the knife.*

The symbolic array could be:

spoon	knife	plate
fork	cup	

while the corresponding representation in long-term memory could be composed by the following frame structure:

```
FRAMENAME: place-setting
AKO: image
PARTS/LOCATION: spoon (1,1) knife (1,2) plate (2,1) fork (2,1) cup (2,2)
```

Moreover, a multi-dimensional array is able to depict spatial (e.g. *left-of, north-of*) and topological (e.g. *adjacent-to, inside*) relations among image parts and can also denote non-spatial dimensions such as speed. Sometimes, a part occupies more than one element in the array, to capture all the spatial relationships involved. Hence, an image frame is related to an individual image or to a class of images sharing the same features. Images can be represented in a viewer- or an object-centered frame of reference: unlike Kosslyn's approach, in Glasgow and Papadias' computational model[18] image representations are typically three-dimensional and object-centered (viewer-independent). Finally, in a nested symbolic array each symbol in the array corresponds to a frame in long-term memory: we can extract a subimage and express it by its symbolic array parts.

Glasgow and Papadias discuss the concept of *interpretation* of a symbolic array as follows: "For a given domain with associated spatial relations R, the *interpretation* of a symbolic array representation depends on a set of primitive functions defined in array theory.

For example, in a map domain with corresponding relation R = {*north-of, south-of, east-of, west-of, borders-on*}, we can define a domain function that takes as parameters a symbolic array representation of a map containing several countries and a symbol for a country in the map and returns a list of all the countries that *border-on* the specified country. Similarly, in a block world, functions might be defined for a set R = {*on-top-of, below, left-of, right-of, in-between*}. Symbolic arrays can also be used to denote more complex spatial and topological arrangements between parts of an image, such as the concept of inside/outside or the molecular concept of a 6-member ring cycle. In addition, arrays can be used to express other comparative dimensions such as time, size, speed or age."[18].

The functions implemented express simple operations and reasoning techniques based on the transformation and inspection of image representations (see figure 1): 1) *generating* spatial representations from long-term memory; 2) *composing and superimposing* image representations; 3) *adding, deleting* and *moving* parts of an image; 4) *focusing* and *unfocusing* attention on parts of an image; 5) *rotating* and *scanning* a spatial array; and 6) *inspecting* and *retrieving* spatial relations from an image representation. Some of these functions have been implemented into the functional programming language Nested Interactive Array Language (NIAL)[19].

Glasgow and Papadias' scheme seems to have some of the features that Johnson-Laird[20] has advocated as desirable characteristics of mental representations: it achieves conclusions without laborious chains of inferences. It is also consistent with Johnson-Laird's multiple model of imagery according to which we have to consider a propositional representation, a mental model (as the structural - spatial - layout), and a visual image corresponding to the perceptual viewer-centered representation. Glasgow and Papadias discuss Johnson-Laird's influence as follows: "The motivation for the proposed multiple representations was not expressibility; no new information is gained from the visual and spatial representations. Rather, the distinction is analogous to the issue of high-level programming languages: Why do we have languages with abstract data types such as arrays when all computation can be carried out by manipulating strings of symbols on a Turing machine? Obviously, such abstract structures exist to aid in the process of programming. In the development of our representation scheme for computational imagery, we were motivated by the same distinction. This scheme includes a descriptive representation, corresponding to Johnson-Laird's propositional representation, and two depictive representations, corresponding to the mental model and the visual image. The depictive representations do not necessarily increase the expressive power of the scheme; they do, however, provide a high-level representation for the spatial and visual properties of an image and imply greater programming ease and efficiency when implementing the processes of imagery"[18].

Finally, we should remember that the proposed multiple representation for mental imagery can account for the conflicting experimental data that cannot be explained by either a descriptive, visual or spatial representation.

Certainly, there are other computational approaches to spatial reasoning, that may have many advantages: the symbolic array representation may be too restrictive for certain purposes.

3. IMAGERY AND PROBLEM SOLVING

In accordance with the computational model previously illustrated, we can consider spatial representations as descriptive. Thus, they are expressed by propositions containing predicates such as spatial relationships and arguments as imaginable objects.

Is there a difference between descriptive and depictive representations?

The spatial representation does not add information that cannot be expressed by propositions; notwithstanding this, the spatial representation is not computationally equivalent to a descriptive one. In several imagery-related tasks (e.g. inspecting) spatial representation may reduce the computational complexity of the solution: the symbolic array adds more constraints to the search. The advantage in processing syllogisms by inspecting a symbolic spatial array is well-known.

Certainly, the symbolic arrays do not have the expressiveness of first-order logic and consequently they cannot represent many situations of vagueness or indeterminacy. Nevertheless they "explicitly" depict the present parts of an image and where they are located relative to one another, as well as depicting what is not present: the full representation is always complete.

As the spatial representations are depictive, and denote the important spatial relations among parts of the image, they are useful in the development of problem solving devices related to the inspection and transformation of images. Moreover, in processing visual information, two kinds of parallelism are involved: spatial, corresponding to the same operations applied concurrently to different spatial locations in an image; and functional, which occurs when different operations are applied to the same location.

Cognitive maps can be mentally inspected and used by an agent for route planning; imagery is used in planning and design; as stated above, the use of imagery in scientific discovery illustrates a mechanism of thinking relevant to hypothesis generation; the role of imagery in problem solving that involves scanning and inspection at multiple levels of structural hierarchy is shown in the computational model by the use of nested arrays of varying granularity, where attention can be focused to retrieve details; finding spatial similarities and equivalence is involved in many problem solving activities and is related to the problem of visual analogical mapping[21, 22]; imagery also involves the simulation of image transformations in order to anticipate the consequences of an action or event; constructing novel images through operations such as compose, superimpose, and put, allows us to detect information not previously seen.

The main objective is now to consider how the use of visual mental imagery in thinking may be relevant to hypothesis generation. We plan to explore whether the computational tool illustrated above can be modified to delineate the first features of what we call *visual abduction*.

What is abduction? The following section deals with the concept of abduction from an epistemological point of view and section 5 proposes an image-based hypothesis generation. It is necessary to show the meaning of visual abduction to illustrate how an *imagery hypothesis* can perform an explanation of a problem presented in an observed initial visual image.

4. THE EPISTEMOLOGY OF ABDUCTION

Abduction is becoming an increasingly popular term in AI[23-31], especially in the field of medical Knowledge Base Systems (KBSs)[32-35].

Let us consider the following interesting passage, from an article by Simon in 1965, published in the *British Journal for the Philosophy of Science*[36] and dealing with the logic of normative theories: "The problem solving process is not a process of 'deducing' one set of imperatives (the performance programme) from another set (the goals). Instead, it is a process of selective trial and error, using heuristic rules derived from previous experience, that is sometimes successful in *discovering* means that are more or less efficacious in attaining some end. If we want a name for it, we can appropriately use the name coined by Peirce and revived recently by Norwood Hanson[37]: it is a *retroductive* process. The nature of this process - which has been sketched roughly here - is the main subject of the theory of problem solving in both its positive and normative versions"[36].

The word "retroduction" used by Simon is the Hansonian neopositivistic one replacing the Peircian classical word *abduction*: they have the same epistemological and philosophical meaning.

As Fetzer has recently stressed, from a philosophical point of view the main modes of argumentation for reasoning from premises to conclusions are expressed by these three general attributes: *deductive* (demonstrative, non ampliative, additive), *inductive* (non-demonstrative, ampliative, non additive), *fallacious* (neither, irrelevant, ambiguous). *Abduction*, which expresses likelihood in reasoning, is a typical form of fallacious inference: "it is a matter of utilizing the principle of maximum likelihood in order to formalize a pattern of reasoning known as 'inference to the best explanation'"[38]. A hundred years ago, Peirce[39] was also studying and debating these three main *inference types* of reasoning.

Peirce[40] interpreted abduction essentially as a *creative process* of generating a new hypothesis. Abduction and induction, viewed together as processes of production and gener-alization of new hypotheses, are sometimes called *reduction*, that is ἀπαγωγή. As Łukasiewicz[41] makes clear, "Reasoning which starts from reasons and looks for consequences is called *deduction*; that which starts from consequences and looks for reasons is called *reduction*".

To illustrate from the field of medical knowledge, the discovery of a new disease and the definition of the manifestations it causes can be considered as the result of the creative abductive inference previously described. Therefore, creative abduction deals with the whole field of the growth of scientific knowledge. However, this is irrelevant in medical diagnosis where instead the task is *to select* from an encyclopedia of pre-stored diagnostic entities, diseases, and pathophysiologic states, which can be made to account for the patient's condition. On the other hand, diagnostic reasoning also involves abductive steps, but its creativity is much weaker: it usually requires the selection of a diagnostic hypothesis from a set of pre-enumerated hypotheses provided from established medical knowledge. Thus, this type of abduction can be called *selective abduction*[42]: it deals with a kind of *rediscovery*, instead of a genuine discovery.

Induction in its widest sense is an ampliative process of the generalization of knowledge. Peirce distinguished three types of induction and the first was further divided into three sub-types. A common feature of all kinds of induction is the ability to compare individual statements: using induction it is possible to synthesize individual statements into general laws (types I and II), but it is also possible to confirm or discount hypotheses (type III). Clearly we are referring here to the latter type of induction, that in our model is used as the process of reducing the uncertainty of established hypotheses by comparing their consequences with observed facts.

Deduction is an inference that refers to a logical implication. Deduction may be distinguished from abduction and induction on the grounds that only in deduction is the truth of inference guaranteed by the truth of the premises on which it is based.

Thus, *selective abduction* is the making of a preliminary guess that introduces a set of plausible hypotheses, followed by deduction to explore their consequences, and by induction to test them with available data: 1) to increase the likelihood of a hypothesis by noting evidence explained by that one, rather than by competing hypotheses; or 2) to refute all but one (see figure 3).

If during this first cycle new information emerges, hypotheses not previously considered can be suggested and a new cycle takes place: in this case the *nonmonotonic* character of abductive reasoning is clear.

There are two main epistemological meanings (see figure 4) of the word abduction: 1) abduction that only generates plausible hypotheses (*selective or creative*) - and this is the meaning of abduction accepted in our epistemological model - and 2) abduction considered as *inference to the best explanation*, that also evaluates hypotheses. In the latter sense the classical meaning of abduction as *inference to the best explanation* (for instance in medicine, to the best diagnosis) is described in our epistemological model by the complete abduction-deduction-induction cycle. All we can expect of our "selective" abduction, is that it tends to produce hypotheses that have some chance of turning out to be the best explanation. Selective

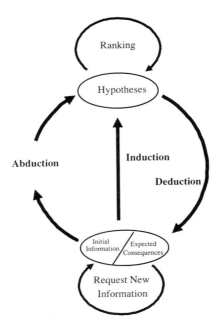

Figure 3. Abduction-deduction-induction cycle.

abduction will always produce hypotheses that give at least a partial explanation and therefore have a small amount of initial plausibility. In this respect abduction is more efficacious than the blind generation of hypotheses.

Indeed, our epistemological model should be considered as an illustration of *scientific theory change*: in this case selective abduction is replaced by creative abduction and there is a set of competing theories instead of diagnostic hypotheses. Furthermore the language of background scientific knowledge is to be regarded as open: in the case of competing theories, as they are studied by the epistemology of theory change, we cannot - contrary to Popper's viewpoint - reject a theory merely because it fails occasionally. If it is simpler and explains more significant data than its competitors, such a theory can be acceptable as the best explanation.

"Visual abduction" (see figure 4), a special form of abduction, occurs when hypotheses are instantly derived from a stored series of previous similar experiences. In this case there is no uncertainty. It covers a mental procedure that tapers into a non-inferential one, and falls into the category called "perception"[43]. Philosophically, *perception* is viewed by Peirce as a fast and uncontrolled knowledge-production procedure. Perception, in fact, is a vehicle for the instantaneous retrieval of knowledge that was previously structured in our mind through inferential processes. By perception, knowledge constructions are so instantly reorganized that they become habitual and diffuse and do not need any further testing. Finally we should remember, as Peirce noted, that abduction plays a role even in relatively simple *visual phenomena*. Many visual stimuli are ambiguous, yet people are adept at imposing order on them: "We readily form such hypotheses as that an obscurely seen face belongs to a friend of ours, because we can thereby explain what has been observed"[30]. This kind of image-based hypothesis formation can be considered as a form of *visual abduction*.

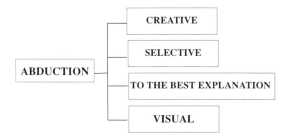

Figure 4. The four meanings of the word abduction.

5. VISUAL ABDUCTION

5.1. Image-based Explanation

Having illustrated many issues concerning the phenomenon of imagery, which is commonly and consciously experienced as the ability to form, transform and inspect an image-like representation of a scene, and having indicated that such representations play a role in problem solving strategies involving visual or spatial properties of an image, we will now discuss, from a computational perspective, a *visual abductive problem solving strategy*. To provide manageable bounds to our very general objective, i.e. to analyze the role of visual hypothesis generation, which is so crucial to scientific discovery, we have initially limited ourselves to the subtask of illustrating some structurally similar examples from the field of common sense reasoning, where it is very easy to find many cases dealing with what we have just called visual abductive problem solving. Moreover, we have limited ourselves to the spatial representation.

Although there is considerable agreement concerning the existence of a high-level visual and spatial medium of thought as a mechanism relevant to abductive (selective and creative) hypothesis generation, the underlying cognitive processes involved are still not well understood. Notwithstanding this, we will attempt to work around this gap in our understanding: although describing a computational model able to "imitate" the real ways the human brain works when it makes visual abductions would be best, our primary concern is its expressiveness, efficiency and inferential adequacy, rather than its explanatory and predictive power as regards psychological research.

Let us consider the following preliminary cognitive case: many visual stimuli are ambiguous, yet people are adept at imposing order on them. As stated above, this is the case when we readily form hypotheses such as that an obscurely seen face belongs to a friend of ours, because we can *explain* what has been observed (Peirce's visual abduction is related to this example).

More generally, we can face an *initial* (eventually)* observed image in which we recognize a problem to solve. For example, given a visual or imagery datum, we may have to explain: 1) the absence of an object; 2) why an object is in a particular position; or 3) how an object can achieve a given task moving itself and/or interacting with the remaining objects in the scene/image. How can "visual" reasoning perform this explanation? To answer this question it is necessary to show how visual abduction may be relevant to hypothesis generation, that is, how an *image-based explanation* is able to solve the problem given in the initial image.

_____

* Of course the initial spatial image can be the representation of a real case.

Faced with the initial image, in which we have previously recognized a problem to solve, as stated above, we have to work out an *imagery hypothesis* that can explain the problem-data. Thus, the formed image acquires a hypothetical status in the inferential abductive process at hand.

The generation of a "new" imagery hypothesis can be considered the result of the *creative* abductive inference previously described; in this respect we can consider how the imagery representations of new hypotheses lead to scientific discovery. The selection of an imagery hypothesis from a set of pre-enumerated imagery hypotheses, stored in long-term memory, also involves abductive steps, but its creativity is much weaker: this type of visual abduction can be called *selective*[42] (see section 4).

All we can expect of visual abduction is that it tends to produce imagery hypotheses that have some chances of turning out to be the best explanation. Visual abduction will always produce hypotheses that give at least a partial explanation, and therefore have a small amount of initial plausibility. In this respect abduction is more effective than the blind generation of hypotheses.

How can we computationally perform the generation of imagery hypotheses which are able to explain problem-data? This complex task can be achieved in an environment supplied with suitable levels of expressivity and adequate inferential systems.

5.2. Visual Abduction System (VASt) Structure

We are developing a system that focuses on spatial reasoning tasks and is hierarchically organized into three levels (see figure 5):
1) a *universe* of spatial worlds
2) *spatial worlds*
3) *imagery objects* that are included as elements in each spatial world (an imagery object may be included in one or many worlds).

At the computational level, the universe is composed of a collection of *representations* that describe each spatial world, and of a *navigation device* that allows an imagery object to move from the originary world to a different world. Thus, the different worlds can communicate with each other by means of the navigation device. Indeed, each world (represented, according to Glasgow and Papadias' computational model, by a nested symbolic array) is composed of a domain of suitable imagery objects (represented by the symbols in the array) and by a collection of rules identifying it (e.g. Newtonian mechanics).

To represent quantitative relations explicitly among the objects of the array we have to modify Glasgow and Papadias' knowledge representation scheme: it is necessary to add the representation of empty entries and of suitable distance functions. The imagery objects are of two kinds: *active* and *passive*. Both kinds of objects "know" the rules of the world where they are included (or to which they move); moreover, they can be individually characterized by a family of functions that imply some definite kinds of action they are able to perform. Finally, 1) each active object may interact with the remaining objects in a world; and 2) each passive object in a world may only act after interacting with an active object.

5.3. Imagery Hypotheses

From a functional point of view the computational system is extendable in several ways - each related to different capabilities. First, we may limit ourselves to solving problems in a unique spatial world; this approach naturally leads to qualitative physics reasoning.

To illustrate this situation we can consider the well-known monkey-banana problem. In a room there is a banana, a box, and a monkey. The monkey cannot reach the banana because

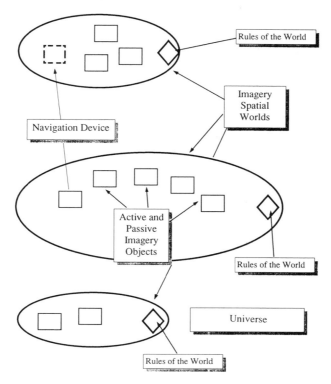

Figure 5. VASt structure.

it is on the ceiling, but it can push the box to a point below the banana, climb on top of it and so reach the banana. In VASt the room is represented by an array where all the objects are indicated by symbols on the array. If we describe a primitive distance function, say dist(x,y), we are immediately transported to a new formulation of the goal, that is, dist(monkey,banana)=0. Let us suppose the "object" monkey is supplied with a set of primitive functions, such as *walk-right, walk-left, push-left, push-right, climb, grasp*. It is possible to describe visually the effect of every function in the array. A feasible procedure could be 1) choose an action 2) if this action leads to a decrease in the distance then 2.1) do it 3) else go to 1). Such an algorithm clearly does not always lead to the goal. It has been improved with a device that is capable of detecting stationary states which are not the solution of the problem. We can also avoid this impasse by adding to the distance function a penalty function which evaluates the "distance" of the actual state from the set of stationary states.

If we consider each *new* step leading to the goal (i.e. each new configuration of the array generated by the new positions of the objects) as an *imagery state*, we can say that the "monkey" (or "people" faced with the monkey-banana problem) forms different *imagery hypotheses* devoted to achieving the task. Thus, each step represents a particular imagery world. The configuration of the array in which the goal is achieved performs the image-based best explanation, i.e. the visual abduction: this generated abductive imagery hypothesis is the best explanation of the problem-data, and hence able to perform the planning task (see figure 6).

Finally, if our spatial world represents a room, it can be supplied with a collection of rules

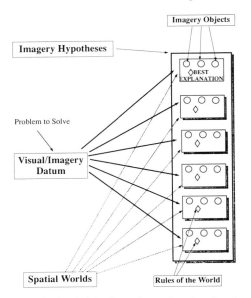

Figure 6. Visual abduction as image-based explanation.

of varying detail (e.g. Newtonian mechanics), yet it can also be supplied with completely different rules. Of course, using less rules results in the objects of the world being less constrained. For example, weakening the rules of Newtonian mechanics can lead to a new kind of spatial world, which is very abstract and considered as being a "virtual" one[44].

In the example above, the initial and final states were slightly different: we can also consider strategies that involve a greater number of different spatial worlds.

The previous example satisfies cases in which we have *to explain*, in an initial (eventually) observed image, how an object can achieve a given task, moving itself and interacting with the remaining passive objects. Nevertheless, it is possible to build new imagery worlds by moving objects in the universe using the navigation device and then map the generated worlds: we are extending VASt in order to map worlds onto worlds, that is, imagery hypotheses onto imagery hypotheses. In the first two simple examples stated above we were faced with an initial image in which we had to explain 1) why an object is in a particular position; 2) the absence of a particular object.

We shall clarify the first case with the following example, which is based on common sense reasoning: we see a broken horizontal glass on the floor, near the table. On the floor there are also some leaves and we see that the window is open. If we retrieve from long-term memory another *visual* (imagery) description still containing the glass (intact), the table, and the window, and we recognize this new representation as being a slightly different version of the previous one, we have to explain the presence of the leaves and broken glass in the initial image. They constitute an anomaly that needs to be solved (explained). If VASt is able to link the leaves to the presence, say, of wind, we are in turn transported to a new imagery explanatory hypothesis. Of course such a performance needs a preliminary knowledge base and a suitable form of connections between data. We plan to perform these connections by means of a neural network that links objects in the spatial world universe. In particular we may be able to establish excitatory links between objects belonging to the same arrays that are stored in long-term memory: this mechanism will enable the detection of the anomaly represented by the leaves and broken glass in the example above. A further link established between the leaves and wind will lead to the completion of the explanatory task.

The second case deals with the capacity to justify the absence of a given object in a scene. Let us consider the following example: one of our friends is accustomed to travel the same route every day. The road passes near to a little bridge, under which ducks can usually be *seen* swimming. On a particularly cold day our friend does not *see* the ducks. He asks himself where the ducks could be, but, since he has never seen any ducks in a different setting, while he is able to detect the anomaly he is unable to explain it. The imagery explanatory reasoning is impossible: therefore, it is stopped. On the contrary, if our friend had previously seen the ducks, say, under the roof of a farmhouse, once he has detected the absence of the ducks he can retrieve from long-term memory the image of the ducks sleeping under the roof. The imagery explanatory hypothesis is immediately achieved. The computational mechanism is the same as in the first example. By comparing an image of a place with a similar one stored in long-term memory it is possible to detect an anomaly (such as the absence of the ducks). A link is established between different objects (in the example, between the ducks and the roof) which allows the retrieval from long-term memory of another image (spatial world) that constitutes an hypothesis that explains the absence of the ducks.

6. CONCLUSION

This paper has examined a knowledge representation scheme together with certain cognitive results in order to explore whether a kind of hybrid imagery-linguistic representation can be used to model image-based hypothesis generation. Performing a system that delineates the first features of what we call *visual abduction*[45] we have provided evidence to support the position that imagery constitutes an alternative reasoning paradigm: faced to an initial image, in which we have recognized a problem to solve, we have to generate an imagery hypothesis able to explain problem-data.

To provide manageable bounds to our very general objective, i.e. to analyze the role of visual hypothesis generation, we have initially limited ourselves to the subtask of illustrating some structurally similar examples from the field of common sense reasoning, where it is very easy to find many cases dealing with visual abductive problem solving. Moreover, we have limited ourselves to spatial representation.

This paper has described a computational system, VASt, dedicated to show how an image-based explanation is able to solve the problem given in an initial image. The system's design clearly reflects our observation that the existence of a high-level spatial medium of thought is a mechanism relevant to abductive hypothesis generation.

Given a visual or imagery datum, the system is able to perform an image-based explanation of how an object can achieve a given task moving itself and/or interacting with the remaining objects in the scene/image. From a functional point of view the system is extendable in several ways - each related to different capabilities: given a visual or imagery datum, it would be able to provide an image-based explanation of: 1) the absence of an object in a scene; 2) why an object is in a particular position in a scene.

Finally, although describing a computational model able to "imitate" the real ways the human brain works when it makes visual abductions would be best, our primary concern has been expressiveness, efficiency and inferential adequacy, rather than its explanatory and predictive power as regards psychological research. Notwithstanding this, our system can contribute to discussions in the psychological area of the imagery debate.

ACKNOWLEDGEMENTS

The first author is indebted to Paul Thagard and Janice I. Glasgow for helpful discussions on visual thinking and abduction. His visit to Paul Thagard's Computational Epistemology Laboratory, Waterloo University, Ontario, Canada, in April 1993, helped him with developing the idea of visual abduction. The work described in this paper was supported by CNR (Consiglio Nazionale delle Ricerche, Italy).

REFERENCES

1. A. Paivio, Perceptual comparisons through the mind's eye, *Memory and Cognition,* Vol.3, No.6, pp.635-674 (1975).
2. M.J. Farah, The neuropsychology of mental imagery: converging evidence from brain-damaged and normal subjects, in *Spatial Cognition - Brain Bases and Development,* J. Stiles-Davis, M. Kritchevsky, and U. Bellugi eds., Erlbaum, Hillsdale, NJ, pp.33-56 (1988).
3. R.A. Finke and K. Slayton, Explorations of creative visual synthesis in mental imagery, *Memory and Cognition,* Vol.16, pp.252-257 (1988).
4. S.M. Kosslyn, *Image and Mind,* Harvard University Press, Cambridge, MA (1980).
5. M. Tye, *The Imagery Debate*, MIT Press, Cambridge, MA (1991).
6. S.M. Kosslyn and O. Koenig, *Wet Mind. The New Cognitive Neuroscience,* Free Press, New York, NY, p.128 (1992).
7. G. Holton, On trying to understand scientific genius, *American Scholar,* Vol.41, pp.95-110 (1972).
8. N.J. Nersessian, Opening the black box: cognitive science and history of science, *CSL Report 53*, Cognitive Science Laboratory, Princeton University, Princeton, NJ (1993).
9. R.D. Tweney, Fields of enterprise: on Michael Faraday's thought, in *Creative People at Work. Twelve Cognitive Case Studies*, D.B. Wallace and H.E. Gruber eds., Oxford University Press, Oxford, UK, pp.91-106 (1989).
10. A.I. Miller, Imagery and intuition in creative scientific thinking: Albert Einstein's invention of the special theory of relativity, in *Creative People at Work. Twelve Cognitive Case Studies*, D.B. Wallace and H.E. Gruber eds., Oxford U.P., Oxford, UK, pp.171-187 (1989).
11. J.I. Glasgow and D. Papadias, Computational imagery, *Cognitive Science,* Vol.16, No.3, pp.355-394 (1992).
12. S.M. Kosslyn, *Ghosts in the Mind's Machine: Creating and Using Images in the Brain,* W.W. Norton, New York, NY (1983).
13. Z.W. Pylyshyn, The imagery debate: analogue media versus tacit knowledge, *Psychological Review,* Vol.88, pp.16-45 (1981).
14. G. Hinton, Some demonstrations of the effects of structural descriptions in mental imagery, *Cognitive Science,* Vol.3, pp.231-250 (1979).
15. R.A. Finke, *Principles of Mental Imagery,* MIT Press, Cambridge, MA (1989).
16. D. Marr and H.K. Nishihara, Representation and recognition of the spatial organization of three-dimensional shapes, in *Proceedings of the Royal Society of London,* B200, pp.269-294 (1978).
17. T. More, Notes on the diagrams, logic and operations of array theory, in *Structures and Operations in Engineering and Management Systems,* P. Bjorke and O. Franksen eds., Tapir Pub., N (1981).
18. J.I. Glasgow, The imagery debate revisited: a computational perspective, *Computational Intelligence,* Vol.9, No.4, pp.309-333 (1993).
19. M.A. Jenkins, J.I. Glasgow, and C. McCrosky, Programming styles in NIAL, *IEEE Software,* Vol.86, pp.46-55 (1986).
20. P.N. Johnson-Laird, *Mental Models,* Harvard University Press, Cambridge, MA (1983).
21. P. Thagard, D. Gochfeld, and S. Hardy, Visual analogical mapping, in *Proceedings of the Fourteenth Annual Conference of the Cognitive Science Society,* Erlbaum, Hillsdale, UK, pp.522-527 (1992).
22. P. Thagard and S. Hardy, Visual thinking and the development of Dalton's atomic theory, in *Proceedings of the Ninth Canadian Conference on Artificial Intelligence,* Vancouver, CDN, pp.30-37 (1992).
23. H.E. Pople, On the mechanization of abductive logic, in *Proceedings of the International Joint Conference on Artificial Intelligence,* Vol.8, pp.147-152 (1973).
24. H.E. Pople, The formation of composite hypotheses in diagnostic problem solving, in *Proceedings of the Fifth IJCAI,* Morgan Kaufmann, Los Altos, CA, pp.1030-1037 (1977).
25. J.A. Reggia, S.N. Dana, and Y.W. Pearl, Expert systems based on set covering model, *International Journal on Man-Machine Studies,* Vol.19, pp.443-460 (1983).

26. J.A. Reggia and D.S. Nau, An abductive non monotonic logic, in *Proceedings of the Workshop on Non-Monotonic Reasoning*, pp.385-385 (1983).

27. I. Peng and J.A. Reggia, A probabilistic causal model for diagnostic problem solving I: integrating symbolic causal inference with numeric probabilistic inference, *IEEE Transactions on Systems, Man, and Cybernetics,* Vol.17, pp.146-162 (1987).

28. I. Peng and J.A. Reggia, A probabilistic causal model for diagnostic problem solving II: diagnostic strategy, *IEEE Transactions on Systems, Man, and Cybernetics,* Vol.17, pp.395-406 (1987).

29. J.F. Sowa, *Conceptual Structures. Information Processing in Mind and Machine,* Addison-Wesley, Reading, MA (1984).

30. P. Thagard, *Computational Philosophy of Science,* MIT Press, Cambridge, MA (1988).

31. P. Thagard, *Conceptual Revolutions,* Princeton University Press, Princeton, NJ (1992).

32. J.R. Josephson, B. Chandrasekaran, J.W. Jr Smith, and M.C. Tanner, Abduction by classification and assembly, in *PSA 1986, Philosophy of Science Association,* Vol.1, pp.458-470 (1986).

33. L. Johnson and E.T. Keravnou, *Expert Systems Architectures,* Kogan Page, London, UK (1988).

34. M. Ramoni, M. Stefanelli, L. Magnani, and G. Barosi, An epistemological framework for medical knowledge-based systems, *IEEE Transactions on Systems, Man, and Cybernetics,* Vol.22, No.6, pp.1361-1375 (1992).

35. L. Magnani, Abductive reasoning: philosophical and educational perspectives in medicine, in *Advanced Models of Cognition in Medical Training and Practice,* D.A. Evans and V.L. Patel eds., Springer, Berlin, D, pp.21-41 (1992).

36. H.A. Simon, The logic of rational decision, *British Journal for the Philosophy of Science,* Vol.16, pp.169-186 (1965), reprinted in *Models of Discovery and Other Topics in the Methods of Science,* H.A. Simon, Reidel, Dordrecht, NL, pp.137-153 (1977).

37. N.R. Hanson, *Patterns of Discovery. An Inquiry into the Conceptual Foundations of Science,* Cambridge University Press, Cambridge (1958).

38. J.H. Fetzer, *Artificial Intelligence: its Scope and Limits,* Kluwer Academic Publishers, Dordrecht, NL, p.103 (1990).

39. C.S. Peirce, Abduction and induction, in *Philosophical Writings of Peirce,* Dover, New York, NY, pp.150-156 (1955).

40. C.S. Peirce, *Collected Papers,* C. Harstone, P. Weiss and A. Burks eds., Harvard University Press, Cambridge, MA (1931-1958).

41. J. Lukasiewicz, Creative elements in science [1912], in *Selected Works,* J. Lukasiewicz, North Holland, Amsterdam, NL, p.7 (1970).

42. L. Magnani, Epistémologie de l'invention scientifique, *Communication and Cognition,* Vol.21, pp.273-291 (1988).

43. D.R. Anderson, *Creativity and the Philosophy of Charles Sanders Peirce,* Clarendon Press, Oxford, UK, pp.38-44 (1987).

44. G. Degli Antoni and R. Pizzi, Virtuality as a basis for problem solving?, *AI and Society,* Vol.5, pp.239-254 (1991).

45. P. Thagard and C. Shelley, *Limitations of Current Formal Models of Abductive Reasoning,* University of Waterloo, Ontario (in press).

MODELS AND DESCRIPTIONS IN MACHINE VISION

Vito Roberto

Dipartimento di Informatica, Università di Udine
Via Zanon 6, I - 33100 Udine, Italy

ABSTRACT

The meaning of *model* and *description* is discussed in the framework of machine
perception. An artificial intelligence perspective is adopted, viewing a model as a *Knowledge
Representation Language (KRL),* and a description as a construct of a KRL. *Generalised
inference* is a key feature, inextricably connected to all kinds of models. The latter are grouped
into three basic schemes: *relational, propositional* and *procedural,* each of which motivated
by the need to capture relevant aspects of perceptual tasks. The distinguishing features of the
schemes are outlined. Profound links are emphasized between relational/structural and
propositional/linguistic schemes, in analogy with mental representations in human beings. The
need to further combine the schemes into unified, hybrid representation languages emerges as
a trend in present and future research.

1. INTRODUCTION AND MOTIVATIONS

The words 'model' and 'description' are ubiquitous in the literature on both human and
machine perception, although not many papers address the issues and the questions raised by
the meaning of such words (see for example[1-7]).

Initially, some part of the real world is to be delimited *(the system);* relevant objects and
relations abstracted, concerning both the system and the environment; tasks to be performed,
clearly defined. This effort of *conceptualisation* is a part of the modelling activity. But, more
than this, a model is the outcome of some *translation* process, in that elements of the
conceptualisation are coded by means of a representation scheme, known to the perceiving
entity, that generates internal descriptions of the world. In automated systems the representative
scheme is formally defined in computational terms.

The complexity, variability and ambiguities of the real world are reflected onto modelling,
which is inherently an ill-defined and expensive activity in terms of resources involved.
Therefore, does a system really need an internal model of the world, in order to behave
reasonably in it? There is an open debate on this point. Some Authors[2], in the tradition of the

research work by the psychologist J.J. Gibson[8], claim that models are indeed not necessary: what is needed is a set of simple mechanisms acting/reacting against the environment in a continuous stream of parallel, multiple feedback cycles.

Going deeper into the debate on internal representations for an intelligent behaviour is beyond the scope of the present paper. We claim that explicit representation schemes are necessary to accomplish a number of selected perceptual tasks. In particular, models are suited to those tasks in which *perceptual* and *cognitive* functionalities coexist and cooperate: access/retrieval from internal memories/databases; reasoning with perceptual cues (associative, spatial, visual reasoning); sensor data fusion and interpretation; perceptual learning and automated model synthesis; planning motion and communication acts; managing man-machine multi-modal dialogue; managing virtual worlds.

The aim of this paper is to outline the main representation schemes used to provide an automated system with internal descriptions of the world; to discuss the features of each scheme, as well as the links among them; to give an overview on possible applications. An introductory bibliography is provided, for the readers who are not familiar with the argument.

The paper is organized as follows. The next section contains an introduction to the terminology and the classification we shall use in the sequel; sections 3 and 4 discuss the large class of relational schemes. Propositional schemes are discussed in section 5, while the next section contains the procedural ones. Our conclusions are reported in section 7.

2. TERMINOLOGY

2.1. Models, Representations, Processes

On general grounds, by *model* we mean *a set of symbols encoding specific aspects of the world, relevant to achieve a goal.*

Intuitively, a model involves static and dynamic aspects: in an attempt to keep them separate, it is useful to distinguish between *representations* - dealing with the former aspects - and *processes,* accounting for the dynamics of a problem.

A *representation* is an abstract term denoting a pair of schemes: a *descriptive* and an *applicative* one. The former is a pair of sets: a *vocabulary,* or *set of primitives,* and a *set of prescriptions,* that may be procedures, or rules, or composition laws combining primitives into well-formed *descriptions (configurations, states)* of the world.

A *process* is an abstract machine connecting an initial state to a final state via a set of *state transitions.* We shall use the term *generalised inference (reasoning)* to indicate those processes manipulating descriptions and providing significant steps towards the solution of a problem.

Although it is a generic term which needs formal specifications in more restricted contexts, generalised inference is a key concept in the design of Intelligent Perceptual Systems (IPSs), since it allows for a unified view of both low-level (purely perceptual, behavioural) and high-level (cognitive) processing steps. Well-known examples are *inferring shape from shading, texture, motion; inferring shape by grouping perceptual cues* (both are kinds of perceptual inference); *inferring new sets of phrases from the existing ones* (kinds of logical inference). Recently, two special forms of inference have been investigated by the psychologists (see for example Glasgow and Papadias[9]): *visual* and *spatial* reasoning. The former deals with the physical/geometrical appearance of objects in a scene (e.g., their shape and size), the latter with the spatial relations among the objects themselves (e.g., their relative positions).

2.2. Models as Knowledge Representation Languages

The starting point of the modelling activity is *abstracting (conceptualising)* those components of the universe and the perceiving system which are relevant to achieve a goal. Such components are physical objects, concepts, relations, actions, and tasks. The outcome of conceptualisation is some *universe of discourse,* which the designer converts (totally or partially) into a computational model.

The problem of building such a model may be addressed by adopting a *formal scheme based on syntactic/semantic prescriptions:* in this way a model is more precisely defined as a *Knowledge Representation Language (KRL).*

Basic elements of a KRL are: an *alphabet of symbols (vocabulary);* a set of *syntactic prescriptions;* a set of *semantic prescriptions;* a set of *pragmatic prescriptions,* concerning how to apply the scheme effectively and efficiently.

There are several motivations to such an approach, which is peculiar of Artificial Intelligence (AI). Firstly, knowledge items (objects, relations, facts, actions) are given an explicit account: they are put forward and expressed, in fact. Secondly, syntactic prescriptions provide the glue to integrate (fuse) the same items into constructs, whose correctness (i.e. internal consistency) can be checked. Semantic prescriptions provide *meaning* to constructs, either by mapping them onto elements of the universe of discourse, or by specifying the operations to be done in order to define them; in these ways external consistency of the model is ensured.

In the AI literature, KRLs are discussed at three distinct levels:
- ontology: basic elements and inference mechanisms assumed to exist;
- design: formal assessment (syntax, semantics, pragmatics);
- realisation: computational complexity; implementation languages.

Defining a KRL suited to a given problem is not yet supported by a methodology; at present, a number of adequacy criteria are widely accepted,[6, 7] providing constraints to the designer. In particular, according to the terminology introduced above, *descriptive adequacy* entails *capacity,* the ability to actually generate descriptions; simple, meaningful *primitives* and *association laws* should be available, as well as *refinement mechanisms,* and *partitions* into more homogeneous subsets of symbols; finally, the representation scheme should provide specifications of how objects are mapped onto signals, and clearly define goals in order to individuate measures of success and effectiveness. On the other hand, *procedural adequacy* addresses *completeness* and *soundness* of a model, i.e. the ability to generate all and only consistent descriptions; *flexibility* in accessing different databases; *efficiency,* supported by complexity estimates and/or performance evaluations.

2.3. A Classification

We shall group the knowledge representation schemes into three main classes[10]: *relational, propositional* and *procedural.*

Relational schemes[1, 11] consider the world as composed of entities and relations, to be expressed via network formalisms; within this large class, it is useful to individuate *analogical* and *conceptual* schemes.

The former directly reflect some of the physical/geometrical features present in the world. Examples are the *world-centered* and the *viewer-centered* representations. On the contrary, conceptual schemes provide abstract descriptions: *semantic networks, frame-based* and *network-based* schemes are here included.

Propositional schemes[11, 12] describe objects and relations in the world in terms of phrases of logic-based languages: *First-Order Predicate Logic (FOPL)* is the reference in this class, which includes several variations on the theme: *higher-order logics, multi-valued logics, fuzzy logics, nonmonotonic logics, modal logics, logics of time and action.*

Procedural schemes naturally emphasize the procedural component of knowledge, i.e. that representing processes to manipulate entities of the domain. They include the rich variety of *physical, geometrical, statistical schemes* and that of *distributed schemes.* The former include, for example, *reconstructive visual models*[4]; *multi-sensor fusion*[13, 14]; *dynamic world models*[15, 16], although most of them are not models in the sense specified in section 2.2. The latter schemes include *pattern-directed inference*[17, 18] and *multi-agent models*[19, 20].

3. RELATIONAL SCHEMES - I

3.1. Ontology and Inference

The evolution provides the physical universe with regularity and structure. Intelligent systems capture the environmental regularities by organising (grouping) perceptual cues[21]: relational schemes play a fundamental role to this aim.

The world is made of objects, concepts, actions, and interrelations among them; descriptions are in terms of networks. This offers the possibility to organise (structure) the whole knowledge by exploiting several kinds of relational links. Let us review a few organisational criteria[10]. *Set-membership* associates specific objects (instances, tokens) to a class (generic type). The *inclusion* relation (also called *is-a*) joins sets to supersets: it is a partial order, and organises generic types into abstraction/generalisation hierarchies. The *aggregation* link *part-of* provides structural hierarchies of objects in terms of their components. The *partition* link allows clustering network subsets into a unique element.

All of these criteria provide ways to index knowledge elements, thus making easier the management of large object databases.

The ontology of network schemes enables a large class of inference mechanisms, mainly concerning *visual reasoning.*

Classification is assigning an object to a set (i.e., activating a set-membership link). *Pattern association* is mapping pattern sets onto pattern sets; interestingly enough, such mappings may be learned by suitable training procedures, as in connectionist schemes. *Instantiation* is specifying an object by assigning values to its attributes; in this way, for example, a segmented pattern from an input image may be recognised as a primitive of a model, either in a data-driven or in a model-driven fashion. Perhaps the most important inference mechanism in relational schemes is *matching*[22], i.e. a quantitative measure of similarity among objects, which can be full or partial, exact or inexact. Other inference mechanisms are closer to logical reasoning: *property inheritance, default reasoning; analogical reasoning* (see section 5.2). Specific kinds of inference are related to the propagation of uncertain beliefs, as in *probabilistic network inference*[23]; *procedural attachment*[11] is a way to let a system react in response to instantiations or other inference steps.

3.2. Analogical Representations

An *Analogical Structure* is a pair {P,R}, where:

$$P = \{P_1, P_2, ..., P_n\} \tag{1}$$

is the set of primitives, being n the number of primitive parts in an object. Each P_i is a k_i-tuple of binary relations *(attribute-value pairs)* over $(A \times V)^{k_i}$.

R is a set of named N-ary relations over P:

$$R = \{PR_1, PR_2, ..., PR_k\} \tag{2}$$

Each element of R, PR_j, is a pair (NR_j, R_j): NR_j is a name for R_j, an m_j-ary relation over the set P.

Primitives may be geometrical, either *local* or *non-local*. They may also be *statistical* and/or *time-dependent*.

As examples of analogical representations we shall briefly introduce the world-centered and the viewer-centered representations.

3.2.1. World-centered representations.

An object or any point in the external space is assigned a reference frame; the appearance of the object is not explicit. Geometric or linguistic primitives are used.

The application scheme makes use of composition/decomposition laws, which may be syntactic (e.g., production rules in a grammar), or algebraic (i.e. boolean operations), or geometric (continuity, proximity, similarity,...). The latter, in some cases, perform *grouping of primitives* in the same way as the Gestalt laws are supposed to operate in human visual perception[24].

Object-centered representations have been extensively investigated and applied in the past, and are currently popular because of their descriptive adequacy in modelling natural forms[21]. *CAD models, Superquadrics, Generalised Cylinders, Constructive Solid Geometry (CSG)* are known examples.

We stress here the role of *syntactic pattern representations*[25]. Coded patterns are strings of a language *(Pattern Description Language, PDL)*, generated by a phrase structure grammar:

$$G = \{V_n, V_t, R, S\} \tag{3}$$

where V_t is the set of terminal symbols (primitives); V_n the set of non-terminal symbols; $V_n \cup V_t$ is the *vocabulary* of the representation scheme.

A pattern/phrase of the PDL is generated by applying recursively the *production rules (relations)* from a finite set R, starting with a root symbol S.

Such a linguistic pattern coding allows the *structural analysis* of a pattern in terms of its constituents, as well as classification/recognition as a result of syntactic analysis *(parsing)*: in our terminology, structural analysis and parsing are inferential mechanisms.

A PDL is naturally a knowledge representation language; it has been successfully applied to several perceptual tasks, such as signal understanding, object recognition, texture analysis and synthesis[25].

3.2.2. Viewer-centered representations.

The role of the observer is emphasized in an active vision[26] approach: suitable schemes have been designed with the main reference frame centered on the viewer; the appearance of an object is made explicit.

An *Aspect Graph*[27] AG is a pair:

$$AG = \{V, E\} \tag{4}$$

with V set of *general views* of an object, as seen from a *viewpoint space;* E is a set of arcs, for each possible transition between two neighbouring general views *(visual event)*.

As a viewpoint space we may consider the *viewsphere,* in which viewpoints are computed via the orthographic projection, or the full *3D space,* requiring the perspective projection to compute the viewpoints. The general views are sets of equivalent, stable object descriptions as seen from a viewpoint (for example, 2-D edges or edge-vertex representations). In figure 1 we report examples of descriptions originated by a viewer-centered representation.

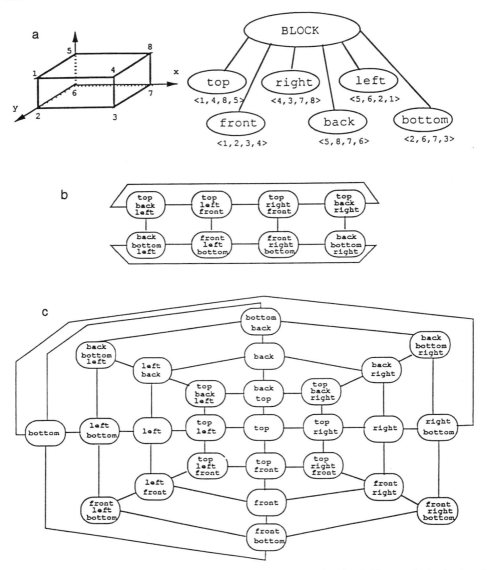

Figure 1. The rectangular block a) is described by two aspect graphs. Graph b) was obtained using the viewsphere; graph c) was obtained using the 3D space (reproduced from Bowyer et al.[27], with kind permission from John Wiley & Sons Ltd).

4. RELATIONAL SCHEMES - II

4.1. Conceptual Representations

Graph-like schemes are used to generate abstract (conceptual) descriptions, in which little or no similarity with the real world is maintained.

The basic formalism is a graph-like specification of an analogical structure:

$$G = \{N, A\} \tag{5}$$

being N the set of labelled nodes describing entities, and A the set of labelled arcs describing relations among entities.

4.2. Semantic Networks

This class of KRLs has been originated by the research in natural language representation and understanding[28]: it is interesting to observe that the same formalism has been applied to perceptual problems, thus capturing profound links between seemingly different functionalities.

Semantic networks are *directed graphs, with labelled arcs connecting labeled node pairs.*

Syntactic constraints are loose, so that potentially flexible and powerful languages may be defined. Node labels represent objects, classes, concepts, facts, events, actions; arc labels denote any kind of n-ary relations among such entities.

It is not surprising that the unconstrained use of semantic networks may result in semantic faults, e.g. ambiguous mappings onto real-world structures. An enlightening discussion on a special arc label, the *is-a*, is reported by Brachman[29]. The opportunity emerged to keep separate, in a knowledge representation language, the *structural/descriptive (concept-forming)* component from the *linguistic/assertional (sentence-forming)* one.

4.3. Frames

Frames were motivated by the need of modelling an agent's *point of view* about a given subworld[30]: the conceptual counterpart of a viewer-centered representation (section 3.2.2). In addition, it is useful to cluster the information concerning specific entity classes into abstraction/structure levels, thus allowing hierarchies to be suitably defined, in contrast to the flat descriptions of semantic networks.

A frame is an expression describing an entity type: *a collection of named slots, each of which can be filled with a value, or a pointer to another frame.*

In this way common sense, default, and stereotypical knowledge can be modelled in an natural fashion[11].

In figure 2 is shown a semantic network and a frame-like description of geological profiles, which are stereotypes in the stratigraphic interpretation of underground images[31].

Frame-based languages have been designed and realised[32, 33], and have proved to be both procedurally and descriptively adequate for those perceptual tasks involving a-priori knowledge, and using structured descriptions of the domain.

4.4. Network-based Languages

Perceptual problems usually require a large amount of procedural knowledge (see also section 6.1). A way to embed such component in a KRL is provided by hybrid network-based schemes, in which relational structures are combined among themselves and complemented with procedures via procedural attachment (section 3.1). The most famous example is KL-ONE[34], which also was designed within a research project in natural language understanding.

KL-ONE applies the quoted distinction between descriptive and assertional structures. The syntax includes a number of constraints on the entity and link types. Basic entity types are the *Concepts* (either primitive, or individual), which are structured objects composed of *Superconcepts* (classes to which the concepts belong), *Roles* (sets of generalised attribute

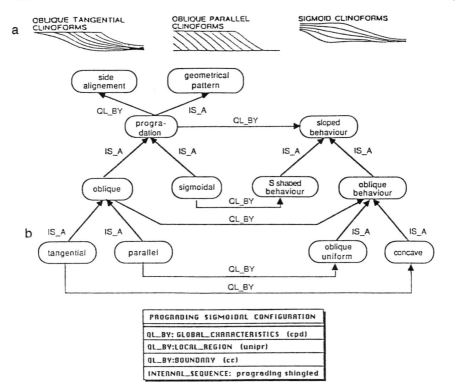

Figure 2. a) Geological underground profiles of the type clinoform; b) the same described by a semantic network; c) one profile (the sigmoid clinoform) described by a frame structure (reproduced from Roberto[31], with kind permission from World Scientific Publishing Ltd).

descriptions) and *Structural Descriptions* (relations among roles). The inference mechanisms are the property inheritance and the procedural attachment.

Examples of hybrid network-based languages motivated by pattern and scene recognition are given by Niemann et al.[6] and Ardizzone et al.[35] Let us briefly review the ideas underlying the latter research work.

The representation of knowledge is performed at two distinct - although integrated - levels: the *geometrical* and the *conceptual* one.

The former adopts a Constructive Solid Geometry (CSG) scheme, with superquadrics as primitives, and generic solids described as results of boolean operations - union, difference, intersection - of simpler ones; of deformation operations; of position-scale operations.

The set of productions of a grammar generating such models is reported here in the Backus-Naur form:

$$C \rightarrow C + C \mid C - C \mid C \cdot C$$
$$C \rightarrow ST(C) \mid EL(C) \mid BN(C) \mid TW(C)$$
$$C \rightarrow PS(C) \mid S \tag{6}$$
$$S \rightarrow (\mathbf{R}; \varepsilon_1, \varepsilon_2; a_1, a_2, a_3)$$

where C is a generic object; S is a representation primitive characterized by a rototranslation matrix $\mathbf{R}$ and the five shape parameters of the superquadric. The deformation operators are also represented: *Stretching* (ST), *Elongation* (EL), *Bending* (BN) and *Twisting* (TW). PS(C) is the Position-Scale operator.

The conceptual model is based on the KL-ONE language.

Input image data are approximated by a composition of superquadrics: it is the *internal description of the perceived scene;* a *conceptual description* is generated by KL-ONE constructs. We report in figure 3 the conceptual description of a primitive shape.

The last observation about this model concerns the semantics: it is fixed by *fuzzy matchings*[36] like the following: *the meaning M(o) of a conceptual description o is a fuzzy subset of the set of geometric descriptions, with a suitable membership function.*

5. PROPOSITIONAL SCHEMES

5.1. Ontology and Inference

Propositional (also called linguistic) schemes naturally provide assertional functionalities; their are largely based on *logic formalisms,* and are appealing for their well-understood formal background.

The world is made of objects, relations and functions[12] and represented in terms of collections of *phrases (predicates, formulas, clauses).*

Inference is *logical inference*, i.e. generating phrases, each one proved (*logically implied*) by applying an axiom or the modus ponens inference rule to a set of phrases (knowledge base). Different kinds of logical inference have been extensively studied: deduction, induction/ abduction, reasoning with uncertain beliefs[12].

In perceptual domains, logical inference has been mainly applied to *spatial reasoning.*

The most popular KRL is the First-Order Predicate Logic (FOPL). The main advantage of FOPL is that a formal semantic theory is available: it is independent of the domain currently described. For this reason the scheme is widely adopted not only to model specific problems, but also as a meta-scheme, a sort of reference to study and assess the properties of other schemes. This happened for frames, for example.

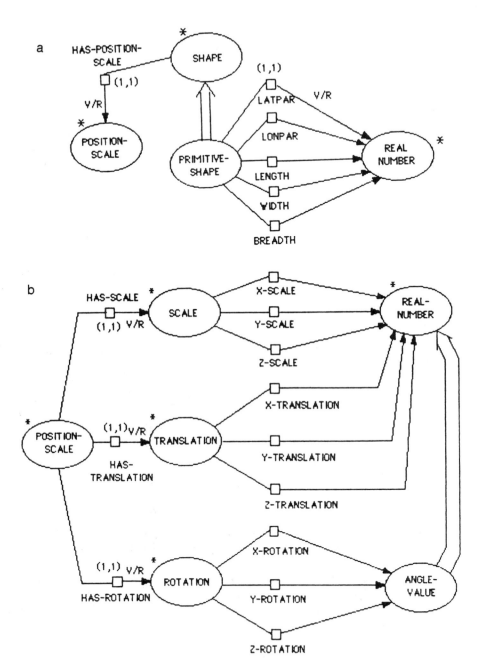

Figure 3. a) Conceptual description of a primitive shape using KL-ONE constructs; b) detailed description of the node *position-scale* (reproduced from Ardizzone et al.[35], with kind permission from Kluwer Academic Publishers).

5.2. From Frames to Predicates

Frames can be readily translated into phrases of FOPL.[37, 38]

A frame E representing entity x, with slot-relationships $R_1, R_2, ...R_n$ is an assertion in first-order predicate logic:

$$\forall x \ (\in (x,E) \Leftrightarrow \exists y_1, y_2, ...y_n \ (R_1(x,y_1) \wedge .... \wedge R_n(x,y_n))) \tag{7}$$

This allows to describe in logical terms a few basic inferential steps, which are typical of frame-based languages (see also section 4.3). For example, considering the frame specified here above, *matching is an inference rule,* as follows:

$$\forall x \ ((R_1(x,y_1) \wedge .... \wedge R_n(x,y_n)) \Rightarrow \in (x,E)) \tag{8}$$

Partial matching can be represented in an obvious way from the latter formula.

In the same way, other basic inference steps may be represented as pairs of sets of phrases: we shall use in the following a semi-formal account, in terms of well-known IF (.........) THEN (.......) pairs.

The basic *inheritance reasoning* step writes:

> IF (X subset of Z)
> AND ($R_i(z,y_i)$ holds for z $\in$ Z)
>
> THEN ($R_i(x,y_i)$ holds for x $\in$ X)

The basic *default reasoning* step writes:

> IF (Entity x matches frame E)
> AND (the value for some slot R_i is not known for x)
>
> THEN
>   IF (there is a default value y_i for R_i in E)
>   THEN ($R_i(x,y_i)$ holds)

The basic *analogical reasoning* step writes:

> IF (object x 'is like' object z)
> AND ($R_1(z,y_1) \wedge .... \wedge R_n(z,y_n)$)
>
> THEN (infer the same values y_i for some R_i in x)

It should be observed that FOPL is not adequate to perform inexact matching and compute matching values; to default reasoning; to analogical reasoning. Extensions of FOPL are needed[11, 38] if one wants to maintain a propositional description.

5.3. Logical Representations for Perceptual Tasks

Purely logical schemes have been widely applied to those perceptual tasks which also require linguistic functionalities, such as man-machine multi-modal communication[39], intelligent query/retrieval from image databases[40], verbal scene interpretation[41,7]. The latter contribution is particularly relevant to the scope of this paper, and will be briefly reviewed in the following.

The Authors present the design of a first-order language for visual knowledge representation; their task is interpreting hand-drawn maps of geographical areas, but the main ideas underlying the approach may be applied to different domains.

A basic distinction is operated between *the image domain* and *the scene domain*. Each domain is described in terms of first-order formulas (axioms), and complemented with additional sets of axioms expressing assumptions and constraints. A third set of axioms specifies the semantics, i.e. the *image-scene domain mappings.*

Having done this axiomatisation effort, image interpretation can be formally defined: *an interpretation is a model (in the logical sense)*[12] *of the full set of axioms (image, scene, mapping).*

In the same way, ambiguities can be given a formal account, in that some of the axioms (assumptions, constraints) are violated.

In figure 4 a few results of the quoted approach are reported.

The purely logical representation scheme appears both descriptively and procedurally adequate for tasks requiring spatial reasoning, although in the presence of restricted *(closed)* worlds, i.e. in which all axioms are supposed to be known. Key questions are left open - such as how to reason with ambiguities; nevertheless, the efforts to formalize perceptual domains are promising and encourage further investigations.

6. PROCEDURAL SCHEMES

6.1. Ontology and Inference

Procedural schemes naturally emphasize the large procedural component of perceptual knowledge.

The focus is on processes (also called *agents)*, i.e. how to manipulate descriptions and use them to achieve goals. The world is left open, dynamic with its intrinsic variability. It is a set of time-dependent states, which might be created or destroyed, together with a set of processes/agents responsible of the state transitions.

In such a world, generalised inference appears as a special process and might even include physical action[3]. We recall *perceptual processes,* such as feature extraction; grouping; segmentation; inferring shape from shading, texture, motion; fusing perceptual data; and *cognitive processes,* like read/write actions on memories; visual, spatial reasoning; exchange of messages; planning.

Most of the procedural systems do not implement knowledge representation languages: quite often they address and solve selected, well-defined tasks with little or no need of explicit representation schemes.

Nevertheless, the literature of machine perception contains a number of very interesting research contributions in which procedural schemes are close to the basic viewpoint expressed in this paper (section 2); we shall review some of these schemes, which represent knowledge in an integrated and comprehensive fashion.

The semantics is formulated in two basic ways. One consists in matching dynamically the elements of the internal description with their counterparts in the environment. Another way is peculiar of distributed models: specifying the semantics consists in specifying the actions an agent performs in a multi-agent world, in order to achieve a definite state transition[19, 20].

6.2. Dynamic World Modeling

Open worlds, such as those occurring in robotic perception, require a dynamic scheme in order to build and maintain descriptions. Dickmanns[16], Crowley and Demazeau[15] and many

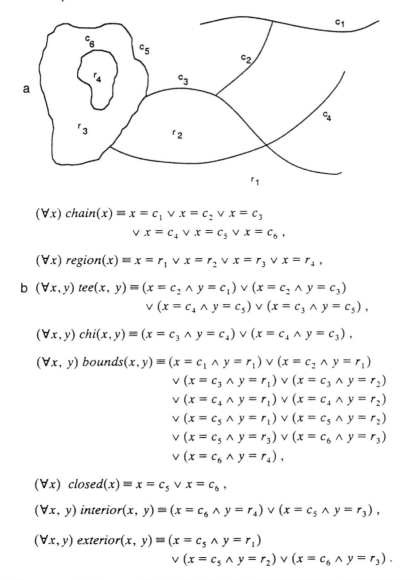

$(\forall x)\ chain(x) \equiv x = c_1 \lor x = c_2 \lor x = c_3$
$$\lor\ x = c_4 \lor x = c_5 \lor x = c_6\ ,$$

$(\forall x)\ region(x) \equiv x = r_1 \lor x = r_2 \lor x = r_3 \lor x = r_4\ ,$

b $(\forall x, y)\ tee(x,\ y) \equiv (x = c_2 \land y = c_1) \lor (x = c_2 \land y = c_3)$
$$\lor\ (x = c_4 \land y = c_5) \lor (x = c_3 \land y = c_5)\ ,$$

$(\forall x, y)\ chi(x, y) \equiv (x = c_3 \land y = c_4) \lor (x = c_4 \land y = c_3)\ ,$

$(\forall x,\ y)\ bounds(x, y) \equiv (x = c_1 \land y = r_1) \lor (x = c_2 \land y = r_1)$
$$\lor\ (x = c_3 \land y = r_1) \lor (x = c_3 \land y = r_2)$$
$$\lor\ (x = c_4 \land y = r_1) \lor (x = c_4 \land y = r_2)$$
$$\lor\ (x = c_5 \land y = r_1) \lor (x = c_5 \land y = r_2)$$
$$\lor\ (x = c_5 \land y = r_3) \lor (x = c_6 \land y = r_3)$$
$$\lor\ (x = c_6 \land y = r_4)\ ,$$

$(\forall x)\ closed(x) \equiv x = c_5 \lor x = c_6\ ,$

$(\forall x,\ y)\ interior(x,\ y) \equiv (x = c_6 \land y = r_4) \lor (x = c_5 \land y = r_3)\ ,$

$(\forall x, y)\ exterior(x,\ y) \equiv (x = c_5 \land y = r_1)$
$$\lor\ (x = c_5 \land y = r_2) \lor (x = c_6 \land y = r_3)\ .$$

		Predicate				
Extension	ROAD	RIVER	SHORE	LAND	WATER	
1	$C_1, C_2, C_3, C_4, C_5, C_6$			R_1, R_2, R_3, R_4		
2	C_1, C_2, C_3, C_4, C_5		C_6	R_1, R_2, R_3	R_4	
3	C_1, C_2, C_3, C_4		C_5, C_6	R_1, R_2, R_4	R_3	
4	C_1, C_2, C_3	C_4	C_5, C_6	R_1, R_2, R_4	R_3	
5	C_1, C_2, C_4	C_3	C_5, C_6	R_1, R_2, R_4	R_3	
6	C_1, C_4	C_2, C_3	C_5, C_6	R_1, R_2, R_4	R_3	

Figure 4. a) A hand-drawn sketch map; b) the set of closure axioms constraining the image domain knowledge; c) the set of models (interpretations) for the map in a) (reproduced from Reiter et al.[7], with kind permission from Elsevier Science Publishers).

others suggest to adopt statistical, time-dependent primitives. In the latter paper, a dynamic world representation is defined as a finite set of primitives:

$$M(t) = \{P_1(t), P_2(t),...,P_m(t)\} \tag{9}$$

where

$$P_i(t) = \{NP_i, X_i(t), CF_i(t)\} \tag{10}$$

where NP_i is a unique identifier for the i-th primitive (part i of the world);

$$X_i(t) = \{x_{i1}(t), x_{i2}(t), ... , x_{ik}(t)\} \tag{11}$$

is a vector, an *estimate* of the state of the i-th primitive; and $CF_i(t)$ is some dynamic estimate of uncertainty.

Independent observations are transformed into a common coordinate space and vocabulary. Then, they are integrated (fused) into an internal description by a cyclic process. The basic cycle is composed of three inference steps (phases): *Predict, Match and Update*. The framework is sketched in figure 5.

In the prediction phase, the current state of the model is used to predict the state of the external world at the subsequent observation time; in the match phase, a correspondence is established between predictions and transformed observations; the update phase integrates the observed information with the predicted state of the representation, thus creating an updated description: numerical techniques such as the *Kalman filtering* are used to this purpose.

6.3. Pattern-Directed Schemes

Perceptual tasks like speech understanding or signal and image interpretation motivated the introduction of highly modular schemes, in which the tasks are distributed among loosely coupled units[18]. These units are the primitives of the procedural knowledge, and are called *Pattern-Directed Modules (PDMs)*. Basically, they are *Condition-Action pairs*, i.e. expressing in form of a pair the set of conditions *(antecedents)* requested to activate some set of actions *(consequents)*.

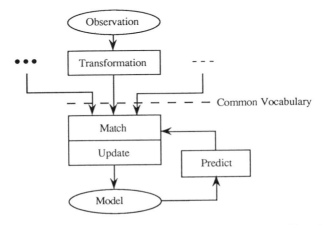

Figure 5. The Predict-Match-Update cycle for dynamic world modeling (reproduced from Crowley et al.[15], with kind permission from Elsevier Science Publishers).

According to the grain size of the actions involved, the structure of a PDM ranges from the well-known *rule* (in this case the PDM is a pair of sets of phrases of a formal language) to the *knowledge source* (the PDM is a set of activation conditions and a procedure to accomplish a task).

In this scheme, the computation proceeds through successive activations of PDMs (generalised inference steps), each of which modifying the internal status of the system (e.g., by writing on a working memory).

Rule-Based (RBSs)[18] and *Blackboard-Based (BBSs)*[17] systems are the most relevant realizations of the pattern-directed approach. RBSs can be adopted to combine procedural and logic-based formalisms, in that rules can assert phrases by performing inference steps as well as drive the computation by means of search strategies. RBSs have been experimented in perceptual problems: their use is generally within more complex architectures including a domain knowledge base and libraries of signal/image processing algorithms. In particular, rule-based modules are charged to accomplish high-level steps: controlling the activation of procedures on selected pieces of information (i.e. *focusing the attention* of the system[42]); integrating/fusing elements of evidence (e.g. doing perceptual grouping; performing associative, visual, spatial reasoning steps); propagating uncertainty through all the abstraction levels, from data to hypotheses. Remarkable results have been reported also in Brooks[43], McKeown et al.[44].

BBSs are hybrid architectures composed of a set of knowledge sources, plus a shared data structure, the *blackboard*. The former have a free internal format, thereby allowing to encode any kind of inference (e.g., asserting phrases, manipulating descriptions). The latter includes the relevant databases in a permanent sector *(long-term memory),* and the status of the computation in a volatile sector *(short-term* or *working memory).* The resulting framework embeds descriptions from any representation scheme, while, in principle, preserving a modular layout[45].

In figure 6 we report the internal levels of the blackboard of a system discussed in Nagao et al.[46] for the interpretation of aerial photographs of the ground surface.

7. CONCLUSIONS

This paper has given a brief overview on the main classes of models, and their role in intelligent systems with perceptual functionalities.

An AI perspective has been adopted, viewing a model as a *language* for knowledge representation (KRL), and a description as a *construct* of such a language.

This approach emphasizes some key aspects of IPSs: integration (fusion) of information in a unique scheme; use of generalised inference to manipulate descriptions; adoption of syntactic constraints to combine (glue) primitive patterns into constructs, and to ensure internal consistency to the latter; use of semantic constraints to maintain external consistency with a dynamic, unpredictable world.

In this view, deep connections become apparent among representation scheme, generalised inference, and action.

In addition, two basic functionalities of a KRL emerge: the *structural (concept-forming)* and the *assertional (sentence-forming)* ones, in such a way that a unique formal scheme can generate both relational (perceptual) and propositional (linguistic) descriptions. This captures important features of the human cognitive processes, as described by several psychologists in the debate on mental images and imagery[9, 47].

However, as any language, a KRL cannot fully capture the complexity of the world; a number of criteria to evaluate the *descriptive* and *procedural adequacy* of a model can be found in the literature. For this reason, no language is a priori preferable to any other: what matters is how *directly* and *naturally* a KRL describes aspects of the universe of discourse.

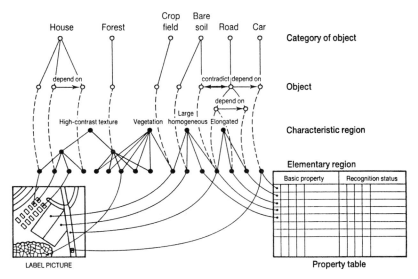

Figure 6. Internal structure of the blackboard of a system for the interpretation of aerial images (reproduced from Nagao et al.[46], with kind permission from Addison Wesley Publishers Ltd).

Semantics is essential in establishing adequacy. One way to specify semantics is using a domain-independent mathematical theory, like in logic-based schemes. As an alternative, robust, flexible mappings can be used from world elements to KRL elements: such mappings do not need to be independent of the specific domain. Finally, operational semantics has proved to be very useful in procedural, distributed schemes.

Relational schemes (both in analogical and conceptual form) are adequate for a large class of perceptual tasks. Knowledge is given a structure, which allows to group objects into classes, clusters, hierarchies; to better define defaults, a-priori elements and stereotypes; to index object databases efficiently; to adopt a number of non-trivial inference techniques. On the other hand, they have shown semantic ambiguities and faults, so that additional syntactic constraints must be imposed to the basic scheme.

Logical schemes are the most convincing ones from the point of view of a formal assessment, because syntax, semantics and inference are related in a consistent mathematical theory. An important drawback is the lack of organisational principles for the elements of knowledge. In addition, in perceptual domains the impact of logic-based formalisms is limited by their scarce flexibility in dealing with open, dynamic, uncertain worlds[5].

Procedural schemes have the advantage to allow the specification of direct interactions among knowledge elements, thus avoiding expensive search techniques, and meeting effectiveness and efficiency requirements. Their major drawback is the often implicit coding of knowledge items, which causes serious problems when reading and modifying them.

Hybrid languages, combining the advantages of more than one scheme have been investigated, and have provided convincing responses to some perceptual problems. Network-based models combine relational with procedural techniques, and have shown their adequacy to scene recognition and image interpretation in the presence of structured a-priori knowledge. The pattern-directed approach offers flexible architectures, fusing in principle any representation scheme, and providing distributed frameworks for pattern understanding. Such frameworks evolved into multi-agent and distributed-AI systems[20].

The search for hybrid schemes and languages is currently a central problem in AI, and a remarkable progress is expected in the near future on both theoretical and applicative sides.

ACKNOWLEDGMENTS

I thank Virginio Cantoni for organising the Third International Workshop on Perception in Pavia: the discussions I had with many colleagues during the workshop helped me to clarify a few of the countless, intriguing aspects of perception. In particular, I am indebted to L. Cordella, S. Gaglio, W. Gerbino, S. Levialdi, R. Luccio, S.L. Tanimoto and B. Zavidovique for their interesting suggestions.

REFERENCES

1. D. Ballard and C.M. Brown, *Computer Vision*, Prentice Hall, Englewood Cliffs, NY (1982).
2. R.A. Brooks, Intelligence without Representation, *Artificial Intelligence*, Vol.47, pp.139-159 (1991).
3. R. Davis, H. Shrobe, and P. Szolovits, What Is a Knowledge Representation?, *The AI Magazine*, pp.17-33 (1993).
4. D. Marr, *Vision*, Freeman, San Francisco, CA (1982).
5. M. Minsky, Logical Versus Analogical or Symbolic Versus Connectionist or Neat Versus Scruffy, *The AI Magazine*, pp.34-51 (1991).
6. H. Niemann, G.F. Sagerer, S. Schröder, and F. Kummert, ERNEST: A Semantic Network System for Pattern Understanding, *IEEE Transactions on Pattern Analysis and Machine Intelligence*, Vol.12, No.9, pp.883-905 (1990).
7. R. Reiter and A.K. Mackworth, A Logical Framework for Depiction and Image Interpretation, *Artificial Intelligence*, Elsevier Science Publishers, Vol.41, pp.125-155 (1989/90).
8. J.J. Gibson, *The Ecological Approach to Visual Perception*, Houghton Mifflin, Boston, MA (1979).
9. J. Glasgow and D. Papadias, Computational Imagery, *Cognitive Science*, Vol.16, pp.355-394 (1992).
10. J. Mylopoulos and H. Levesque, An Overview on Knowledge Representation, in *On Conceptual Modelling: Perspectives from Artificial Intelligence, Databases and Programming Languages*, M. Brodie, J. Mylopoulos, and J.V. Schmidt eds., Springer Verlag, New York, NY, pp.3-17 (1983).
11. R.A. Frost, *Introduction to Knowledge Base Systems*, Collins, London, UK (1986).
12. M.R. Genesereth and N.J. Nilsson, *Logical Foundations of Artificial Intelligence*, Morgan Kaufmann, Los Altos, CA (1987).
13. G.D. Hager, *Task-Directed Sensor Fusion and Planning*, Kluwer, Boston, MA (1990).
14. R.C. Luo and M.G. Kay, Multisensor Integration and Fusion in Intelligent Systems, *IEEE Transactions on Systems, Man and Cybernetics*, Vol.SMC-19, No.5, pp.901-931 (1989).
15. J.L. Crowley and Y. Demazeau, Principles and Techniques for Sensor Data Fusion, in *Signal Processing, special issue on Intelligent Systems for Signal and Image Understanding*, V. Roberto ed., Elsevier Science Publishers, Vol.32, No.1-2, pp.5-27 (1993).
16. E. Dickmanns, An Integrated Approach to Feature Based Dynamic Vision, *Proceedings of the International Conference on Computer Vision and Pattern Recognition*, Ann Arbor, MI (1988).
17. R.S. Engelmore and A.J. Morgan eds., *Blackboard Systems*, Addison Wesley, Wokingham, UK (1988).
18. F. Hayes-Roth, D.A. Waterman, and D.B. Lenat, Principles of Pattern-Directed Inference Systems, in *Pattern-Directed Inference Systems*, D.A. Waterman and F. Hayes-Roth eds., Academic Press, New York, NY, pp.577-601 (1978).
19. G. Agha, *Actors: a Model of Concurrent Computation in Distributed Systems*, MIT Press, Cambridge, MA (1986).
20. C. Hewitt and J. Inman, DAI Betwixt and Between: from "Intelligent Agents" to Open Systems Science, *IEEE Transactions on Systems, Man , and Cybernetics*, Vol.21, No.6, pp.1409-1419 (1991).
21. A. Pentland, Perceptual Organization and the Representation of Natural Form, *Artificial Intelligence*, Vol.28, pp.293-331 (1986).
22. L.G. Shapiro and R.M. Haralick, Structural Descriptions and Inexact Matching, *IEEE Transactions on Pattern Analysis and Machine Intelligence*, Vol.3, No.5, pp.504-519 (1981).
23. J. Pearl, *Probabilistic Reasoning in Intelligent Systems*, Morgan Kaufmann, San Mateo, CA (1988).

24. R. Luccio, Gestalt Problems in Cognitive Psychology: Field Theory, Invariance and Auto-Organisation, in *Intelligent Perceptual Systems,* V. Roberto ed., Lecture Notes in AI, Springer-Verlag, Berlin, D, Vol.745, pp.2-19 (1993).

25. K.S. Fu, *Syntactic Pattern Recognition and Applications,* Prentice Hall, Englewood Cliffs, NY (1982).

26. Y. Aloimonos, I. Weiss, and A. Bandyopadhyay, Active Vision, *Proceedings of the First IEEE Conference on Computer Vision,* pp.35-54 (1987).

27. K. Bowyer and C.R. Dyer, Aspect Graphs: an Introduction and Survey of Recent Results, *International Journal of Imaging Systems and Technology,* John Wiley & Sons Ltd, Vol.2, pp.315-328 (1990).

28. R. Quillian, Semantic Memory, in *Semantic Information Processing,* M. Minsky ed., MIT Press, Cambridge, MA (1968).

29. R.J. Brachman, What IS-A Is and Isn't: An Analysis of Taxonomic Links in Semantic Networks, *IEEE Computer,* Vol.16, No.10, pp.30-36 (1983).

30. M. Minsky, A Framework for Representing Knowledge, in *The Psychology of Computer Vision,* P.H. Winston ed., McGraw Hill, New York, NY, pp.211-277 (1985).

31. V. Roberto, Perceptual and Conceptual Representations in a Geophysical Image-Understanding System, in *Proceedings of the 6th International Conference on Image Analysis and Processing,* V. Cantoni, M. Ferretti, S. Levialdi, R. Negrini, and R. Stefanelli eds., World Scientific Publishing Ltd, Singapore, pp.713-720 (1992).

32. D.G. Bobrow and T. Winograd, An Overview of KRL, a Knowledge Representation Language, *Cognitive Science,* Vol.1, No.1, pp.3-46 (1977).

33. R. Fikes and T. Kehler, The Role of Frame-Based Representation in Reasoning, *Communications of the ACM,* Vol.28, No.9, pp.904-920 (1985).

34. R.J. Brachman and J.G. Schmolze, An Overview of the KL-ONE Knowledge Representation System, *Cognitive Science,* Vol.9, pp.171-216 (1985).

35. E. Ardizzone, S. Gaglio, and F. Sorbello, Geometric and Conceptual Knowledge Representation within a Generative Model of Visual Perception, *Journal of Intelligent and Robotic Systems,* Kluwer Academic Publishers, Vol.2, pp.381-409 (1989).

36. L. Zadeh, *Fuzzy Sets and Systems,* Elsevier, Amsterdam, NL (1983).

37. A. Deliyanni and R.A. Kowalski, Logic and Semantic Networks, *Communications of the ACM,* Vol.22, No.3, pp.184-192 (1979).

38. P.J. Hayes, The Logic of Frames, in *Frame Conceptions and Text Understanding,* D. Metzing, W. De Gruyter and Co. ed., Berlin, D, pp.46-61 (1979).

39. W. Wahlster, E. Andrè, W. Graf, and T. Rist, Knowledge-Based Media Coordination in Intelligent User Interfaces, in *Trends in Artificial Intelligence,* E. Ardizzone, S. Gaglio, and F. Sorbello eds., Springer-Verlag, Berlin, D, pp.2-16 (1991).

40. S.K. Chang, *Principles of Pictorial Information System Design,* Prentice Hall, Englewood Cliffs, NY (1989).

41. J.R.J. Shirra, G. Bosch, C.K. Sung, and G. Zimmermann, From Image Sequences to Natural Language: A First Step Toward Automatic Perception and Description of Motions, *Applied Artificial Intelligence,* Vol.1, pp.287-305 (1987).

42. V. Roberto and C. Chiaruttini, Seismic Signal Understanding: a Knowledge-Based Recognition System, *IEEE Transactions on Signal Processing,* Vol.40, No.7, pp.1787-1806 (1992).

43. R.A. Brooks, Model-Based Three-Dimensional Interpretations of Two-Dimensional Images, *IEEE Transactions on Pattern Analysis and Machine Intelligence,* Vol.5, No.2, pp.140-150 (1983).

44. D.M. McKeown, W.A. Harvey, and J. McDermott, Rule-Based Interpretation of Aerial Imagery, *IEEE Transactions on Pattern Analysis and Machine Intelligence,* Vol.7, No.5, pp.570-585 (1985).

45. V. Roberto, Knowledge-Based Understanding of Signals: an Introduction, in *Signal Processing, special issue on Intelligent Systems for Signal and Image Understanding,* V. Roberto ed., Vol.32, No.1-2, pp.5-27 (1993).

46. M. Nagao, T. Matsuyama, and H. Mori, Structural Analysis of Complex Aerial Photographs, in *Blackboard Systems,* R.S. Engelmore and A.J. Morgan eds., Addison Wesley Publishers Ltd, Wokingham, UK, pp.219-230 (1988).

47. P.N. Johnson-Laird, Semantic Primitives Or Meaning Postulates: Mental Models Or Propositional Representations?, in *Computational Models of Natural Language Processing,* B. Bara and G. Guida eds., Elsevier, Amsterdam, NL, pp.227-243 (1984).

ICONS AND WORDS: METAPHORS AND SYMBOLISMS

Stefano Levialdi[1] *(chairman),* Jörg Becker[2], and Santo Sciacca[3]

[1] Dipartimento di Scienze dell'Informazione, Università degli Studi "La Sapienza"
Via Salaria 113, I - 00198 Roma, Italy
[2] Universität der Bundeswehr München
ET/Physik, D-85577
Neubiberg and ICAS
D-82319 Starnberg, Germany
[3] Assessorato all'Istruzione
Comune di Pavia, I - 27100 Pavia, Italy

ABSTRACT

This panel discussed the role of icons as "meaning containers" in a number of different contexts: as graphical solutions of scientific problems (J. Becker), as gestures in training actors or written communication in poetical style (S. Sciacca) and, finally, as visual tools for computation. No conclusion can be drawn - and this reflects the dialectic nature of the topic - but we can certainly ponder on the concept of a "correct use" of icons as visual metaphors and on the multiplicity of their semantics.

TOPIC INTRODUCTION (S. Levialdi)

Our panel had a very ambitious goal: namely that of capturing the semantics of icons when used as visual metaphors and compare it, if possible, with that of words! The power of metaphors, both visual and linguistic, has been widely recognised not only in everyday communication but also in the context of new visual computational tools. Such tools may be employed when new programs must be used by nonskilled persons; in this case, iconic metaphors help in building mental models for conceptualising the program from a functional point of view.

The first author, J. Becker, shows that there are multiple meanings attached to an icon, generally dependent on the observer and the environment which impinge on the user, he also claims that only about 20% of useful information may be gathered from a text (or picture) which is read (or seen) by an observer. Moreover, the formalization of meaning, in the context of its linkage to an icon, is extremely difficult since reality itself is a fuzzy concept. In the remaining part of the presentation an interesting example of a complete description of reality is taken from quantum physics. Finally, some considerations on socially "positive" and "negative" icons are made to show that their selection and evaluation by us may determine which is the pragmatic information that may be conveniently used for real meaning disambiguation.

The second author, S. Sciacca considers the problem of creating an imaginary reality, i.e. that one of actors in a play. In order to do this all our senses must be committed so that three different forms of memory are evoked: affective, sensory and emotional; to feel by means of all the senses and to achieve completeness is the goal of an artificial reproduction of a given state of mind which, in turn, will produce a corresponding behavious on the stage. As an example of an intense and complete form of communication, the Japanese Haiku poem, contains both a linguistic and an evoked image in its brief textual form.

The third author, S. Levialdi explores visual metaphors keeping in mind their potential for expressing actions - execute queries or, more generally, instructions - either in a visual interface or within an iconic language for a given domain.

There are no conclusions to these different views on reality as described by icons or, at a different level of abstraction, by means of visual metaphors: we may only hope that through studies which come from a number of different fields like psychology and computer science a clearer picture emerges on the role and efficiency that visual tools may have in aiding the process conceptualisation, particularly for new and unforeseen applications.

ICONS, RHIZOMES AND MEANING (J. Becker)

"Most contemporary pictures, videos, paintings, the audiovisual, and the synthetic pictures are literally objects in which there isn't anything to be seen. We can only guess that behind each of them something has disappeared. They are no more than that: Traces of something that has disappeared."

These are words of the French philosopher Jean Baudrillard[1]. Is there any meaning in icons? Or has the meaning disappeared behind them?

Well, Baudrillard is right in 80% of all cases, but there are some 20% left about which it's worthwhile to think. This 80% rule has been formulated by the author on the ground of many experiences, and it has been confirmed in many cases, not only for pictures. (For instance, 80% of the text of any newspaper is completely irrelevant.). Let us start with a popular example of an icon: the owl. The most common meaning of this icon is "wisdom". However, in the context of a conference on vision it may also symbolize "sharp sight", and in the context of cities it may be a symbol for "Athens". Here we see that the meaning of a symbol may be (and usually is) context-dependent. A traffic sign on a road has a certain meaning; put it into a design museum, and the meaning is completely changed.

Apart from the icon and the context (which make part of the environment) there are two other instances involved in the icon: the icon user and the icon observer. Usually the user uses an icon to invoke a reaction in the observer, and we hope that behind the use of the icon there is some meaning and some sense. The observer first has to decode the icon; it then causes some reaction in him (an action, an emotion, etc.). Furthermore, for the user the intended reaction of the observer has some value, but also the observer has his values which need not necessarily coincide with those of the user; so he may react in a different way. It is sufficient to consider traffic signs as an example. The situation is depicted in figure 1.

A similar structure has been discussed as a general model of learning systems where a "robot" replaces the observer and a "teacher" replaces the user[2]; just that in the case of artificial

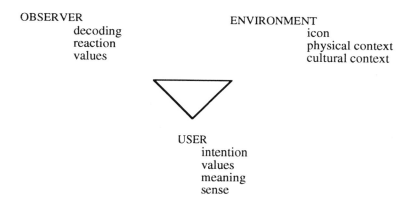

Figure 1. The "trinity" Observer - Environment - User.

learning systems the robot does not possess values of his (her? its?) own. If we represent values, or ethical norms, as mathematical norms of behaviour where behaviour is an element of some normed space we may define pragmatic information S from the point of view of the user and from the point of view of the observer as

$$S \text{ (user)} = \log \text{ (usernorm (reaction))}$$
$$S \text{ (observer)} = \log \text{ (observernorm (reaction))}$$

In general, the pragmatic informations for the user and the observer are different. The term "pragmatic" means action-oriented: whether the observer has "understood" the icon from the point of view of the user may be judged from S (user), but the observer may prefer to act in a different way to increase his own pragmatic information S (observer). Thus, we may imagine a game-theoretical situation in which both the user and the observer try to maximize their pragmatic information. In some cases (e.g., traffic signs) the user may have the option of sanctions to reinforce the "correct" reactions; in other cases (e.g., commercials) he can only try to invoke emotions in the observer that may cause the wanted behaviour.

In any case a number of requirements have to be met such that a sign may become an icon:
• the icon must easily be decoded;
• its form must be distinguished from other forms that occur in the given environment;
• the icon must belong to a family of icons which must not be too large;
• the icon must appear frequently - it must be a "clichè" in the sense of Eco[3, 4];
• the icon must invoke some reaction (attention, behaviour, ...).

It is useful to think about the mathematical structures to which the concepts belong which we are talking about. This is important because only then we are sure which operations may be performed, which phenomena may occur, or which questions make sense.

First we observe that icons, at least in the strict sense, are pictures. The reason for this may be that pictures are perceived more easily than words, are more easily distinguished from other objects, and have a close relationship with the emotional side. Pictures may be superimposed; they may be compared with one another, and there is a continuous transformation from one picture to another one. Hence we may argue that pictures are elements of some Banach space, Hilbert space, normed semi-ring or the like.

Decoding an icon is an act which maps the picture into a verbal interpretation. Language, however, has a structure completely different from that of pictures. For instance, words cannot be superimposed; instead we have different operations on words, like composition according to some syntactical rules, or inference according to some inference rules; thus, language may rather be described by algebraic or lattice structures.

Action, or behaviour, belongs again to a different mathematical structure. In some cases one only has a binary choice to react to an icon, but usually there is again a continuum of possible actions. At any rate, there must be a norm (or at least a metric) defined on the space of actions because it must be possible to judge an action with respect to a desired action. This norm represents an evaluation function. It has been stressed in Becker[2] that it is important that this evaluation function be continuous; therefore it must have the mathematical properties of a norm. Again we may consider actions to be elements of some Banach space or something similar.

So we may view an icon not only as a picture but as a tricky combination of objects belonging to different mathematical structures and maps between them. In a very brief form we may depict it in the following way:

$$\{ \text{set of pictures} \} \xrightarrow{\text{decoding}} \{ \text{set of verbal interpretations} \} \xrightarrow{\text{evaluation}} \{ \text{set of actions} \}$$

Thus, in the language of object-oriented thinking, an icon is not just an image: it is rather an object consisting of various data structures and methods. In a very simplified, symbolic form we may write it like this:

```
type
    icon = object
        form: image;
        context: environment;
        interpretation: string_of_sentences;
        action: behaviour;
        information: float;
        procedure perceive (form, context);
        procedure interpret (form, context, interpretation);
        procedure evaluate (interpretation, action, information);
        procedure perform (action);
    end;
```

This looks like a rather concise, complete, and self-consistent description. But having demonstrated the power of object oriented thinking we shall now destroy the myth that this is the ultima ratio. As e.g. Eco[5] and Vester[6] have pointed out reality is neither concise, nor complete, nor self-consistent. It is rather a kind of rhizome with an ever changing topology, with no clear boundaries, and with circular clauses that are rarely self-consistent. The choice of an object-oriented description is only a convenient and economic local measure to describe certain aspects of reality, but this choice must not be confused with reality. Actually, if we want to talk about the essence of processes, about meaning and relevance, we have to go beyond object-oriented thinking.

There are many similarities of this situation with quantum physics. An object would thus correspond to a measurement described by some self-adjoint operator, but this measurement provides only a partial description of the system under consideration, quite apart from the fact that a measurement usually changes the state of the system. Many scientists have tried to reduce quantum physics to classical physics. They not only failed, but in the end they had to face the fact that quantum physics is a more precise description of reality than classical physics. In the same way it seems to emerge that the rhizome structure of systems is a more precise description of reality than the object-oriented one, in spite of (or because of!) the inevitable uncertainty relations. Pirandello[7] has discovered this effect eight years before the discovery of quantum mechanics.

However, in quantum physics there exists a description of the system which contains all possible information about it: it is the state vector which is an element of a Hilbert space. We may say that this vector is the holistic description of the system whereas a measurement (or its classical interpretation) is an extraction of some information about it. We shall now give an example which shows that such holistic descriptions may be available also in other cases. It is taken from semiconductor physics, but this background does not really matter.

Consider a delta doping structure in a semiconducting material which generates a potential well. The well V may be calculated from the doping concentration N_D via the Poisson equation. We may now put an electron into the well, and if its energy is not too large it will stay there. The Schrödinger equation will tell us about its state ψ depending on the shape of the well. Thus we have the following problem structure (N_D, V, and ψ are functions, or forms):

$N_D(x)$ doping concentration
↓

$V(x)$ potential well
↓

$\psi(x)$ electron state

In figure 2 the graphs give an idea of possible shapes of ψ and V.

This is a task which is straight-forward. But now imagine that we put more electrons into the well, say 100 or 10000. Since the electrons are charged they will deform the potential. In other words, the potential determines the state of the electrons but the state of the electrons changes the potential. In figure 3, we have the unpleasant but realistic rhizome structure which contains a closed loop.

(In reality the system is still more complicated because you have to take into account the Fermi energy and several energy eigenstates, but for our considerations this form of the equations is sufficient. Just for fun we could assume that electrons follow Bose-Einstein statistics in which case the equations would be rather correct).

There is always a way to resolve the loop by a formal resolution: start with an empty well, calculate its shape; put one electron inside, calculate its state and its charge distribution; calculate the change of the potential due to the electron; ecc. During this iteration, gradually increase the number of electrons (N1, N2, ...). In a graph theoretical representation this formal solution looks like the one given in figure 4, in which the rhizome structure is replaced by a tree structure.

We call this a formal solution because nobody can guarantee the convergence of this enterprise. Actually, in practice it turns out that the scheme sometimes converges and sometimes doesn't. (A similar situation may be found in perturbation theory: even if all terms of the series exist the radius of convergence may be zero in which case the series is called a formal one).

Now we do some magics and replace the original rhizome graph by a holistic one as shown in figure 5.

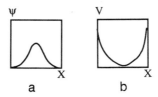

Figure 2. Possible shape for, respectively: a) the ψ function; b) the V function.

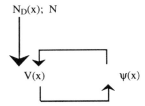

Figure 3. Rhizome structure containing a closed loop. $N_D(x)$ represents the doping concentration; N is the number of electrons in the well.

This is the holistic solution - but is there a way of obtaining it? Imagine a child playing around with two pieces of wire, deforming the shape of each of them, until both shapes simultaneously fit both equations. This kind of Monte Carlo approach can be replaced by a method which is much more elegant and efficient - the method of evolutionary strategies. (A detailed description can be found in Becker[2]; here we just want to state that holistic solutions may exist and may be found. The graph arguments stem from the method of interaction patterns. A graph theoretical foundation may be found in Merzenich[8]).

Let us now come back to the question of relevance and meaning, and to a possible explanation of the 80% rule. For this question it is not sufficient to consider the structure of a single icon, not even of a family of icons. Icons, users, contexts, and observers form an ecosystem of the rhizome type the essence of which cannot be discovered if one resolves the rhizome into local trees.

In recent years the ecological niche for icons has been increased dramatically by innovations in the computer and telecommunication sector. There is no doubt that the population of icons will grow into the niche until it is full. Such logistic growth processes have been observed in many other cultural and economic sectors (see Marchetti[9]). Since this is an evolutionary process we must not be surprised that only a limited fraction of icons really has

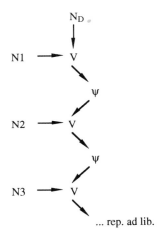

Figure 4. Tree structure representing the iterative potential well computation.

Figure 5. Representation adopting an holistic solution.

a meaning. Here we must focus for a moment on the word "meaning": What does it mean? On a short time scale we would say that an icon has a meaning for us if we have an advantage from it - a useful information, a pleasant feeling, ecc. On a longer time scale meaning must always be considered from a biological point of view - ultimately it can be measured as the influence of an icon on the number of surviving descendants of the observer (see Dawkins[10]).

Now, as icons grow into the niche, there are born all sorts of mutations of icons some of which don't have any meaning at all. But there are also mutations which serve a certain aim of their users: they may induce us to buy things which we don't need, they may be designed to prevent us from doing what we want, they may be created to distract us from truth. There are surrogate icons, parasitic icons, deceptive icons, mimicry icons. Nobody can stop this process. But yet we have an influence on it: which icons will survive is a matter of their evaluation and selection, and it is ours to evaluate and select in order to maximize our own pragmatic information.

HUMAN MACHINE: AN APPROACH TO TRAINING ACTORS (S. Sciacca)

Why actors? Because actors have to create an imaginary reality.

The main point I feel is that in some way we need to create a biography. A sort of reconstruction of the life of a man or, for analogy, the life of a machine. According to science we are completely conscious that our growing up is bound to experience and experience is bound to survival. What is interesting is that we make discoveries through our senses from the first days of life. So, for instance, when we wanted to drink at that time we experienced that our lips were dry, our tongue needed to be wet and we cried because we had already discovered that if we cried some one could come and we could satisfy our need. We used the voice in a particular way to get a result. So our daily voice is the result of that experience and obviously of many others. So according to me in my experience of training actors we have already written every thing in our body at three different levels.

Three different sorts of memory: affective memory, sensory memory, emotional memory. We could call the entire process: psycho-physical-sensorial experience. We need to be more precise: it is like a lot of cells that organize themselves according to their individual function. The way that they choose to organize themselves is strictly related to the senses. That is why actors are trained to increase their sensitivity through smell, taste, sound, sight...through sensorial exercises. In this voyage of discovery they need to work to the smallest details looking for a macrocosmos in a microcosmos, looking for feedback in the "right" memory through a sensorial detail for evoking a previous experience.

If I could synthesize: to feel by being complete. For me being complete means that every sense interacts with others. It is a way of amplifying perception. So sometimes, to find a character, I need to see him in a virtual reality. I am sure that he is not going to move like me, so I have to look carefully with my imagination so I can get into a different body, find new energy, a different rhythm. To do this I have to visualize him, the character. I have to know what he looks like, from what kind of experience he emerges, what happens in his life, somewhat to discover what has been his own experience of physical reality. To get this result we use different techniques.

Portrait. In this experience the actor should become the picture he chooses. He should try to understand the origin that creates expression and through careful observation discover the movement inside the picture and so to move and to speak from that particular point in a virtual reality.

We use the same approach in a different exercise that we call: the animal. The actor studies an animal, especially the way the animal moves, where the movement starts, where the voice comes from. It is a kind of research that aims to discover or recall to our memory how to discover...to have a path, a path based on experience that we can follow. What we discover through these exercises is that each experience changes according to the principle sense being used.

Choice bound to the desire for survival. The point is that for an actor the need for a different reality, the reality of a theatrical piece or of a character, obliges him to discover the structure of an experience in order to recall it when required. If in the morning you concentrate on drinking a cup of coffee, you discover the unbelievable number of sensations that are involved. The cold spoon, the hot cup, the smell, the weight of the cup, the shape...the different way each part of our hand reacts. Like a human machine I need a lot of information to do this job. What I realize is the fact that I need to structure an experience so that I can recall it any time I like. I have to discover a point of departure. The point of departure is where I am. What kind of space, what kind of light, what kind of reality psycho-physic sensorial affects me. Then I try to discover how I feel in this space, larger or thinner, and then how to move in this space. If I am open enough in a certain sense it is the space that create me. It is a conscious preparation for an uncoscious result. It is an interaction of senses which evoke previous childhood experiences. After this first approach I activate the energy of my body to adjust the senses so that I can use the right ammount of energy I need. After this, a second adjustment is that towards the audience.

This adjustment comes from the need to check the feedback with the public. Categories particularly involved are those of rhythm, intensity, movement, energy, distance in other words those that create a common immaginative space where it is possible to meet. It is still a space. Sometimes the actor faces some problem that he has never met in his life, or he is unable to get into a particular atmosphere or unusual situations and so we come to the last adjustment: the metaphor.

When completely involved in feeling life you don't need a lot of words, so life seems to be more similar to poetry, to a different language, closer to a creative process. The metaphor I am speaking about is still sense\related but in a dynamic way. For instance suppose you have to rappresent a particular relationship with someone that you don't understand. In these circumstances it is easier for the actor to use a dynamic methaphor like "to be in fog that rises and falls moved by a soft breeze" so that you never know what you see and what you do not see, so that you never know what to do. An image is more effective than any number of words. To give an example. In Japanese poetry called Haiku the first line is always an image:

> "Old pond
>   frog jump in
>   water sound"

Basho

If you allow yourself to stay for a while in this image you easily realise how the image seems to amplify the original apparently simple meaning and can include a lot of different meanings. A hidden meaning enriched by the symbol where the frog means the animal inside of us, the old pond, our stagnation of the spirit and then... something happens.

...And if you look at the poetry in the original language the calligraphy and the sound of the vowels suggest even more. It seems to me that this short poem in some way is a more complete communication, bound to the essence of life and really simple.

What is the structure?

The first line is an *image*.

The second line shows the *movement* from one polarity to another, the last line is where the poet finds "*the discovery*".

What is amazing is that in Japanese two different symbols for instance "sun" and "moon" when combined give a new different meaning: "shining". This meaning can be changed in a different context by the smallest of strokes. In a very simple way. It is a dynamic process.

> "On a withered branch
> a crow has settled
> Autumn night fall"
>
> Basho

ICONS ET AL (S. Levialdi)

Natural language, by means of words, conveys a number of features exploiting the basic link between themselves (words) and things (objects)[11]. The meaning can be grasped through the construction of a mental image in the listener whenever a linguistic description (a chain of words) is provided by the speaker. The semantic link is obviously not unique and strongly depends on context, speaker's intention, listener's motivation and time of speech. Yet an interesting and important feature of the natural language communication is its *generality* in the sense of understanding a "prototype" of the described object. For instance a sentence like "A flower pot is on the table" generates an image of a middle-sized flower pot, most likely at the center of a rectangular table; no particular colour is associated to the pot nor to the table.

Conversely, we may consider a pictorial communication based on icons, drawings or other sketches, as having details and a wide number of relevant features which determine, at a much finer resolution, the mental image which is formed through visual perception[12].

Globally, both communication vessels may cooperate, provided the conveyed descriptions are compatible, so that what is now perceived resembles a complete picture more than an abstract model of reality. The symbol comprehension and perceptual consciousness is based on the experience of sensorimotor imagery interacting with objects in space, specifically based on a number of interesting and typical features of perception in humans. These features are: the unity of perceptual objects (objects are perceived as wholes, size-colour-texture) which is known as the *binding* property; the *externality* of perception (where the input source is not distinguishable); the *self-awareness* (the bodily image is always present and builds a constant background on which the perception is mapped) and the standard *awareness* where there is a temporal blending of perceptual phenomena as they are absorbed by the human senses. But symbols are one of the three different forms of signs as classified by Pierce[13] at the end of last century; there were other two, namely index and icon: these last two represent a trace of a natural footprint and an abstracted version of a more complex image (or concept) of which it is a substitution.

A recent definition of icon[14] is "...a visually segmented object which tells the viewer about an inside message or information (concept, function, state, mode, etc) assigned by the designer". Examples of icons are the petrol pump in the car control panel board showing, when lit, the need to refuel and a fork and knife on a poster on the road to show the presence of a restaurant at a given distance on the same road. Icons may play a representative function, an abstract function or an arbitrary function.

Taking into account the evolution of icon design in terms of a given meaning and the role played by a represented concept, a more recent definition is[15, 16] "... an *icon* is a perceptible (physical) pattern to which a community of users agrees to assign a meaning so that it can be exploited for human communication and reasoning. This agreement is however reached through an *evolutive* process and does not necessarily reach a stable end." As may be seen by the dynamical description included in the last sentence it is very difficult to assign, at any given time, a unique meaning to an icon, independently from the observer. Yet, the role of icons in visual interfaces for human-task communication is crucial: either they are far better than reading a thick handbook for using a new application or they fail in providing the simplicity and naturalness that is required to start using a new application, machine or complex system.

To this end, the icon must contain a "metaphorical value", in other words, be a straightforward metaphor that naturally leads the observer to the construction of a mental model of the system enabling him/her to "guess" the commands functions, the help options, the way to obtain partial/final results during the execution of a novel application.

The metaphor has been recognised and widely studied within the realm of linguistics (by Pierce) as "a sign being an explicit reference by means of which conclusions may be extracted about a latent object (abstract or concrete)" and, in a standard, everyday concept in the many ways described by Lakoff et al.[17]. More recently, metaphor has also been defined as "the process of borrowing between an intercourse of thoughts" and Chandrasekaran[18] in his introduction to the book written by J.H. Martin, says "...There is persuasive argument that human cultures share a number of basic metaphors, i.e. metaphors are not arbitrarily created entities from individual to individual, or even from culture to culture. This suggests that perhaps we have a storehouse of metaphors that are part of our implicit knowledge of language, and that all actual metaphors are applications of this basic set...". Finally, Borges[19] considers that "...it is perhaps a mistake to imagine that metaphors may be invented. The true ones, those that establish intimate connections between one image and another one, have always existed; the metaphors we may still create are the false ones, those which are not worthwhile creating".

Some fiction stories describe situations which may be contradictory: in terms of time, of place, of characters. This happens, for example, whenever an "impossible" world is described as in some of Kafka's novels where a person enters a building positioned in a certain lown and exits from the same building in a completely different place. Such texts belong fall into the so-called *self-voiding fiction*[20] which have, as a counterpart, a visual metaphor such as the one shown in Figure 6, which displays the "impossibility" of a three dimensional object such as that suggested by the picture. On other hand, if a two dimensional object is considered, since the drawing has been made, it is obviously a possible object. The metaphor maps self-voiding into impossibility: what can we learn from this example? that a complex message can be conveyed by means of a visual metaphor, that such indirect description may be easier to understand than a direct one[21] communicating it to a wider audience, having common cultural background.

Metaphors have been widely used in the programming area, particularly for the development of applications which are based on everyday working life; to show this in a compact form, Carroll[22] compiled Table I.

This table lists, along four columns, the application area, the existing commercial systems, the used metaphor and the knowledge exploited for making the application easier to understand, use and learn. As may be seen in Table II, an extension to Table I (mapped into Carroll's Key Metaphors[23]) may be added with the appearance of Artificial Reality on the arena. Some new games take advantage of personal experience to provide that "true" feeling which is necessary for the construction of a "make believe" world.

The metaphors may also carry denotational information since a rough picture, sometimes

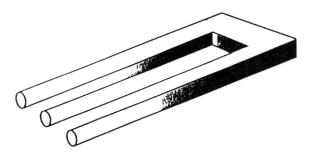

Figure 6. A well known example of a visual metaphor.

called an icon, may represent (for instance) the presence of a variable, of an input device or of storage capacity when a bucket, a ticket office or a file cabinet are respectively displayed on a screen. Moreover, abstract knowledge may also be visualized in figure 7, and therefore grasped, by means of an analogical representation showing the increasing annual gross income of a country by a set of symbols (like the yen symbol) each one being larger than the previous one.

Cognitivists for a long time have been discussing how people learn and have suggested many models for describing this process both to build a general theory and for improving teaching through the analysis of learning from a practical standpoint. A recent conjecture[24] is that, basically, all learning is performed by means of a psicomotorial experience, i.e. by *doing* and that in very few cases, an abstract model of the object, (system, function, etc) is built *before* understanding it (or them). If this is true then the possibility of grasping, by iconic means, the object about which we are reasoning using a metaphor, may greatly enhance the understandability of new, alien concepts.

A great number of authors have studied and analysed the nature of metaphor[25], there is a consensus on the fact that a basic set of metaphors exists, that this set is independent from culture, sex and race and, still more surprising, that even if it is new metaphors, the really important ones are already "built in " our human, personal, knowledge base[19].

It may be argued that, perhaps, our working environment is typical of a certain class of users and that if this environment is considered in the metaphor, a learning path towards the understanding of an application program may be easily figured out and therefore followed by any candidate user of such an application. In fact, the well known desktop metaphor used in the Macintosh interface relies on the above fact, the developers of spreadsheet programs as well as for pen-based software have tried to capture both the sequence of human actions and of their

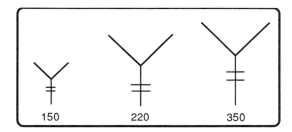

Figure 7. Abstract knowledge by means of analogical representation.

Table 1. Key Metaphors in commercially available software (reproduced from Carroll et al.[23], with kind permission from Elsevier Science Publishers)

Application Area	Systems	Metaphor	Exploits Knowledge of
Word & text processing	Wordstar, WriteNow	Typewriting	typewriting, typing paper, keyboard
Advanced document comp. (formatting languages, desk-publishing, composite editors, structured editors)	Interleaf, Star, GML, PageMaker, Etude	Document	types of graphical objects and their attributes (document components), logical structure of documents
Idea organisers and outline systems	Framework, Thinktank, Maxthink, NoteCards	Outline (as a plan of structuture of knowledge)	structure/decomposition of ideas often using further metaphors: index cards, frames.
Large electronic workspaces	Whiteboard, Chalkboard	Chalkboard	chalkboard attributes: group interaction, freeform text and graphics
Operating environments for accessories	Star, Lisa, Macintosh, Sidekick, Sketchpad	Desktop	office organisation and personal desktop tools workstations, procedures office tools: notepads, calculators, Rolodex
Forms-based business applications	SBA, OBE, Intuitive Solutions, ALL-in-1, Personal Decision Series Formanager	Business forms	codification of business activities in forms, organization of information, information items, report generation
Database management	QBE, OBE, Personal Decision Series, DBIII All-in-1	Table of data	matrix-structured: rows and columns
Spreadsheets	VisiCalc, Lotus 1-2-3	Ledger sheet	matrix-structured numerical data
Object-oriented programming environments	Boxer, Rehearsal World, Alternate Reality Kit	Physical world	physical objects and systems, their attributes, appropriate actions

Table 2. Extension to Table I

Artificial Reality	Voyeur, Mortal Kombat	True 3-D Space	Personal Experience

spatial organization when calculating a partial result or scribbling down some notes in order to simplify such tasks by an electronic notebook.

The basic problem still remains; what is a good metaphor?, which are its basic features? which granularity level should be used? which role can motivation and previous knowledge play?[26] ...The answer to these questions may come both by tentative definitions and by pragmatic work on those metaphors which "work"; this activity may fall under the name of Metaphor Engineering. In fact, a structural description of metaphors could be based on the answers to the following questions:

Which tasks should be solved?, how may they be solved? and what are the visual ingredients that fully exploit human senses? (i.e. the "look and feel" of the screen layout, the keyboard organization, etc).

Perhaps a number of used metaphors could be redesigned if considered after the answers given to the above questions which, in part, touch on new and borderline fields like: models of users and systems, usability issues, interaction strategies. In all these fields the communication problem is the crucial bottleneck since only through an efficient exchange of data between the system and the user can the metaphor play the crucial role of helping in building a mental model of the new program (or application) so as to decrease the semantic gap between program contents and human thoughts.

In order to achieve this goal, new work is going on in the analysis of metaphor design, in the development of visual languages[27, 28] and in the design of visual interfaces, all aiming at an improvement of the human-task communication allowing the user to concentrate on solving his problem instead of forcing him/her to understand the software environment, debug the program, obtain clear displays and printouts and optimize code for a particular computing platform. In short we are waiting for high performance visual metaphors that may greatly help non technical users in their jobs when using new programs or computing systems.

CONCLUSIONS (S. Levialdi)

This panel touched on a wide number of topics and suggested different ways of making use of visual cues as presented in the solution of physical problems, in a gestural language for training actors, in the form of Haiku Japanese poetry or, finally, with icons on a screen for aiding novel users to achieve computerised work. It is obviously difficult to draw conclusions about a wide number of interesting properties that visual objects - whether icons or diagrams as metaphors - have in the conceptualisation of new solutions, systems or programs. We may only say, to stay on the safe side, that a single visual object may have, in general, a number of meanings and therefore, for a correct use it needs disambiguation through syntactic, semantic and pragmatic knowledge, only in this way visualization may really help in wide areas of scientific communication including problem solving, human cooperative work and computer interaction.

REFERENCES

1. J. Baudrillard, quoted from: Patricia Görg, *Der leere Rausch des Vervielfältigens*, Süddeutsche Zeitung, München, D (translated by the author) (1993).
2. J. Becker, Learning as gain of information, in *The paradigm of self-organization II*, G. Dalenoort ed., Gordon and Breach, New York, NY (in press).
3. U. Eco, *Dalla periferia dell'impero*, Fabbri-Bompiani, Milano, I (1977).
4. U. Eco, *Apocalittici e integrati*, Fabbri-Bompiani, Milano, I (1978).
5. U. Eco, *Semiotica e filosofia del linguaggio*, Einaudi, Torino, I (1984).

6. F. Vester, Cybernetic principles of organization, in *WOPPLOT 83*, J. Becker and I. Eisele eds., Springer-Verlag, Berlin, D (1984).

7. L. Pirandello, *Così è se vi pare*, Le muse (1918).

8. W. Merzenich, Algebraische Semantik für Parallele Flubdiagramme, paper presented at *ICoLe 93*, Lessach, A (1993).

9. C. Marchetti, The future, in *Sinergetica ed instabilità dinamiche*, G. Caglioti and H. Haken eds., Società Italiana di Fisica, Bologna, I (1988).

10. R. Dawkins, *The Selfish Gene*, Oxford University Press (1989).

11. M. Massironi, What words see: levels of compatibility between verbal descriptions and visual images, in *On Representation: Relationship between Language and Image*, C.E. Bernardelli and S. Levialdi eds., World Scientific Press, Singapore (in press).

12. N. Newton, How do we understand a symbol? in *On Representation: Relationship between Language and Image*, C.E. Bernardelli and S. Levialdi eds., World Scientific Press, Singapore (in press).

13. C.S. Pierce, *Collected Papers*, Harvard University Press, Cambridge, MA (1958).

14. H. Fuji and R.R. Korfhage, Features and a model for icon morphological transformation, *IEEE Workshop on Visual Languages*, Kobe, Japan, pp.240-245 (1991).

15. P. Mussio, M. Finadri, P. Gentini, and F. Colombo, A bootstrap approach to visual user interface design and development, *The Visual Computer*, Vol.8, No.2, pp.75-93 (1992).

16. S. Levialdi, P. Mussio, M. Protti, and L. Tosoni, Reflections on icons, *IEEE Symposium on Visual Languages*, Bergen, pp.249-254 (1993).

17. G. Lakoff and M. Johnson, *Metaphors we live by*, University of Chicago Press (1980).

18. B. Chandrasekaran, in *A Computational Model of Metaphor Interpretation*, James H. Martin, Editorial comment (1990).

19. J.L. Borges, *Otras Inquisiciones*, Biblioteca di Babele, F. Maria Ricci ed., Parma, I (1989).

20. U. Eco, *Sei passeggiate nei boschi narrativi - Harvard University, Norton Lectures 1992-93*, Bompiani, Milano, I (1994).

21. S. Levialdi, Cognition, Models and Metaphors, Keynote address, *IEEE Workshop on Visual Languages*, Chicago, IL, pp.69-77 (1990).

22. J.M. Carroll and J.R. Olson, Mental models in human-computer interaction, *Handbook of Human-Computer Interaction*, M. Helander ed., Elsevier Science Publishers B.V., North-Holland, pp.45-66 (1988).

23. J.M. Carroll, R.L. Mack, and W.A. Kellogg, Interface metaphors and user interface design, *Handbook of Human-Computer Interaction*, M. Helander ed., Elsevier Science Publishers B.V., North-Holland, pp.67-85 (1988).

24. F. Antinucci, Communication during the seminar at our University of Rome, June 17 (1993).

25. U. Eco, *Metafora*, Enciclopedia Einaudi (1982).

26. D.D. Woods and E.M. Roth, Cognitive systems engineering, *Handbook of Human-Computer Interaction*, M. Helander ed., Elsevier Science Publishers B.V., North-Holland, pp.3-43 (1988).

27. Shi-Kuo Chang, *Visual Languages*, Plenum Press (1986).

28. M.M. Burnett and M.J. Baker, A Classification System for Visual Programming Languages, *TR 93-60-14*, Dep. of Comp. Science, Oregon State University, Corvallis, OR 97331 (1993).

VISUAL THINKING: STABILITY
AND SELF-ORGANISATION

Riccardo Luccio

Dipartimento di Psicologia, Università di Trieste
Via dell'Università 7, I - 34123 Trieste, Italy

1. INTRODUCTION

As Ashby and Lee[1] recently pointed out, there is a great deal of trial-by-trial variability in all perceptual representations. However, *stability* is the first apparent visual concept of the world to which we have adjusted. Although proximal stimulation is continuously changing, our phenomenal world is usually *stable*, made up of objects which usually remain the same size, shape, colour and identity. The second aspect is *harmony*. Very often, objects in nature keep a specific regular and harmonic structure. It is interesting to note that we are particularly pleased when we find such regularity and *harmony*. Almost perfect beautiful examples of axial or central symmetry can be found in the inanimate world as well as in the biological world. These are often considered conclusive evidence that natural phenomena conform to natural laws. At the same time, the pleasure that we experience in perceiving regular and harmonic configurations (and the tension that we feel when we face configurations that depart from this regularity and harmony), is considered conclusive evidence of the fact that perceptual organisation is dominated by the *tendency to Prägnanz*. Therefore, this tendency can be considered the leading principle that governs perception.

We know that the Prägnanz tendency is an assumption, not only of classical Gestalt psychology, but is widely accepted by non-Gestalt followers too. Kanizsa and Luccio[2, 3] insist that in Gestalt tradition, Wertheimer[4-6] and other researchers use this term with two distinct meanings: i) Prägnanz is a *phenomeno logical* characteristic, ii) and Prägnanz is a property of *process*. "Good" Gestalten or Prägnanz Gastalten in the *first* instance, are configurations that are built according to a sufficiently transparent mathematical function, of which symmetry is the most apparent regularity of this kind. We can call this kind of Prägnanz, *singularity*[7]. This is because these structures are experienced as unique examples, or prototypes of a family of configurations. Singularity or figural goodness is a phenomenal *property* of certain forms that are "better" than others. The *second* description of Prägnanz refers to the characteristics of the perceptual process. A phenomenal event is Prägnant when it is governed by a minimum principle, finally achieving maximum stability and resistance to change. There is a general *tendency* to achieve this result. Notice that stability and singularity are not equivalent concepts and that *tendency towards stability* and *towards singularity* are two different things. To assume that the visual system tends towards order and prefers to produce maximum regularity and maximum possible symmetry is misleading. As Kanizsa and Luccio[8] pointed out, only a few natural objects have a regular structure. The majority are ill-shaped, therefore few phenomena or events have a "good" form, and are therefore "better" than others. A percept is always the result of a lawful process, but it is very seldom a singular phenomenon. Kanizsa and Luccio[2] deny that a tendency towards Prägnanz exists at all, when with this we understand a tendency towards singularity. The visual system is not governed by a *tendency towards singularity* but rather a *tendency towards stability*.

In this paper, we put forward the thesis that stability is the result of the visual systems capacity of self-organisation. The principles to which the system self-regulates, are essentially those that Wertheimer[5,6] pointed out (i.e. proximity, similarity, common fate and so on). These principles tend towards a perceptual result that is better only in the sense of maximum stability. The possibility of a phenomenal "singularity" is only a by-product.

2. AUTO-ORGANISATION AND PATTERN FORMATION

A model of self-organisation that appears to be particularly apt, at least to describe such perceptual phenomena is *synergetics*. This has been developed over the last few decades by Hermann Haken[9, 10]. As Stadler and Kruse[11] pointed out, there is a continuity between Gestalt theory on autonomous order formation and the currently fast developing theory of self-organising non-equilibrium dynamic systems. Let us consider this approach briefly (see also Kanizsa and Luccio[12]). Subsequently we will see how this model could be applied in three different circumstances: i.e. perception of movement, perception of multistable configurations and perception of rhythmical patterns.

In synergetics, pattern formation can be described in terms of evolution of state vectors in the following form:

$$q(x,t) = (q_1, q_2, q_3, \ldots q_n) \tag{1}$$

The evolution is described in terms of time derivative:

$$\frac{dg}{dt} = N(q, \alpha) + F(t) \tag{2}$$

N is a non-linear function that governs the temporal changes that occur. This function depends on a *control parameter*, while $F(t)$ is a function that describes internal and external fluctuations. The dynamics of the whole system is governed by *order parameters* only. This means that if $\dot{q}$ describes the system at a micro-level, the high-dimensional equation could be reduced at a macro-level to equations of the ξ_u parameter only:

$$\dot{\xi}(t) = \tilde{N}(\xi_u) + \tilde{F} \tag{3}$$

where $\tilde{N}$ is a non-linear function, $\tilde{F}$ are internal or external fluctuations, and the dot · indicates the time derivative. Haken states that the reduction of the behaviour of this complex system is ruled by the "slaving principle". Near instability, the macroscopic behaviour of the system is characterised by only a few modes, which are enough for its description. When the control parameter change, the old status is replaced by a new one in the form:

$$q = q_0 + ve^{\lambda t} \tag{4}$$

Note that λ is an eigenvalue which can assume either positive or negative values. Therefore the solution for q from a starting point q_0 can be broken down into two parts. The first part is for positive eigenvalues which amplitude is the ξ_u order parameter. The second is for negative eigenvalues, i.e. the stable mode which amplitude is the ξ_s order parameter:

$$q(x,t) = q_0 + \sum_u \xi_u(t)v_u(x) + \sum_s \xi_s(t)v_s(x) \tag{5}$$

What happens in such a system is represented in figure 1. To begin with, when the control parameter is under the critical value, there are fluctuations in the system that cause a slight increase in the order parameter. This tends to relax towards a stable state (rest state in figure 1a). Therefore ξ_u tends towards 0, and according to the slaving principle ξ_s also tends towards 0 so that $q = q_0$. When the control parameter exceeds the critical value, the first state is replaced by two possible ones. Symmetry and bifurcation are broken in these two possible states, of which only one is chosen (see figure 1b).

A very important consequence of this is hysteresis. When the initial state is one of the two possible shown in figure 1b, there is a tendency to remain in the current state, which lasts longer

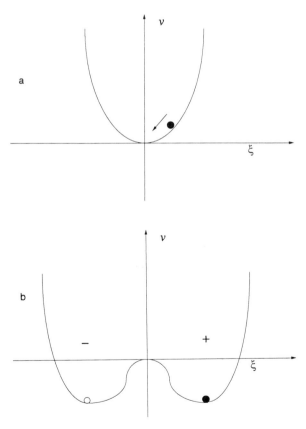

Figure 1. a) Behaviour of a complex system which relaxes towards a stable state; b) behaviour of the system when exceeding the critical value, with the appearance of two possible states.

than when the system relaxed for the first time. The application of this model of a *transitional phase of non-equilibrium* is quite straight forward both for behavioural problems and above all for perceptual problems. As Kruse and Stadler[13, 14] and Kanizsa and Luccio[12] pointed out, multistability has previously been a privileged field of research. More generally, this approach offers a very powerful means to interpret the processes of self-organisation, according to Köhler[15].

3. AUTO-ORGANISATION AND THE PERCEPTION OF MOVEMENT

Let us introduce the problem of self organisation in perception of movement with the well known example of the cycloid. If asked, anyone will describe what they see when looking at a rotating wheel very simply. The wheel performs a linear translation while its parts perform circular movements around the axis of the wheel itself. Indeed, only one part of the wheel, its hub goes on a "physical" path that corresponds to the phenomenal path. All the other parts perform motions that are different from what we see. To be able to see the actual motion, we need to isolate a single part from all the other parts, as Rubin[16] and Dunker[17] first observed. For instance, we can fix a small lamp or a phosphorescent dot somewhere on the wheel, and we can see it making a series of loops called cycloids. If we add a second light to the periphery of the wheel, phenomenally a rotary movement of each point around the other prevails[18]. This phenomenal result becomes coercive if we increase the number of illuminated points on the perimeter. In reality, each point traces a cycloid, but we see a phenomenal breakdown of the actual cycloid motion into a rotary and translatory component. The points rotate around an invisible centre and displace themselves together along another invisible plane. This has often been considered proof of the *tendency towards the Prägnanz*. A circular movement is "better" than a cycloid movement. In any case, the number of points is the crucial factor to be able to perceive the "good" motion.

This fact could be demonstrated in a very striking way in the following case (see Kanizsa et al.[19]). If three dots move along three circular paths which are partially overlapping (see figure 2) we are unable to see the actual paths, but rather an elastic triangle rotating and twisting in space. If we increase the number of dots on each path from one to five, the patterns are still invisible. In the area where the circular paths overlap, the dots form continuously transforming and disrupting groups. An observer can detect the circular movements only when there are more than six points on each path. Note that the observer is quite aware of the three distinct circular paths.

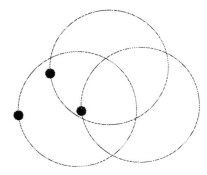

Figure 2. Set of three dots moving on circular paths partially overlapping.

More precisely, with one-two dots per path subjects (Ss) report that they see the vertices of one (or respectively two) virtual triangle that rotates on the screen. With three dots the dots move perceptually inward and outward, with reference to the centre of the figure. With 4-5 dots Ss see a chaotic springing up of the dots from the centre. From 6 dots on, the circular paths appear more and more clearly.

Cutting and Proffitt[18] have stressed the importance of the distinction between absolute, common, and relative motion. When it is very clear what absolute motion is [16,17,20,21], common motion is the apparent motion of the whole configuration relative to the observer, and relative motion is the apparent motion of each element relative to other configural ones. According to our results, in the situation with 1-2 dots there is a very clear prevalence of common motion, while with 6 or more dots, relative motion clearly prevails. Here we clearly see the dots moving along circular paths. These dots, however, are seen *relatively* to the other dots, forming three independent rotating circumferences.

Notice that in all conditions the dots occupy in each path the vertices of regular polygons inscribed in a circumference (equilateral triangle, square, regular pentagon, and so on). Notice also that when DS (the average distance of each dot from the other dots of the same path) is different, according to the class of the polygon, DO (the average distance of each dot from the dots of the other paths) is always the same, leaving the relative position of the three paths constant. If the centres of the paths are on the vertices of an equilateral triangle, as in our case, DO is always the same, with 3, 4, 5... 15 dots per path. (We have discovered this geometric regularity with some surprise *a posteriori*).

In the figures that we used in the experiment, DO was constant, while DSs were shorter than DO in conditions from 15 dots to 6 dots, 4.65 cm (5 dots), 4.83 cm (4 dots), longer with 3 or 2 dots, and about equal with 6 or 5 dots. So, we hypothesised that the difference between DO and DS could be a crucial factor for the prevalence of either common or relative motion. When DO is clearly longer than DS common motion prevails. In the opposite case, relative motion prevails. This hypothesis was tested in a second experiment, varying the distance between paths and leaving the number of dots per path constant. The results clearly supported this hypothesis. When DOs are less than DS, it is very difficult to detect the circular paths. A perceptual organisation in which the dots are phenomenally moving along these paths is possible only if the average distance between dots on a path is less than the average distance of each of them to the dots of the other paths.

Thus, the proximity of dots is a crucial factor in perceptual organisation of phenomenal motion. The relevant control parameter is the difference between the average distance from the dots of other paths (DO) and the average distance from the dots of the same path (DS). If DO is clearly less than DS the order parameter of common motion emerges and the system is in a stable attractor state. If DS is clearly less than DO the order parameter of relative motion emerges and the system is in a totally different stable attractor state.

3. MULTISTABLE DISPLAYS

Much of the same could be said about multistable patterns, like Rubin's faces or Necker's cube. In this case, we are in the situation of the figure 1b, where there are two attractors, corresponding to relatively stable states. A classical example of the use of multistable situations is in the research on stroboscopic alternatives carried out by von Schiller[22] sixty years ago. This example has been recently extensively re-investigated by Kanizsa and Luccio[12].

Von Schiller's analysis begins with the simple tachistoscopic device illustrated in figure 3. In the first phase he projects the two dots in A simultaneously on a screen, and then, after turning

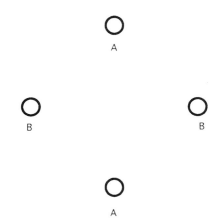

Figure 3. The tachistoscopic device concerning von Schiller's analysis.

them off, the two dots in B. The four stimuli are located at the vertexes of a square, thus the distance between each pair of contiguous vertexes is equal. In this case, the phenomenal displacements occur indifferently either clockwise or anticlockwise. However, the results become immediately clearer when we introduce a differentiating element, for instance a difference in colour. When in the first phase the upper dot is black and the lower is white and in the second phase the right dot is black and the left one is white, the prevailing motion seen will be clockwise. Similar results are obtained when, instead of chromatic equality, visual objects similar in shape, size, or brightness are used. In all these cases the alternative disappears definitively according to the principle of *similarity*. The tendency of visual objects to maintain their own identity and to persist in time without changing as far as possible their aspect, takes over the situation.

Von Schiller then contrasts the strength of many factors and found that similarity (of shape, size, colour, or brightness) is decisive in determining an alternative direction of motion. All these similarities succeed in counteracting the tendency to prefer the shorter path. He demonstrated that *colour* always prevails over *size*, and that *brightness* is more influential than either *colour* or *size* in determining the direction of the motion. Besides similarity and proximity, closeness and good continuation are other factors that affect the direction of the motion and that can emerge through stroboscopic alternatives. In conclusion, it can be said that factors favouring grouping in static situations also condition the successive belonging in kinematoscopic situations.

Further research carried out by P. Kolers[23], F. Hoeth[24, 25], H. Erke and H. Gräser[26], and V. Ramachandran[27] has essentially confirmed the results obtained by von Schiller. In cooperation with Kanizsa I have also performed two experiments. Our aim was not to find something particularly new; but rather to focus on two aspects of the results that we felt had been somewhat neglected by previous authors. The two aspects on which we focused our attention were: a) the very first motion the subjects see; b) the ease with which they can pass from one kind of motion to an alternative one. The first aspect is not particularly relevant in this context. The second one is: it is the problem of hysteresis.

In the first experiment, we contrasted stimuli for size, shape, and colour. The results substantially supported those obtained by von Schiller. In the second experiment we added the factor of proximity. The proximity factor turned out to be dramatically effective. For all Ss the very first motion perceived is towards the nearest point, and it proves very hard for them to voluntarily invert the direction of the motion; in fact, for about half the Ss this was almost impossible. Spontaneous inversions are less frequent and last a very short time before the Ss returned to the first motion seen. This is not the case, however, if we contrast proximity with directionality of the figure. Here the two alternative directions are equiprobable.

The results of the two experiments were hardly surprising, and essentially confirm all previous findings. What is important to stress is the dramatic difference between proximity and other factors like colour, shape, and size (with the exception of directionality, which, however, is a tertiary quality). Proximity proved to be an autochthonous factor, preceding all other factors in organising movement and highly resistant to *Einstellung* or other cognitive factors. The other factors prevail only when they are recognised. So, they require a more precise definition of what the order parameter needs to take in account such data. But the strong effect of hysteresis that we can detect is enough to indicate that this line of interpretation is highly promising.

We could similarly argue for another line of research on multistability, with the use of the figure-ground articulation. Also in this case we have many factors that we can contrast with each other: the *relative size* of the parts, their *topological* relationship and the kind of their *contours*. The area that is *smaller, included*, with *convex* contours, will tend to emerge as a figure. In figure 4 all these factors concur to favour the central zone as a figure, so the situation is neither ambiguous nor is there any spontaneous reversibility. When no single factor prevails over the others, a typical ambiguous situation occurs.

After the classical research by Rubin[28], other factors were discovered. In 1928 P. Bahnsen[29], a pupil of Rubin, pointed out the role played by *symmetry* in figure-ground articulation. A factor analogous to symmetry, identified and studied by S. Morinaga[30], is *Ebenbreite* or *constancy of width* of a region of the field. In 1975 Kanizsa[31] contrasted convexity with symmetry. It seemed to him that symmetry could be overcome or at least counterbalanced

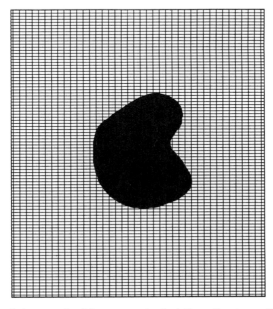

Figure 4. An example of figure-ground articulation without reversibility.

by a preference for a convex form. According to Biederman, Hilton, and Hummel[32], the union of symmetrical regions in a unitary ground should be favoured in Kanizsa's displays by the fact that collinearity of the top and bottom edges is stronger for the symmetrical sections than for the asymmetrical ones. When we interrupt collinearity of contours of symmetrical regions, these become more visible as figures.

Also in these cases, we must stress the great importance of the phenomenon of hysteresis, in switching from one configuration to another.

4. RHYTHMICAL PATTERNS

This approach is particularly interesting, because it allows us to consider in a unitary perspective data coming from processes that are apparently very different. Among them are the rhythmical patterns, both in generation of rhythmical patterns and in recognition of rhythmical sequences, as in (amongst others) psychology of music.

First, notice that almost every motor task requires an harmonisation of a multiplicity of actions. A classical example is walking: there is a great number of muscular groups involved, and almost the whole body is involved. However, both the walker and the observer do not notice this complexity. The cooperation of all muscular groups reduces itself to a unitarian, steady, fluent, overt motion, that suggests that this situation is similar to the ones considered above.

Notice that in movement the tendencies toward simplicity, regularity, and so forth are particularly apparent. Vogt[33] has carried out some elegant experiments on reproduction of movements, that clearly indicate this tendency toward Prägnanz (as singularity), that charac-terises the cognitive secondary processes[2, 8].

Let us examine some examples. We could ask a subject[34], to perform a rhythmical task with a finger, paced by a metronome, but with different phases or[35] with two arms with an increasing tempo again with different phases or to perform a bimanual task with the two hands out of phase[36], again paced by a metronome. We noted in all of these cases, a sharp and spontaneous phase transition from one motor behaviour to another. Referring to the last example, i.e. the prototype of this kind of research (see also Schöner and Kelso[37, 38], for a fuller theoretical treatment), the Ss had to perform a co-ordinated rhythmical bimanual task with a tempo increasingly faster. At a certain (critical) point we can observe a spontaneous change in the pattern of response. At first the subject could perform the task both in phase or in counter phase, subsequently only in counter phase. The problem is to determine the order parameter, the only variable on which the phase transition depends. In this case the order parameter is the phase φ. The time derivative of the phase is given by:

$$\dot{\varphi} = -\frac{dV}{d\varphi} + \sqrt{Q}\zeta \tag{6}$$

with

$$V(\varphi) = -a\cos(\varphi) - b\cos(2\varphi) \tag{7}$$

a and b are constant. ζ is a white noise with Q strength, which accounts for random fluctuations. When

$$\frac{a}{4b} < 1 \tag{8}$$

the system has two attractors, with a relative phase equal to 0 e π. When however there is

$$\frac{a}{4b} > 1 \tag{9}$$

then there is a single stable phase with φ equal to 0.

Kelso and Pandya[35] developed this model, changing from production to recognition of rhythmical patterns. In this light, the two processes are one thing only.

5. GLOBAL AND LOCAL FACTORS

As stated at the beginning, the phenomena world is usually a highly *stable* one. Therefore we can say that there is an autonomous tendency towards *stability*. A *tendency* towards Prägnanz in the perceptual field *does not exist*, when we mean that Prägnanz is a tendency towards singularity. Note that nearly any stimulus situation, although always *plurivoque* and therefore theoretically capable of giving rise to many phenomenological outcomes, tends to give a *perceptual unique outcome*. This is only with the exception of some very special multistability cases. Generally this is not towards the most singular solution, but rather towards the most stable one. This occurs despite by trial by trial variability in perceptual representations as stressed by Ashby and Lee[1]: trial-by-trial variability, but in the same time trial-by-trial stability of outcome.

Gestalt psychology has since long taught us that organisation factors are both numerous and often in antagonism with each other (proximity vs. closure vs. continuity of direction, etc.). Therefore, the more stable outcome must be that which corresponds to a maximum equilibrium between opposing factors. Of course it does not necessary mean that the resulting configurations also has figural "goodness". We can assume that the tendency towards stability coincides with the tendency towards Prägnanz only in special cases, in particular those cases in which there is only one factor. However, as Kanizsa and Luccio[8] point out, the more numerous the interacting factors are, the more rarely the stable solution coincides with the prägnant one. This does not mean that stability is simply the result of *local* factors. In this case, local factors should be more important than *global* factors, categorically contradicting a fundamental principal of Gestalt psychology, which I must say I consider to be still valid. If the segmentation of a perceptual field is due to the autonomous process (i.e. through a dynamic self-distribution of the interacting forces in the field), such segmentation must be the result of a global process, in which all elements interact with all other elements. However the effects of such organisation are different in different regions, and acquire a particular importance locally as interaction of elements vary in intensity as a function of the reciprocal relationships between the elements and the anysothropies of the field[39].

In my opinion, the interpretation of interactions between organisational factors in terms of order parameter is an elegant means to solve the complexities involved in so many interacting variables. In this synergetic perspective, the local vs. global controversy takes on a different meaning, and many apparent paradoxes become clear.

REFERENCES

1. F.G. Ashby and W.W. Lee, Perceptual variability as a fundamental axiom of perceptual science, in *Foundations of Perceptual Theory*, S. Masin ed., Elsevier, Amsterdam, NL, pp.369-399 (1993).
2. G. Kanizsa and R. Luccio, Die Doppeldeutigkeiten der Prägnanz, *Gestalt Theory*, Vol.8, pp.99-135 (1986).

3. G. Kanizsa and R. Luccio, Formation and categorization of visual objects: Höffding's never confuted bus always forgotten argument, *Gestalt Theory*, Vol.9, pp.111-127 (1987).

4. M. Wertheimer, Experimentelle Studien über das Sehen von Bewegung, *Zeitschrift für Psychologie*, Vol.62, pp.371-394 (1912).

5. M. Wertheimer, Untersuchungen zur Lehre von der Gestalt I, *Psychologische Forschung*, Vol.1, pp.47-58 (1922).

6. M. Wertheimer, Untersuchungen zur lehre der Gestalt II, *Psychologische Forschung*, Vol.4, pp.301-350 (1923).

7. E. Goldmeier, *The Memory Trace: Its Formation and Its Fate*, Lawrence Erlbaum Ass., Hillsdale, NJ (1982).

8. G. Kanizsa and R. Luccio, The Phenomenology of Autonomous Order Formation in Perception, in *Synergetics of Cognition*, H. Haken and M. Stadler eds., Springer, Frankfurt, D, pp.186-200 (1990).

9. H. Haken, *Synergetics: An Introduction*, Springer, Berlin, D (1983).

10. H. Haken, *Advanced Synergetics: Instability Hierarchies of Self-Organizing Systems and Device*, Springer, Berlin, D (1983).

11. M. Stadler and P. Kruse, The self-organization perspective in cognition research: historical remarks and new experimental approaches, in *Synergetics of Cognition*, H. Haken and M. Stadler eds., Springer, Frankfurt, D, pp.32-52 (1990).

12. G. Kanizsa and R. Luccio, Multistability as a research tool in experimental phenomenology, in *Perceptual Multistability and Semantic Ambiguity*, M. Stadler and P. Kruse eds., Springer, Berlin, D (in press).

13. P. Kruse and M. Stadler, Stability and instability in cognive systems: multistability, suggestion and psychosomatic interaction, in *Synergetics of Cognition*, H. Haken and M. Stadler eds., Springer, Frankfurt, D, pp.201-217 (1990).

14. P. Kruse and M. Stadler, The relevance of instability and nonequilibrium phase transition in the research on cognitive systems, in *Perceptual Multistability and Semantic Ambiguity*, P. Haken and M. Stadler eds., Springer, Berlin, D (in press).

15. W. Köhler, *Die Physischen Gestalten in Ruhe und im Stationren*, Zustand, Braunschweig, Vieweg (1920).

16. E. Rubin, Visuell wahrgenommene wirkliche Bewegungen, *Zeitschrift für Psychologie*, Vol.103, pp.384-392 (1927).

17. K. Duncker, Über induzierte Bewegung, *Psychologische Forschung*, Vol.12, pp.180-259 (1929).

18. J.E. Cutting and D.R. Proffitt, The minimum principle and the perception of absolute, common, and relative motion, *Psychological Journal*, pp.211-212 (1982).

19. G. Kanizsa, P. Kruse, R. Luccio, and M. Stadler, Conditions of visibility of actual paths, *Japanese Psychological research* (in press).

20. G. Johansson, *Configurations in Event Perception*, Almkvist Wilsell, Uppsala, S (1950).

21. G. Johansson, Visual perception of biological motion and a model for its analysis, *Perception and Psychopsysics*, Vol.14, pp.201-211 (1973).

22. P. von Schiller, Stroboskopische Alternativversuche, *Psychologische Forschung*, Vol.17, pp.179-214 (1933).

23. P. Kolers, The illusion of movement, *Scientific American*, Vol.211, pp.98-106 (1964).

24. F. Hoeth, *Gesetzlichkeit bei stroboskopischen Alternativbewegungen*, Kramer, Frankfurt, D (1966).

25. F. Hoeth, Bevorzugte Richtungen bei stroboskopischen Alternativbewegungen, *Psychologische Beiträge*, Vol.10, pp.494-527 (1968).

26. H. Erke and H. Graeser, Reversibility of perceived motion: selective adaption of the human visual system to speed, size and orientation, *Vision Research*, Vol.12, pp.69-87 (1972).

27. V. Ramachandran, The neurobiology of perception, *Perception*, Vol.14, pp.97-103 (1985).

28. E. Rubin, *Visuell wahrgenommene Figuren. Teil 1*, Gyldenalske, Kopenhagen, DK (1921).

29. P. Bahnsen, Symmetrie und Asymmetrie bei visuellen Wahrnehmungen, *Zeitschrift für Psychologie*, Vol.108, pp.129-154 (1928).

30. S. Morinaga, Beobachtungen über Grundlagen und Wirkungen anschaulich gleichmässiger Breite, in *Archiv für gesamte Psychologie*, Vol.110, pp.310-348 (1942).

31. G. Kanizsa, The role of regularity in perceptual organization, in *Studies in Perception*, G. Flores d'Arcais ed., Martello-Giunti, Firenze, I, pp.48-66 (1975).

32. I. Biederman, H.J. Hilton, and J.E. Hummel, Pattern goodness and pattern recognition, in *The Perception of Structure*, G.R. Lockhead and J.R. Pomerantz eds., APA, Washington DC, pp.73-96 (1991).

33. S. Vogt, Einige gestaltpsychologische Aspekte der zeitlichen Organisation zyklischer Bewegungsabläufe, *Bremer Beiträge zur Psychologie*, No.77 (1988).

34. J.A.S. Kelso, J. Del Colle, and G. Schöner, Action-perception as a pattern formation process, in *Attention and Performance XIII*, M. Jeannerod ed., Erbaum, Hillsdale, NJ, pp.139-169 (1990).

35. J.A.S. Kelso and A.S. Pandya, Dynamic pattern generation and recognition, in *Making Them Move*, N.I. Badler, B.A. Barsky, and D. Zelter eds., Morgan Kaufmann, San Mateo, CA, pp.171-190 (1989).

36. H. Haken, J.A.S. Kelso, and H. Bunz, A theoretical model of phase transitions in human hand movements, *Biological Cybernetics*, Vol.51, pp.347-356 (1985).

37. G. Schöner and J.A.S. Kelso, A synergetic theory of environmentally-specified and learned patterns of movement coordination. I. Relative phase dynamics, *Biological Cybernetics*, Vol.58, pp.71-80 (1988).

38. G. Schöner and J.A.S. Kelso, A synergetic theory of environmentally-specified and learned patterns of movement coordination. II. Component oscillator dynamics, *Biological Cybernetics*, Vol.58, pp.81-89 (1988).

39. M. Stadler, P.H. Richter, S. Pfaff, and P. Kruse, Attractors and perceptual field dynamics of homogeneus stimulus area, *Psychological Research*, Vol.53, pp.102-112 (1991).

SPATIAL REASONING AS A TOOL FOR SCENE GENERATION AND RECOGNITION

Giovanni Adorni

Dipartimento di Ingegneria dell'Informazione, Università di Parma
Viale delle Scienze, I - 43100 Parma, Italy

ABSTRACT

Although a great deal of effort has been put into the research and development of artificial intelligence reasoning systems, spatial reasoning is a relatively new independent research area. Up to now spatial reasoning problems have been considered in a variety of areas, including computer graphics, computer vision, robotics, geographical information systems, man-machine interaction, autonomous systems, and expert systems. Spatial reasoning involves spatial task planning, navigation planning for robots, representing and indexing large spatial databases, the integration of symbolic reasoning with geometrical constraints, and multisensor data fusion.

In this paper I focus on the aspects of spatial reasoning that are more closely related to high-level computer vision. More precisely, after a brief review of studies performed by psychologists of perception related to the field, I investigate the problems of: i) the description of objects and space modelling, ii) the representation of spatial relationships, iii) functional aspects of objects and naive reasoning.

1. VISUAL INFORMATION PROCESSING

In everyday life our visual information processing system processes complex arrays of three-dimensional (3-D) objects. We also have the ability to recognize complex scenes and object pictures from them. A picture is a two-dimensional (2-D) surface of limited size that represents or symbolizes 3-D objects that may actually be of quite a different size to their representation in the picture. Sigel[1] points out that pictures may differ along a number of dimensions and still adequately fulfil their symbolic purpose. For example, pictures of an object, such as a glass, may vary depending upon how much detail is present, and whether dimensional cues such as shading are present. Pictures that depict more than one object may vary in their organization of the objects, their complexity, the spatial relationships among the objects, and so on. Yet we identify the presence of a glass whether it is shown as a simple outline on a blank background or as one of many objects in a complex photograph of a kitchen.

That we can easily recognize the conceptual identity of pictures and their corresponding real objects implies a picture processing task. How is this task performed? What can we learn from our visual information processing system in order to build an artificial visual information processing system that performs well in solving tasks that are relatively simple for a human being?

One way to discover how pictures are processed is to examine how people move their eyes when they look at pictures. Yarbus' experiments[2] demonstrate that a very large number of fixations are made on areas of the pictures/parts of the objects that are "highly informative". For example, in scanning a picture of an animal or human face, subjects are most likely to fixate on the eyes, mouth, and nose. Yarbus' records of eye movement also indicate which visual items do not attract many fixations. First, fixations are not necessarily drawn to either the very darkest or the very brightest regions of the picture. Second, fixations are not necessarily drawn to the regions with greatest detail. Details receive relatively few fixations unless they are informative in the sense mentioned earlier. Moreover, contour does not seem to influence fixation patterns; that is subjects do not appear to follow the outlines of figures as they scan. Contours are scanned only in viewing pictures in which informative aspects lie along the contours. Yarbus also reports that scanning patterns are affected by the subject's purpose in looking at the picture.

Cooper[3] reports some experiments performed with subjects who listened to short prose stories that contained the names of several concrete objects (e.g., lion, zebra) and conceptual categories (e.g., safari, Africa). As they listened, subjects were allowed to freely view a display containing pictures of many of the objects mentioned in the story. Cooper found that when subjects heard the name of something pictured on the screen, their eyes quickly moved and fixated on the picture of the named object. They also fixated on pictures of things conceptually related to items in the stories. Mackworth and Morandi[4] show that the above definition of informativeness predicts fixations. They report that subjects make more fixations in regions containing contours (high informative content) than in regions composed largely of unbounded textures (low information content). Moreover, after picking out an important object or person in the picture, subjects explore the characteristics of this region before moving on (see also Antes[5], Berlyne[6], Loftus and Mackworth[7], on these topics).

Studies of eye movements reveal that most of the viewer's information comes from fixations on a few important objects or areas in the picture. The detection experiments performed by Biederman[8] show that for complex pictures, overall organization exerts a top-down influence even when the subject knows in advance what to look for and where to look for it. In standard viewing such prior expectations help a subject to detect the expected objects.

Biederman[9] observes that to comprehend a picture's content requires not only an accurate inventory of the objects present but also an understanding of the functional and spatial relations among them. Viewers must arrive at an overall representation, or schema, that serves to integrate all the separate aspects of the picture. Biederman[9] explains that "the schema specifies both the items appropriate to a given scene and the physical ... relations that should hold among them". In other words, a schema is a mental structuring of data that embodies necessary real-world constraints on the content and organization of the picture and thus contains expectations about what should appear in the picture. Biederman lists five important types of constraint imposed by picture organization: support, interposition, probability, position, and size (see figure 1).

The support constraint requires that objects be supported. For example, picture of a staple floating in the air violates this constraint. Requirements for support are somewhat contingent on the objects depicted; for example, birds and airplanes can float in the air without any visible means of support. Interposition requires that the contours of an object apparently behind another not be visible. The probability constraint posits that some objects are likely and others unlikely in any particular scene. For example, a fire hydrant in a dining room scene violates the

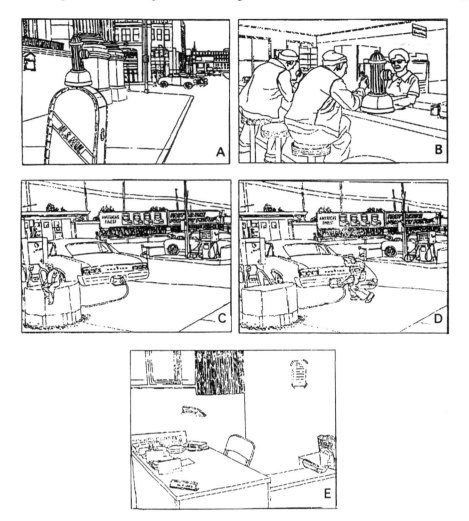

Figure 1. Examples of relational violations according to Biederman, Teitelbaum and Mezzanotte[10]: A) position violation; B) probability violation; C) size violation; D) interposition violation; E) support violation.

probability constraint. The position constraint is violated by a picture of a car on top of a skyscraper. The size constraint requires that objects appear in appropriate relative sizes. The last three constraints are mainly determined by the nature of the individual objects in the picture.

Also the organization imposed by the spatial relationships between objects in a picture affects the perceptual processing of more complex pictures. Picture memory experiments show that such organization affects memory as well as perceptual processing (see, for example, Mandler and Parker[11] and Mandler and Johnson[12]).

Mandler and Parker reason that there are at least three types of memory-relevant information a subject could extract from a picture, all of which might be influenced by the organization of the picture. These are i) an inventory of the main items in the picture, ii) the physical appearance of these items, and iii) the locations of the items, including the relative spatial relationships between them. Physical details are one form of pictorial organization: thematic content is another (see experiments performed by Bransford and Franks[13]).

An important question about picture memory is whether pictures are integrated into thematic wholes (as linguistic material is) or whether picture memory is specific enough for individual pictures to be remembered. Pezdek[14] studied this question, and the results of his experiments indicate that memory for pictures may be quite similar to memory for linguistic material. We apparently process the pictures for theme and gist, and we do not necessarily remember the exact visual details that are necessary to distinguish between conceptually similar pictures.

Up to now we have described several studies that show that the encoding of specific details is important for picture memory, but we have not considered whether such details are verbally labelled or are stored in another fashion.

Our everyday experience tells us that we need not apply labels to parts of pictures in order to remember them. When asked to recall the appearance of a very familiar object, we often form a "mental picture" or "mental image", from which we can obtain more specific information. Psychologists make a distinction between two qualitative different forms of representing information in memory (see, for example, Anderson[15]). In a *propositional representation*, information is stored in terms of propositions (predicates) about the relationship between one or more objects, or an object or a property. All this information could also be stored in a memory as an "imaginal" (or analog) representation. The mental image would like include many more visual details and characteristics than the few preceding propositions. An *imaginal representation* contains a great deal of detail, much of which cannot be easily expressed in propositions. We therefore can code concepts such as table in terms of both visual, sensory imagery and in terms of verbal, semantic characteristics (e.g., a piece of furniture, a noun, etc.). This distinction is the essence of the dual coding hypothesis[16]. Concrete nouns can be encoded into both a verbal and an imaginal memory system at the time of presentation and therefore, offer more opportunities for meaningful associations. Abstract nouns can only be encoded verbally.

The demonstrations by Atwood[17] and by Brooks[18] of selective, modality-specific interference for visual and verbal memory systems show that the information in the imaginal representation system can be disrupted by incoming visual stimulation. This finding suggests a strong similarity between the structural representations used in imagery and those used in visual perception. Shepard and his colleagues have developed a set of experimental findings that suggest a relationship between the two that Shepard[19] calls a second-order isomorphism. A simple isomorphic relationship implies that there is a direct, one-to-one correspondence between the parts of two entities. However, according to Shepard, "the proposed equivalence between perception and imagination implies a more abstract or "second-order" isomorphism in which the functional relations among objects as imagined must to some degree mirror the functional relations among those same objects as actually perceived[19]".

In the following sections I investigate how some of the previously discussed issues can be used in systems able to carry out reasoning on 3-D scenes, and how spatial reasoning can be a useful tool for object recognition.

2. OBJECT DESCRIPTION AND SPACE MODELING

A common approach to the problem of scene interpretation is to generate tentative hypothesis about the position and size of objects and try to use these expectations to guide the search for picture areas which exhibit the expected features. But where do these expectations come from? If a robot operates in a known environment, expectations can be self-generated on the basis of built-in knowledge and previously experienced situations.

Another very common source of information can be some kind of external input, often based on natural language communication. A piece of information like: "look for the pencil", "where?", "on the table", conveys a lot of information about the presence of a reference object (a table) and the characteristics of a surface (a table top), which must be located in order to restrict the search for the target-object (a pencil).

To take advantage of these linguistic information sources we must be able to extract from a qualitative expression like "on the table" all the quantitative constraints that are relevant from a geometric modelling point of view. These problems could seem much more related to the generation of visual analog representation than to understanding of a scene.

But what does *understanding* a scene mean exactly? When we analyse a scene, we use a lot of geometric and non geometric knowledge. We are not surprised to find cigarette butts in an ashtray and a glance is enough to classify them, but we could have some trouble recognizing that it contains goldfish, and this is surely not only due to geometric constraints! Therefore, as discussed in the previous paragraph, the processing of visual knowledge must be based on cognitive models that are able to handle different sources and kinds of information, and in this sense I feel that there is not a clear cut divide between scene analysis and scene generation. In the rest of this section, I mainly deal with the representation of objects, trying to point out how linguistic information can be related to visual information.

Knowledge of the structure of an object is often intimately related to our capability of understanding the meaning of a spatial relationship; for instance, the meaning of the sentence "the cat is under the car" is clear, even if it may depend on the state of the car, moving or parked; in contrast, the sentence "the cat is under the wall" is not clear, unless the wall has fallen down or it has a very particular shape. Moreover, in some cases only a rough definition of object dimensions may be sufficient, while in other cases more sophisticated knowledge about the structure of objects is required. If we consider, for example, the sentences "open the door" and "open the drawer", these sentences have different geometric meanings because the movements of doors and drawers usually follow different rules. In some cases, the recognition of a feature allows the generation of hypotheses about the presence of an object; in other cases, when we look for an object like, for example, a pencil, we do not start analysing the whole scene, but we first look for places in which or on which it is reasonable to find the object being searched for, like, for example, a table or some other piece of furniture.

From the previous discussion it follows that all object modeling techniques must at least deal with the following issues[20]:
• objects must be described in several levels of detail;
• the articulation of a movable object part must be properly described;
• characteristic features of objects must be pointed out;
• typical relationships between objects must be described.

Some artificial intelligence researchers suggest organizing the knowledge about objects, facts or concepts using special structures called "frames" and "scripts". *Frames*, based on the concepts of schema introduced in the first section, were originally proposed by Minsky[21] as a basis for understanding visual perception, natural language dialogues, and other complex behaviours. *Scripts*, frame-like structures specifically designed for representing sequences of events, have been developed by Schank and Abelson[22] and their colleagues. Both refer to methods of organizing knowledge representation in a way that directs attention and facilitates recall and inference.

In the following I use a frame-like structure, called THOUGHT[23], to describe objects, in which solid objects are described by means of "generalized cones"[24] at several levels of detail.

Cones can be interconnected by means of fixed or movable points, with arbitrary constraints on rotation and shifting. Specific joint elements are defined to properly describe the surface of an articulated object.

A THOUGHT is a frame like structure within which new data is interpreted in terms of concepts acquired throughout previous experience[22]. A THOUGHT supports propositional representation as well as imaginal representation through a TYPE definition. A type determines a set of operations applicable to thoughts. It is possible have the following predefinited types:
- DESCRIPTIVE: defines the complete description of a physical, abstract, animate or not, object, in terms of Conceptual Dependency theory[25];
- PROTOTYPE: defines the structural part of a physical object;
- JOINT: defines the connection element among physical objects with the goal of building more complex objects or scenes;
- SPATIALREL: defines spatial relationships among objects.

A descriptive type can contain every type of thought; a prototype type can contain joint or prototype thoughts; a joint type can contain only joint thoughts and a spatialrel type can contain only spatialrel thoughts.

The structural part of an object is described by a prototype thought, in terms of generalized cones[24] at several levels of detail. A *generalized cone* is defined as a simple, monolithic object defined by a Cartesian triple, a parametric description of the boundary of a generic section of the cone and a law which gives the variations of the section along the axes. To describe the cone axis and the cone boundary it is possible to connect "lines" and "arcs of a circle". The element of connection is completely defined by its spatial angles ϕ, φ, ω, which give the standard position of the connected parts. See Adorni[26] for more details on the syntax of thoughts.

An example of using these types of thought is given in figure 2, in which the body of the prototype-chair is described in terms of ((prototype joint prototype) and (prototype joint prototype) and (....)), where prototypes and joints are described according to the syntax informally introduced above.

A descriptive thought is specified by its body, which contains some fields. Every field gathers information on a feature or a property of the object. Physical features and relations with other object have been pointed out. The fields are:
◊ LEVELS field contains a hierarchical description of the object shape at different levels of detail.
◊ DESCR field contains all the basic non specialized knowledge about the object.
◊ USE field contains the description of the most common activities which involve the utilization of the object. These descriptions can be simple conceptualizations or pointers to script like knowledge representation.
◊ POSITION field gives the most common spatial relations between the described object and other objects in standard scenes. For instance, an object can be a part of another object, or can be commonly found inside another object.
◊ SUPPORT field is used to point out the capability of the object to give a support to another object in standard situations. For instance, a chair usually supports a human being.
◊ CONTENT field is analogous to support field, and describes the standard use of the object as a container for other objects.
◊ COLOR and MADE fields describe some relevant features like the possible set of colours and materials, while the WEIGHT field contains an expectancy about the range of possible weights.
◊ DYNAMIC field contains the current expectations about the size of the object; it can be dynamically updated every time a new object of the same class enters the knowledge data base of the system.

Any of these fields can be empty. Referring to the example in figure 2, a trivial definition of a chair is given in figure 3, where the full description of the body of the thought is shown.

CHAIR_1 is prototype
    thought
        ((LEG_1 CONNECT_1 SEAT_1)
        (LEG_1 CONNECT_2 SEAT_1)
        (LEG_1 CONNECT_3 SEAT_1)
        (LEG_1 CONNECT_4 SEAT_1)
        (BACK_1 CONNECT_5 SEAT_1))
    end CHAIR_1.

BACK_1 is prototype
    though
        axes_is   line (40.0)
        low_is   constant
        boundary_is rectangle (5.0 40.0)
    end BACK_1.

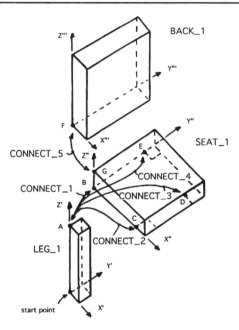

Figure 2. Example of a chair.

The discussion about the spatialrel type involves the analysis and the representation of the spatial relationships. The representation of knowledge and the analysis of the spatial relationships requires the choice of a suitable set of primitive concepts. All the linguistic relationships like "above", "under", "behind", and so on, can be transformed into quantitative geometrical relationships among the coordinates of some points of the objects involved.

```
CHAIR is descriptive
  thought
    descr  (isa chair furniture)
    levels ((parallelepiped_1)
       (chair_1))
    use    ((0.8 (human_being on chair))
       (0.3 (cat on chair)))
    position ((0.9 (chair on floor))
             (0.4 (chair in garden))
             (0.6 (chair near table)))
    support ((0.9 (human_being on chair))
            (0.4 (cat on chair)))
    color  ((0.9 light_brown)
           (0.4 red))
    made_of ((0.9 wood))
    weight (3.5)
    dynamic
       dn  (30)
       dmax (0.5 0.5 1.0)
       dmin (0.3 0.3 0.8)
       dsigma (0.02 0.02 0.01)
  end CHAIR.
```

Figure 3. Definition of the body of the chair-thought.

While the analysis and formalization of spatial relationships is discussed in the next section, in the following I discuss several issues regarding the kind of information necessary for an exhaustive description of spatialrel thought.

From a computational point of view, reducing spatial prepositions to quantitative metric relations seems to hold some promise since it can offer simple and universal inference tools. On the assumption that the use of prepositions could be reduced to merely geometric considerations, the answer to any question about the relative positions of two objects could be obtained by means of simple computations that only involve coordinates[27].

From a psycho-linguistic point of view, there is evidence of privileged directions and planes. For example, an axis oriented according to the direction of the force of gravity appears to be a good reference for concepts like "upper" and "lower"; similarly, "in front" and "behind" usually refer to a vertical plane containing the axis of symmetry of the object, and to a direction normal to that plane[28]. On this basis, Clark[29] supports the idea of using a Cartesian reference frame locating a horizontal plane at ground level, a vertical direction normal to that plane (things above the ground can be perceived more easily) and lying on a second plane which discriminates between "in front" and "behind" for a person in a vertical position, and a horizontal positive direction towards the observer (things "in front" of other things can be perceived more easily than those "behind" them).

Analysis of prepositions emphasizes the existence of other types of reference systems. In the sentence "far from the table", the directional aspect is not relevant; in fact, the same concept could be better expressed by means of spherical coordinates for which the radius value is independent of the angular values. This would also allow us to manage situations in which angular variations are important, regardless of distances (as in the sentence "turn right") in some other cases, we are interested in the height of an object but not in its horizontal position, like, for example, in "the plane is flying at 10000 meters".

A cylindrical coordinate system can better emphasize the axis of symmetry of most objects[30] and this is very useful in describing the structure of the objects themselves.

Finally, there may be situations in which we refer to the position of an object on a trajectory; the trajectory is not very significant in itself (although it cannot be approximated by a straight line), since emphasis is placed on the distance between two points belonging to the trajectory (for example, "the accident occurred two kilometers after the bridge", "after three miles, he was again at the starting-point"). In this case, it is necessary to introduce a curvilinear coordinate that can be represented by functions which can be specified only if the trajectory is completely known; this coordinate can be undefined and kept separate from the coordinates of the basic reference system.

Addressing is never absolute, whatever the actual reference system may be. Usually, the object of a preposition is used as local origin for a particular relation: in "the book on the table", the table is the reference for the book. In some cases, the reference is associated with an implicit observer; in "the bridge behind the tree", the preposition "behind" is referred to an observing direction which cannot be deduced directly from the tree. A human being is able to switch from one system to another, during a conversation, without problems; the same ability should be exhibited by an artificial system, which would exploit, for example, multiple representations and related conversion rules derived from analytic geometry.

Although the perceived environment is generally 3-D, in many cases prepositions refer to one or two-dimensional aspects of the objects. In other cases, spatial interpretations are given for intrinsically two-dimensional objects.

In some cases, the *reference dimensionality* depends on the preposition, which emphasizes some particular aspects of the objects. An example is given by the preposition "on", which usually refers to a functionally significant surface of the object involved in the relation, as in "the book on the table", and "the fly on the wall". Similarly, "along" recalls one-dimensional aspects in both i) objects with a prevalent dimension ("along the river") and ii) structured objects for which attention is focused on boundary lines ("along the frame of the window") or on other significant lines ("along the ridge of the hill"). However, in other cases, the reference dimensionality depends more on the object than on the preposition, because the prototypical interpretation is altered.

Let us consider, for example, the preposition "in". Its basic meaning describes a relation between volumes, in which the subject is partially or entirely contained in the object. Nevertheless, in sentences like "the tree in the garden" or "the caravan in the desert", it is not clear what is the volume defined by the object of the relation. In these cases, the object seems to be used as a conceptual environment rather than a geometric volume, and the inclusion of the preposition "in" is a 2-D projection on a horizontal plane, namely, a "map". A track of the 3-D value is left because the 2-D inclusion remains valid under some constraints on the third dimension; for example, we can say "the bird in the garden", even if the bird is not physically touching any of the objects in the garden, provided that it is not flying too high.

The constraint on the third dimension is sometimes related quite explicitly to the dimensions of the objects usually belonging to the environment (an object that is over the tree-tops is "over the forest" and not "in the forest"); it may be related to scale factors: an airplane can fly "along the borders" of a region.

Another example of 2-D inclusion is given by vertical projections. In the sentence "the man in the doorway", the depth of the doorway is not important; what we mean is that the vertical projection of the man is inside the vertical projection of the doorway. However, also in this case, the three-dimensionality, peculiar to the preposition "in" is still respected, since the sentence is only valid if a vertical plane intersecting both objects exists; in some cases, this three-dimensionality is no longer present, as in "the target in the sight".

Opposite situations, are those in which 2-D objects are interpreted as 3-D ones. We usually say "the field in the picture" or "the image in the mirror", because the picture and the mirror produce perspective images conveying the idea of the third dimension. If the subject of the relation forms no part of the image, the physical two-dimensionality gets the upper hand; as a consequence, we will say "the car in the picture" if the object is part of the image, but "the car is over the picture" if the sentence relates to the picture as a physical object.

The knowledge representation model previously introduced has been designed to support, as much as possible, the inferential activities necessary to build a mental image of a scene as it may be imagined by the reader or the listener of a story. These inferential activities are heavily based on default assumptions, the shapes, structures, physical characteristics, locations, positions of objects, on several standard sequences of actions during which objects are used, on basic or current goals of actors in the story and so on. An example is given by the following sentence:

"Giovanni entered the living room and opened the drawer"

the word "living room" sets the basic environment; when the system reads this word, it creates an instance of living room that inherits from the living room DESCRIPTIVE thought a set of default assumptions about the shape (from the LEVELS field) and the objects contained (from the CONTENT field) in standard positions. Default assumptions must be done flexibly. For instance, the CONTENT field of the living room thought can contain a very long list of objects which, in the experience of the system, can be found in a living room; however only a small subset is probably used to build the standard image of a living room, the choice depending on the likelihood of the presence of each object. Such a likelihood is a value which is associated to each object in the description and is dynamically modified by each new experience made by the system. The objects listed in the CONTENT field of the living room thought may have, in turn, a not empty content field, which is used to create new objects and so on.

Let us now turn to the second part of the sentence. When the drawer is mentioned, the system looks for some relation between the drawer and the living room in the POSITION field of the drawer thought. If the system has no experience about drawers scattered in a living room, no information can be extracted either from the POSITION field of the drawer or from the CONTENT field of the living room. However, in the POSITION field, the drawer is described as a part-of a piece of furniture, with a very high probability; hence, the system will look for such a piece of furniture in the scene. If no instance of this type exists, the presence of a new object is inferred, balancing the probability of being contained in a living room with its probability of having a drawer. Otherwise, if an instance of such a piece of furniture has been already created, its actual structure is checked; in fact, even if its PROTOTYPE does not exclude a drawer, the current instance may have not it. In this case, a "minimum number of objects" principle is followed, in the sense that the system prefers to modify the assumed structure of that existing instance of a piece of furniture rather than creating a second object. This modification of the structure of the existing piece of furniture implies the capability of deleting a chain of default assumptions, because the presence of the drawer can no longer be compatible with the presence of other objects inside that piece of furniture, some probability in the CONTENT field of the PROTOTYPE can be different if a drawer is assumed and so on. Another example can be given by the following sentence:

"Sitting in the armchair, Giovanni called George by telephone".

In this sentence, the relative position of the armchair and the telephone are not explicitly stated. However the system infers the "near" relationship using its knowledge about the current position of Giovanni, the structure of the body of a human being (arm length, articulations and so on) and the activities involved in a standard use of the telephone. A detailed description of these activities is contained in a frame-like knowledge structure connected to the USE field of the telephone DESCRIPTIVE thought. This description includes, for instance, the action of grasping the receiver, which in turns implies a physical contact between a part of the telephone and the actor's hand. The search in the USE field is triggered by a number of events like, for instance, the information, clearly stated or even only supposed, that an object is an instrument for some action. Another triggering event is an effect for which the system must create a causing object. In the sentence:

"Giovanni switched on the light in the living room"

a change in the state of the room is asserted; among the objects listed in the CONTENT field of the living room a chandelier is characterized by a USE field in which such a change of state of the containing environment is described. Therefore the system will infer the presence of a chandelier in the living room.

3. THE MEANING OF SPATIAL PREPOSITION

SPATIALREL thoughts are highly procedural, because a number of conditions must be verified or assumed to be true even if the relation is used in its basic, 3-D sense[31]. For instance, if "A on B", the system has to verify that the objects A and B are not known to be in a position that is incompatible with such a relation (for instance the two objects are far apart); then, checking the physical structure, the size and the weight of A and B, the possibility for B to support A is tested. After the relation has been checked, it is translated into geometrical relations among the points of A and B.

Another reason for the complexity of SPATIALREL thoughts is that the interpretation of the spatial relationships requires a large amount of different types of knowledge, so that it can be regarded as a challenging problem for cognitive modeling. Obviously, the first domain involved is linguistics, since even the simplest form of description, such as,

<object> <preposition> <object>

may give rise to a surprisingly high number of ambiguities and context dependent interpretations. Some basic works in this field are those of Bierwisch[32], Cooper[33], Goguen[34], Boggess[31], Waltz and Boggess[35], Waltz[36, 37], Adorni, Boccalatte and DiManzo[27], Adorni, DiManzo and Giunchiglia[38, 39], DiManzo, Adorni, and Giunchiglia[40], Herskovits[41].

While the problem of the reference system and dimensionality has been discussed in the previous section, in the following section I analyse the meaning of spatial prepositions and related reasoning activities.

When somebody is told that "A is on B", he probably thinks of the object A placed on top of the object B, in physical contact with it and supported by it against the force of gravity. Relations like "the book on the table" or "the chair on the floor" fit this interpretation, which can be regarded as the *prototypical meaning* of the preposition "on"[31]. However, it is easy to find deviations from such a core meaning. For example, in the sentence "the picture on the wall", the vertical constraint is relaxed, while in "the town on the sea" (this sentence has been literally translated from Italian, to stress that, in some languages, ambiguity of spatial relationships is even more emphasized), no supporting action is suggested, and the interpretation is closer to the concept of proximity. Hence, the choice of most prepositions and the selection of their meanings in the current context depend on several factors, which are related to the objects involved or to the general context of the speech[41].

Many geometrical, physical, and functional features of the objects may affect the interpretation of a preposition. For example, in the sentence "the pebble in the shoe" and "the nail in the shoe", the preposition "in" suggests two different images: in the first case, the likeliest interpretation is that the pebble is inside the shoe, while, in the second case, a listener probably prefers to think of a nail embedded in the shoe. The reason for the second interpretation is that the listener has some special knowledge about nails and shoes; he knows that one of the basic functions of a nail is to be embedded in solid objects to connect different parts together, and this function is actually performed in shoes. In the case of a pebble, this functional property does not hold, and therefore the prototypical interpretation of "in" (i.e., 3-D containment) is preferred.

Sometimes the choice of a preposition seems to be due to purely conventional reasons; we usually say "in the car" and "on the bus", using two different prepositions to indicate the same spatial relation between a customer and two different types of means of transport. *Salience* is a typical example of *contextual factor*. As regards figure 4a, we can say that "the small book is on the table", even if it is on top of the stack of books; on the other hand, in figure 4b, "the lid is on the jar" and not "on the table" because the relation between the lid and the jar is salient. This means that some objects may be *transparent* to some prepositions, but this property depends on the current context; it cannot be generalized as a *transitivity* property of a preposition[31].

Another basic contextual factor is *relevance*. When the preposition "in" is used to describe 3-D containment, the contained object is expected to be inside the container and supported by it. However, it is not normally useful to distinguish between the situations described in figure 4c and 4d; then we can say that in figure 4d the ball is in the basket, even if the 3-D enclosure is violated. On the other hand, the ball in figure 4e is not in the basket because lack of direct or indirect support makes relevant the relative spatial positions.

A characteristic of most spatial relations is *indeterminacy*. There is some tolerance in the use of prepositions, and the limits of this tolerance depend on the objects involved. Therefore, the appropriateness of a relation between two objects may depend on the absence of other objects in more favourable positions.

It should be clear from the previous discussion that the use of prepositions involves a large amount of knowledge, specific to many different domains. The most feasible approach to carrying out automatic interpretation seems to lie in the definition of a proper set of non ambiguous spatial primitives. The mapping of prepositions into primitives can be based on a large taxonomy of objects which should try to encode at least the role of the characteristics of objects by the assignment of meanings.

The following examples derive from an analysis of the Italian use of the preposition "in", which is very close to the English one[40].

The preposition "in" differentiates the objects involved in a relation (subject and object) on the basis of the considered volume (concave, full, ...). For this purpose, objects have been

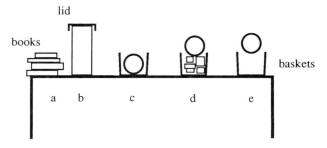

Figure 4. Example of a simple scene.

divided into the following types:

O1:   objects with a cavity able to contain other objects (room, drawer,...);

O2:   objects without internal holes;

O3:   geographical objects (lake, mountain, ...);

O4:   fluids;

O5:   solid objects that can contain other objects and assume the shapes of what is contained (sack, ...).

The classes defined for the subject are similar to those of the object (S1, ... S5), with only one exception: objects like hole, cavity, vein, crack, form a new class (S6), necessary to point out the structural degradation they produce in an object.

By using the above classification, it is possible to represent, for example, the sentences: "hole in the box", "wine in the glass". The first sentence is conceptualized as:

<subject_S.6> <in_punch> <object_O.1>

that is the subject modifies the object's structure, producing an empty space which depends on the subject's structure. The second sentence is conceptualized as:

<subject_S.2-S.5> <in_cont> <object_O.1>

that is, the subject is contained in the object.

When a speaker describes a scene, he usually avoids boring the listener with a number of details which are not important for the story. However, most omitted features are necessary to give consistency to the *scene imaged* by the listener and therefore they must be inferred. For instance, when someone hears of "a branch above the roof", probably he does not think of a branch magically hanging over a floating roof: a more likely image is "a tree near a house", with a branch above the roof. But this seemingly trivial instantiation involves a surprising amount of sophisticated reasoning.

First of all, the listener must use his knowledge about the physical structure of objects to realize that roofs and branches are typically parts of other objects, namely, houses and trees. The reference to a branch and a roof and the involved relation between them allow the listener to think of an outdoor environment, where a tree suggests something like a garden, due to the listener's knowledge about the usual living conditions of trees.

The expected position of a tree in a garden is upright and in physical contact with the ground. From a structural description of a prototypical tree and house, a possible position for the branch and the roof can be derived, which is consistent with the stated relation and the listener's knowledge of typical relations between houses and trees (e.g., tree near house). Hence, the previously resulting image can be accepted as a probable completion of the scene sketched in the input sentence. Even if this example may seem relatively simple, the inference path is quite long and involves several kinds of knowledge. See DiManzo, Adorni, Giunchiglia[38] for a more detailed discussion on these topics, and on the treatment of default knowledge.

We deal now with the *instantiation problem*, that is, the problem of activating expectations about unmentioned objects. Some rules for instantiating objects are suggested in Adorni, DiManzo, Giunchiglia[38]. To some extent, the effect of all these rules is to replace the current objects and/or relations, with new objects and/or relations. An example of this type of replacement rules (a slightly simplified version of the formula discussed in Adorni, DiManzo, Giunchiglia[39]) is:

ON(A,B) and GEO(B) and OPENCONTAINER(B) and not GEO(A) and not FLYING(A)

$\Rightarrow$ ON(A,water) and CONTAINED(water,B).

This rule states that if the input sentence is "A on B", and B is a geographical object (i.e., GEO(B)) and a container with an open top (i.e., OPENCONTAINER(B)), as defined by Hayes[42], while A is not a geographical object and cannot fly (i.e., not FLYING(A)), then it is reasonable to assume that B is full of water, and the relation ON holds between the object A and the water itself.

In fact, geographical open-containers are objects like rivers, lakes, seas, and so on; so if someone says "the boat on the lake", the proper placement of the boat in the imaged scene depends on its capability to be supported by a suitable amount of water. Therefore, the relation ON(A,water) (instead of ON(A,lake)) is the right conceptualization.

Rules of this type operate on a *local context*, that is, they are activated by some features of the objects involved in a single spatial relationship. The activating features are typically related to the physics, structure, and functional characteristics of objects. A first case is the sub-part relation. In the example at the beginning of this section, the two mentioned objects, namely, the branch and the roof, are sub-parts of a tree and a house, respectively. There are strong functional connections between branches and trees and between roofs and houses, and therefore it is reasonable to assume that, if a sub-part is mentioned, the whole object exists.

A sub-part relation can be established between two objects every time their functional connection is strong enough to cause reasonable expectations. For instance, a lock can exist without a door, but it must be connected to a door in order to perform its typical functions; hence, a reference to a lock allows us to imagine a door, if there are no specific reasons for rejecting this hypothesis. Physical properties are the second major source of expectations. For example, supported liquids can be inside a container or spread on a surface. Also, those objects which are assemblies of a large number of small, unconnected parts (e.g., sugar, salt, and so on) often require a suitable container.

When the customer in a restaurant asks for some wine, a very common inference is to imagine the waiter bringing him a bottle which contains the requested wine. So, several types of objects generate a strong expectation for a container, whose kind depends on the current story and environment. This fact can be described in the knowledge base by linking the "fluid" object to a proper set of containers by means of a CONTAINED_IN relation description.

Another meaningful case is provided by the *environmental conditions* for living organisms. When somebody says "a plant in the living-room", a listener probably realizes that a plant, to be in good physical condition, needs some earth, which in turn requires a pot. It is worth noting that this instantiation is based on the assumption of *good physical state*, which can be rejected if the stated relation evokes a specific environment, as in "the fish in the fishmonger's shop". Several *local instantiation rules* of this type can be formalized, but, to use them properly, a number of constraints must be fulfilled.

A first relevant problem is the *relation inheritance*. Given two objects A and B, linked by a spatial relation REL (i.e., A rel B), and assuming that A causes the instantiation of a third object C, the relation REL may or not hold between C and B, depending on the involved objects, the relation, and the adopted instantiation rule. For example, if A is a liquid and C is the related container, the relation inheritance is strongly expected, and, with the exception of a few special cases, it is independent of the actual relation. So, "water on the table" leads to "water in a bottle" and "bottle on the table", and "sugar in the drawer" should be interpreted as "sugar in the sugar-basin" and " sugar-basin in the drawer". In other cases, different rules may be followed when the sub-part mechanism is invoked, the spatial relation describes support or containment, and the relation inheritance is not the privileged rule. So, "the branch on the roof" does not mean "the tree on the roof", and the relation "tree near roof" is to be preferred.

If the inheritance does not hold, and a new relation between two (inferred) objects must be defined, the choice depends basically on our knowledge about the default positions of the objects in the current environment; the initially stated relation can only restrict the set of acceptable defaults. For instance, if one hears of "the paw in the drawer", a cat and a table can be inferred. Since "the cat in the drawer" is not the likeliest interpretation, it is necessary to use default knowledge about the possible positions of a cat in a room. Assuming that a cat can stay "on the floor" or "on the table", the first choice, namely, "on the floor", for geometrical reasons (e.g., typical dimensions of tables and typical positions of drawers) seems to prevent the cat from putting a leg inside the drawer; therefore, the cat is probably "on the table" and "near the drawer".

The choice of relations between inferred and stated objects, by inheritance or by default considerations, is a crucial point, since it is a preliminary step in checking, from a geometrical point of view, the consistency of the resulting scene; impossible or unlikely geometry is a good reason for rejecting otherwise feasible object instantiations. Consider, for example, the sentence "the branch under the car". Even if it can be conceived that, due to an accident, a car has climbed a tree, default knowledge about the geometry of cars suggests that the mentioned branch should be very close to the ground. However, this position is not in good accordance with the known geometry of trees; on the contrary, broken branches are expected to lie on the ground. Hence, the instantiation of a tree must (probably) be rejected for geometrical reasons.

When the presence of an object is suggested by some instantiation rule, its inferred position must be compared with its default positions in the current environment. In each specific environment, objects may or may not be in default positions, giving rise to a wide range of degrees of expectation.

As far as simple spatial relations are concerned, the most important underlying physical phenomenon is gravity, and the basic consequence of gravity is that every object must be properly supported by other objects or by the ground, otherwise it falls down. Hence, a *stability expert* must be invoked whenever a supporting relation is inferred.

The analysis of stability may also result in the instantiation of some specific object. For example, in the analysis of the sentence "the picture on the wall", a nail or a hook are inferred to give support to the picture; the picture can be hung up and a nail or hook can be partially embedded in the wall, so the whole scene is consistent.

When functional links cause the instantiation of unmentioned objects, no contextual or inferred evidence for an improper use of the stated subpart is found. In fact, it is reasonable to imagine a tree when someone says "a bird on the branch", but it is no longer reasonable to instantiate a tree when we hear "Giovanni hits George with a branch".

The definition of the inner functional characteristics of objects and the reasoning about their actual utilization is discussed in the next section.

4. FUNCTIONAL ASPECTS AND NAIVE REASONING

Recognition is basically a matching process, where the 3-D structures inferred from the visual data are compared to a set of prototypes, each describing a class of objects. If the set of prototypes incorporates only structural (i.e. geometric) knowledge, that is, they try to describe *what objects look like*, it is very difficult to cope with all the possible shapes of the objects belonging to the same class, without losing too much discrimination power. Therefore, in this section, recognition is approached from a functional point of view, defining a framework within which the class prototypes can describe *what objects are for*.

When we address the problem of object representations, the question immediately arises whether a description based on purely geometrical features can be powerful enough to capture the inner semantics of the observed objects. Despite the large popularity of descriptions based on primitive volumes, like, for example, in ACRONYM[43], or surfaces, like, for example, in IMAGINE [44], when we try to use these kinds of representations to model even a simple object, like a chair, we quickly get into trouble. Consider, for instance, the set of objects depicted in figure 5.

We describe each of these objects by means of a list of primitive components plus a set of interconnecting relations, all the resulting descriptions are clearly different from each other. However there are some similarities, so that, given some proper restrictions on the overall

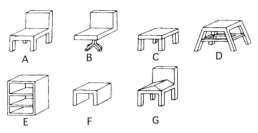

Figure 5. A set of objects with some similarities: A), B) chairs; C), D), F) stools; E) shelf; G) unknown object.

dimensions, nobody should fail in identifying "a" and "b" as two instances of a chair, "c", "d" and "f" as stools, "e" as shelving and "g" as an unknown object which resembles a chair but which could not be easily used this way.

This kind of clustering can also be performed on the basis of the previously suggested descriptions, by relaxing the structural constraints on the matching conditions and putting a threshold on the total allowed mismatching. However, it is difficult to tune the matching procedure so that the obtained clusters correspond to our common sense classifications. In fact, depending on the assumed weights and thresholds, we can obtain, for example, the clusters:

$$\{a, c, d, g\}, \{b\}, \{e, f\} \text{ or } \{a, b, c, g\}, \{d\}, \{e\}, \{f\}.$$

The reason for this difficulty of having truly satisfactory clustering is that common sense classifications are often based on some kind of judgment on the apparent functionality of the observed objects, and, unfortunately, there is no simple way of relating functionality to structure. Very often a lot of even complicated structural details are only due to stylistic reasons, or perform specific tasks, like increasing robustness, (see object "d" versus object "c") or permit rotation (see object "b"), which only have a secondary influence on the object's main functional goal.

What knowledge is then necessary to check the *chairness* of an observed object or, in other words, how can we give a satisfactory intentional representation of it?

If someone is asked to describe a typical chair, a likely and reasonable answer is that a chair is a seat plus a back plus one or more legs. Although this answer can be interpreted as a rough description of the shape it really means that a chair is the assembly of three components having distinct purposes: the first, (seat), is designed to give support to a sitting human body, the second (back) to stop the back-bending of the trunk, and the third one (legs) to allow the chair to maintain a stable position on a supporting surface (typically the floor). Therefore, such a description is an attempt to give the concept of chairness by decomposing it into a set of more *primitive functions*. The idea of defining what objects are for by means of functional decomposition is appealing for several reasons:

• it makes a formal definition of the semantics of each primitive easier, i.e. the set of geometrical constraints the primitive imposes on the object shape. On this basis a set of specialized experts can be designed, each one devoted to check whether an arbitrary 3-D structure is suited to perform a specific function;

• the possible interactions between primitives and the related restrictions can be clearly understood;

• a Functional Description Language can be defined to build object classes as networks of primitives[45];

• descriptions can be used as models to carry out the recognition process; models may include explicit definitions of strategies for handling incomplete matching and default conditions.

The question is now whether it is possible to describe a reasonable set of objects using a limited number of primitives. Looking at many manufactured objects in everyday use, some concepts seem to be repeated. Many objects are designed to support something, such as chairs, tables, shelves and so on. A special case of supporting is hooking, which requires very peculiar shapes, so it is convenient to treat it as a separate function. The purpose of a large class of objects is to contain something, such as wardrobes, drawers or bags. Sometimes they are specifically designed to contain liquids, pots, bottles, glasses, wash-basins; as in the case of hooking, liquid containment imposes some special constraints on shape, and can be considered to be a specific function.

Allowable movements can influence the structure of objects. Some objects must rotate on fixed supports, like wheels, doors or taps; sometimes an object part must rotate relative to another object part, such as the seat and the legs of some types of chairs. Sliding can also constrain shapes, as happens with drawers, push-buttons or keys.

A special case is embedding, which involves the capability of penetrating solid objects, like nails or forks. Sometimes the functional goal is to block movements; for example, the back of a chair is suited to block the bending of the trunk of a sitting human body. Often the main function of objects is to close up some opening, as is the case with doors, windows, lids or caps. Very frequent requirements are graspability and liftability, since many objects are designed to be handled in some way. Hence tens of objects seem to be suited for descriptions based on a small working set of primitives.

Within functional descriptions, the role of geometric knowledge is to relate (naive) physical concepts to observed shapes. Therefore, most of the available geometric knowledge is hidden in the primitive experts, mentioned above. Sometimes, however, a purely functional specification of certain structural details would require an enormous amount of very sophisticated physical knowledge: on the other hand, the performed function may impose very restrictive constraints on the shape of the component involved. In this case, it is much easier and more efficient to maintain some geometrical relations explicitly at the description level. So, some of the primitives, which will be introduced in what follows, do not implement functional concepts, but simply describe geometric restrictions.

We can represent an object through a detailed description of its functions rather than a classification of its shape: in particular we say that the function of an object is the result of concurrent elementary sub functions.

$$\text{obj1} \Leftarrow a \Rightarrow \text{FUNCTION} \Leftarrow o \Rightarrow \text{obj2}$$

In this simple graph, "obj1" and "obj2" are the entities involved in the specified sub function and, more precisely, obj1 is the "actor", while obj2 is the "object". This means that "obj1 (the actor) performs the declared function on obj2 (the object)". Now we consider a chair: its main feature is to support a sitting human body; then we could write:

$$\text{chair} \Leftarrow a \Rightarrow \text{SUPPORT} \Leftarrow o \Rightarrow \text{sitting\_human\_body}$$

This graph is too general because it does not specify which part of the chair is used to support a sitting human body: in fact, we mean that "there exists a part of the chair able to support a sitting human". To evidentiate this situation, we have to introduce the concept of "variable", the purpose of which is to represent an unspecified part of an object. Variables are also useful when we have to link particular functions to unnamed parts of an object.

$$Y \Leftarrow a \Rightarrow \text{SUPPORT} \Leftarrow o \Rightarrow \text{objects}$$

This one means: "in the object under examination there is a part devoted to supporting a generic kind of object". We can find such a graph in the descriptions of a table, a chair (where "objects" will be replaced by "sitting human body"), a shelf, and so on.

Each functional primitive characterizes a single part of an object, and all primitives concur to perform the typical activity of the object itself. There are some cases in which it is useful to represent explicit relations between two or more parts of an object. As mentioned in the previous section, spatial relationships are used to describe relations between objects (or parts of them) and the external world. A bottle has to stand on a table: this means that it has a flat base that can lie "on" a plane; the seat of a chair has to stay at a proper height from the floor, that is, the seat has to stay "over" a plane. This kind of information may be represented as follows:

$$obj1 \Leftarrow a \Rightarrow SPATIAL\_RELATION \Leftarrow o \Rightarrow obj2$$

the meaning of which is: "something maintains a particular relation with something else", and either obj1 or obj2 belongs to the external world. We note that all functional, spatial and structural relations have the same syntactic form: the intrinsic semantics are decisive to distinguish them. The use of a spatial relation allows us to characterize an object's capability of maintaining a specific spatial relationship with another object while performing some primitive function. This information is a sort of boundary condition describing the interaction between an object and its typical environment.

From a general point of view, we can say that a generic function is expressed by some restrictions on proper elementary sub functions (such restrictions are described through structural and spatial relations). For example, we know that "the seat of a chair has to stay at a proper height from the floor"; we can represent that with the graph:

$$seat \Leftarrow a \Rightarrow OVER \Leftarrow o \Rightarrow plane$$

With reference with this last example, a question is raised: how is it possible? In the previous sentence, nothing specifies the way in which a seat is maintained over a plane: if we want to introduce this information, we have to modify our graph by adding a new semantic case (called "instrument case") according to the following rule:

$$obj1 \Leftarrow a \Rightarrow (FUNCTIONAL, STRUCTURAL, SPATIAL RELATION) \Leftarrow o \Rightarrow obj2$$
$$\Uparrow$$
$$i$$
$$\Downarrow$$
$$obj3$$

This means: "obj1 maintains the specified relation with obj2 by means obj3".

Other semantic cases may be introduced in order to declare some modifications or some restrictions on the meaning implied by the graph itself. If we say: "a large quantity of fluid flows through a hole", the graph becomes:

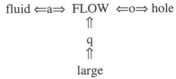

where the arc "q" expresses a "quantity specification". The labelled arc "q" (and each arc "modifiers" one) is related to the functional specification rather than the referred object because we mean that all modifications and restrictions perform a sort of "parametrization" of the semantics of the function involved. So if we wanted to say "a drawer slides in a certain direction", we would use this graph:

$$drawer \Leftarrow a \Rightarrow SLIDE \Leftarrow o \Rightarrow guide$$
$$\Uparrow$$
$$d$$
$$\Uparrow$$
$$Z$$

where the directive arc "d" specifies the direction in which the action is performed. In the previous graphs, we need to know how to specify the relations between the parts or properties of an object and the object itself: we also need to declare (if necessary) what a variable is. We introduce some "declarative arcs", e.g., the "part_of" arc, the "is" arc, and the "orientation" arc. The general form of the "part_of" arc is:

$$obj1 \; \text{—part\_of} \rightarrow obj2$$

and it means that obj1 is a specified part of obj2. Similarly, we represent the "orientation_of" arc as follows:

$$direction \; \text{—orientation\_of} \rightarrow obj$$

and it declares that the obj has that particular direction. We have introduced the concept of variable to represent an unspecified part of an object: to name a particular variable, we use the "is" arc. Its general form is:

$$var \; \text{—is} \rightarrow obj1$$

All the elements previously introduced can be used to represent the functionality of an object, like, for example, the chair described in the following example of a "CHAIR":

◊ A part of the chair, seat, able to support a sitting human body exists.

$$\text{sitting\_human\_body} \Leftarrow a \Rightarrow \text{SUPPORT} \Leftarrow o \Rightarrow X \text{ — is} \rightarrow \text{seat}$$

◊ The "seat surface" has to stay at a proper height over a plane parallel to it; it means that a proper element exists, and this element is "the legs".

$$parallel(plane)$$
$$\Downarrow$$
$$d$$
$$\Downarrow$$
$$\text{seat\_surface} \Leftarrow a \Rightarrow \text{OVER} \Leftarrow o \Rightarrow \text{plane}$$
$$\Uparrow$$
$$i$$
$$\Downarrow$$
$$Y \text{ — is} \rightarrow \text{legs}$$

◊ An element able to block the rotation of a human (sitting on the seat surface) in a certain direction exists: it is "the back".

$$any$$
$$\Downarrow$$
$$d$$
$$\Downarrow$$
$$\text{sitting\_human\_body} \Leftarrow a \Rightarrow \text{BLOCK\_ROTATION} \Leftarrow o \Rightarrow Z \text{ — is} \rightarrow \text{back}$$
$$|$$
$$is\_on$$
$$\downarrow$$
$$\text{seat\_surface}$$

◊ The whole chair must be stable on a plane.

$$perpendicular(plane)$$
$$\Downarrow$$
$$d$$
$$\Downarrow$$
$$\text{chair} \Leftarrow a \Rightarrow \text{ON} \Leftarrow o \Rightarrow \text{plane}$$
$$\Uparrow$$
$$(i)$$
$$\downarrow$$
$$Y \text{ — is} \rightarrow \text{support}$$

◊ Seat, back, legs are parts of the chair.

$$\text{back} \quad —part\_of—_{\downarrow}$$
$$\text{seat} \quad —part\_of\rightarrow \text{ chair}$$
$$\text{legs} \quad —part\_of—^{\uparrow}$$

The functionality of an object can then be used for object recognition, by comparison with the structure under examination. This fact implies that each sub function must be satisfied by at least one (or more) part of "the candidate form", and that a proper set of rules has to be linked to each functional (or structural or spatial) relation to express its intrinsic semantics.

All recognition activity can be performed by a system of cooperating *experts*, one for each relation. It becomes very important to define at that point what information the various experts have to exchange and how they can synchronize each other.

When we analyse a "candidate structure chair", we can start by seeing if there is an element able to support a sitting human body: that will be the "seat". Then we check on other functionality, and finally we can also recognize a chair specifying its constituent elements and their properties.

Each expert works in a similar way and when it finds a structure that is in accordance with its own request, it finishes successfully and its main result is the classification of a part of the involved object.

This fact may be expressed on the graph by introducing the "result" arc: it shows what products are coming from each expert. A "result arc" links a relation to an object or an "is_arc". We label the "result arc" ending in an "is_arc" as "R" when more than one arc comes out of a relation and we want to clearly evidence the most important one.

Let us consider, now, a chair: the results coming from each expert are represented as follows:

$$\underset{\Uparrow \qquad\qquad\qquad\qquad\qquad\qquad \Downarrow}{\overline{\qquad\qquad = R = \qquad\qquad}}$$

sitting_human_body ⇐a⇒ SUPPORT ⇐o⇒ X — is→ seat
                          ⇓
                          r
                          ⇓
                    seat_surface

                    parallel(plane)
                          ⇓
                          d
                          ⇓
seat_surface ⇐a⇒ OVER ⇐o⇒ plane
              ⇑   ⇓
              i   R
              ⇓   ⇓
                 Y — is→ legs

                    any
                     ⇓
                     d
                     ⇓   ⎯⎯⎯ R ⎯⎯⎯
                                              ⇑   ⇓
sitting_human_body ⇐a⇒ BLOCK_ROTATION ⇐o⇒ Z — is→ back
                          |
                       is_on
                          ↓
                    seat_surface

As seen before, object recognition activity is performed by a set of cooperating "active elements" (called experts and denoted by uppercase letters) that express all the elementary sub-functions of the object itself. The main problem is to provide proper rules to synchronize the various experts' activities, and subsequently we need to put such rules into the graph: in fact, we would also like to obtain a description able to guide the activity of each expert. A complete functional description of an object is an oriented, connected graph with one set of nodes called "experts" and another set of nodes called "objects". Figure 6 shows the oriented, connected graph for a chair.

We define "external object node" (and we put it in quotes) as any node that identifies an element not belonging to the structure under examination. The external objects nodes represent the elements of the external world that interact with the object. Each object node may be marked

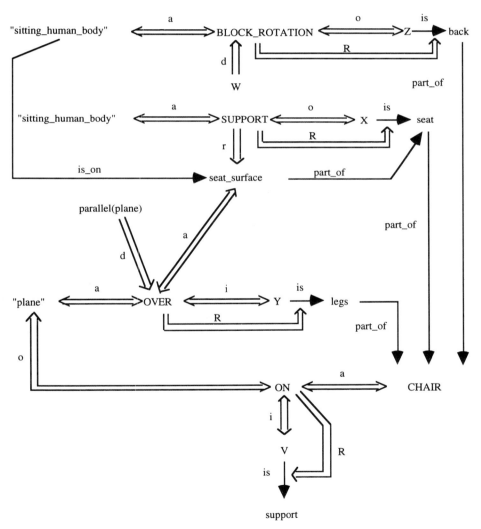

Figure 6. Oriented, connected graph for a chair.

or unmarked: at the beginning all the "external nodes" (not linked to any "is_arc" type) and all the variable nodes are marked while the others are unmarked.

A "state" is associated with each expert: the allowed states are: blocked, active, failed and successful. An expert is "blocked" if at least one of its neighbours is unmarked. An expert is "active" when all its neighbours are marked. When an expert becomes active, it begins its activity: we say that it is active while it is working to check its own functionality. An expert is "failed" when it finishes its activity without satisfying its conditions. An expert is "successful" when it finishes its activity after finding proper "structures" that satisfy its conditions. When an expert passes from the active to the successful state, all the nodes at the end of result arcs, which originate from it, become marked. As a result, we are able to activate other experts until a complete recognition is performed.

We note that, when a variable is involved, the arc "R" ends in a "is_arc": in this case, the object at the end of the "is_arc" becomes marked.

Let us now consider some examples of "recognition activities" based on the above representation. If we want to recognize a chair (see the graph of figure 6), at the beginning, only the SUPPORT expert is active (all its neighbours {Sitting_human_body, X} are marked), while all the others are blocked. In fact the unmarked node "seat_surface" blocks the OVER and the BLOCK_ROTATION experts (through the "is_on" arc): the ON expert is blocked by the unmarked node "chair". When the SUPPORT expert becomes successful, the nodes "seat", "seat_surface" (and consequently, the node "sitting_human_body at the end of the "is_on" arc) become marked, and the OVER and BLOCK_ROTATION experts become active and work in parallel. The ON expert is still blocked: in fact, the nodes "back" and legs" are unmarked. The general rule involving part_of arcs is that "a node at the end of more than one "part_of" arc becomes marked when "all" the nodes at the beginning of such arcs are marked". When both the OVER and the BLOCK_ROTATION experts become successful, "back" and "legs" become marked and the ON expert is allowed to begin its activity. Recognition activity ends when all experts are successful. Each expert could work asynchronously: the marks on the objects are essential for the synchronization of the activity of the various experts[45].

As an example let us now analyse an expert: the support expert. A complete description of the experts introduced in the previous example, can be found in Di Manzo et al.[46]. The support expert aims to verify whether an object may be placed on a pattern of supports so that stability is ensured. The positioning strategies are analysed with the following restrictions:
• objects must have a plane base, which includes the vertical projection of the center of gravity;
• supports must have a plane horizontal top surface; in case of several supports, all their top surfaces must lie on the same horizontal plane;
• only straight edges are handled.

With these restrictions, the stability analysis can be carried out in 2-D space, looking at the shapes of the object base and of the support-top surface(s). The basic concept is that of "supporting area", that is the minimum convex contour including all the intersections between the object and the pattern of supports, like that shown in figure 7.

The minimum requirement for stability is that the (projection of the) object center of gravity is contained in the supporting area; however, when the supporting area is too small in respect to the overall object dimensions, our feeling is that the stability is low. Therefore each object has an its own "minimum supporting area" which must be fully contained in the supporting area, in order to have satisfactory positioning. This requirement may result in the search for more complex positions, possibly involving a number of supports; without such a requirement, the naive strategy of putting the object center of gravity inside the contour of any supporting surface could be always adopted. The last general requirement is that the free space upon the supporting surface is large enough to contain the object.

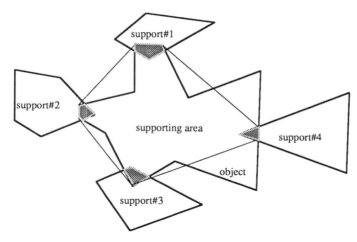

Figure 7. Supporting area.

When the object is very small in respect to the supporting surface, we must simply look for a piece of it large enough to fully contain the object. Hence problems should only arise when the object and the support dimensions are comparable. In such a case, if the contour of the supporting surface is convex, the simple strategy of superimposing the object and the support centers of gravity can be adopted. However, with concave contours, this approach may not work, as shown in figure 8; in this case, the supporting surface must be decomposed into a set of convex pieces, looking for the most convenient one. Note that holes or concavities which are small with respect to the object dimensions can be ignored (see figure 9).

Let us define a couple of perpendicular straight lines x and y, centered on the center of gravity of an object, so that they divide it into four parts, as shown in figure 10. A simple sufficient condition for stability is that each part of the object touches at least one support, since this ensures that the center of gravity is contained in the supporting area. This condition corresponds to the naive strategy of placing a rectangle on two supports so that all the four corners are supported.

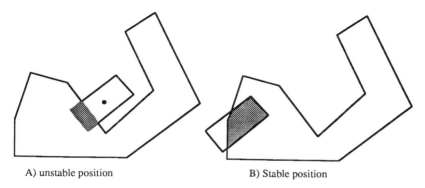

A) unstable position   B) Stable position

Figure 8. Concave support.

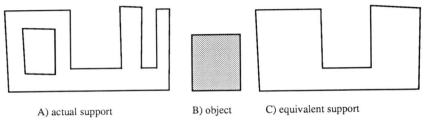

A) actual support   B) object   C) equivalent support

Figure 9. Ignorable holes and concavities.

This strategy leaves two degrees of rotational freedom: the first comes from the orientation of the pair of axes with respect to the object, which results in an arbitrary definition of the object parts; the second comes from the possibility of rotating the object axes relative to the pattern of supports. To reduce this rotational freedom, we can adopt the intuitive strategy of placing the object as "perpendicular" as possible to the pair of supports (see DiManzo et al.[46], for a detailed description of these strategies).

In the rest of this section we shall leave the object shape rather unrestricted. Specific features will be defined whenever they are needed; let us discuss here only some widely used parameters.

We define as UP_RIGHT_TANG the straight line tangent to the UPPER_RIGHT part of the object and parallel to the x axes; similarly we can define the UP_LEFT_TANG, LOW_RIGHT_TANG and LOW_LEFT_TANG, as shown in figure 11. The smallest distance between an UP tangent and a LOW tangent defines the INNER_LENGTH of the object, while the largest distance defines the OUTER_LENGTH (see figure 11). The basic restriction we put on the object shape is the "side continuity", i.e. both the right and the left object parts, considered as separate objects, must be continuous. This condition is satisfied by the object in figure 12a, but is not by the object in figure 12b.

Let us summarize the previous definitions and discussions by listing some useful predicates:

OBJECT(o)   o is a object;
ALIGNED(o)   the x axis of o is parallel to the support reference line;
BARI(p,o)   the point p is the center of gravity of o;

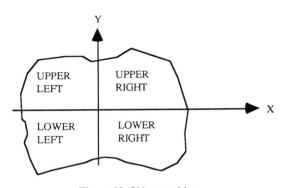

Figure 10. Object partition.

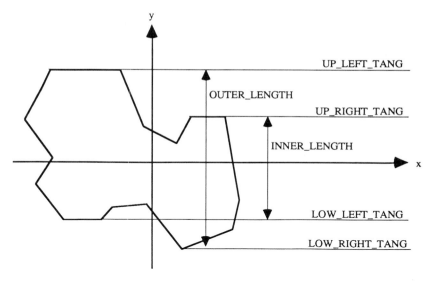

Figure 11. Object parameters.

UPPER_RIGHT(o',o)   o' is the uppermost right part of o;
UPPER_LEFT(o',o)   o' is the uppermost left part of o;
LOWER_RIGHT(o',o)   o' is the lowermost right part of o;
LOWER_LEFT(o',o)   o' is the lowermost left part of o;
UP_RIGHT_TANG(t,o)   t is the uppermost right tangent to o;
UP_LEFT_TANG(t,o)   t is the uppermost left tangent to o;
LOW_RIGHT_TANG(t,o)   t is the lowermost right tangent to o;

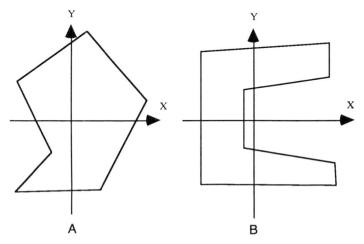

Figure 12. Object side continuity.

LOW_LEFT_TANG(t,o)	t is the lowermost left tangent to o;
INNER_LENGTH((l,o)	l is the inner length of o;
OUTER_LENGTH(l,o)	l is the outer length of o;
L_HIGHER(r,s)	r and s are straight lines parallel to the x axis; the y coordinate value of r is greater than the s one;
P_HIGHER(p,s)	p is a point and s a straight line parallel to the x axis; the y coordinate value of p is greater than the s one;
P_LOWER(p,s)	p is a point and s is a straight line parallel to the x axis; the y coordinate value of p is smaller than the s one.

Note that in the following discussions we shall assume a frame of reference aligned with the x and y axes associated to the object. A support belongs to this class when it consists of two parallel rectangles whose x_length is greater than the x_length of the (aligned) object.

With this type of support we can define the following parameters (see figure 13):

IPR(s)	s is a couple of infinite parallel rectangles;
UP_OUTER_LINE(t,s)	t is the top side of the uppermost support;
UP_INNER_LINE(t,s)	t is the bottom side of the uppermost support;
DOWN_INNER_LINE(t,s)	t is the top side of the lowermost support;
DOWN_OUTER_LINE(t,s)	t is the bottom side of the lowermost support.

A simple sufficient condition for stability can be defined as follows:

$Q\_PR\_SUP(o,s) \Leftarrow IPR(s)$ and $OBJECT(o)$ and $ALIGNED(o)$
    and [$UP\_RIGHT\_TANG(t_1,o)$ and $UP\_LEFT\_TANG(t_2,o)$ and
        $UP\_INNER\_LINE(r_1,s)$ and $L\_HIGHER(t_1,r_1)$ and $L\_HIGHER(t_2,r_1)$]
    and [$LOW\_RIGHT\_TANG(t_3,o)$ and $LOW\_LEFT\_TANG(t_4,o)$ and
        $DOWN\_INNER\_LINE(r_2,s)$ and $L\_HIGHER(r_2,t_3)$ and $L\_HIGHER(r_2,t_4)$]
    and [$BARI(p,o)$ and $UP\_OUTER\_LINE(r_3,s)$ and
        $DOWN\_OUTER\_LINE(r_4,s)$ and $P\_LOWER(p,r_3)$ and $P\_HIGHER(p,r_4)$]

In the previous notes I have discussed some aspects of object positioning (for a complete discussion see Di Manzo et al[46]). When the object has been positioned so that stability is ensured, the 3-D structure of the support must be examined in order to verify whether: i) all its parts are connected, ii) its structure is strong enough to bear the (apparent) object weight. The analysis of connections is rather trivial, even if it may involve some inferences about the possible continuation of those parts that are partially occluded by other parts. The second point is more difficult, since it requires some knowledge about the propagation of forces and moments in rigid structures. An initial examination of this problem has been made in Adorni[47], where causal analysis proposed by De Kleer and Brown[48] has been applied to study the elastic behaviour of plane frames in static conditions.

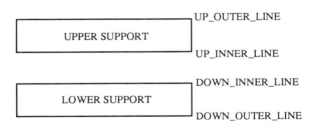

Figure 13. Parallel rectangles parameters.

5. CONCLUDING REMARKS

Let's now turn back to the representation of chairness, and modify the thought of figure 3 in the following way:

...

use ((0.8 (chair horizontal_support sitting_human_body))

...

position ((0.9 (floor horizontal_support chair))

...

The restriction of the supporting section is obviously introduced to distinguish a chair from a hook which can give support by hanging; the introduction of functional conditions "position" asserts that a chair is usually not floating in air, so its structure must be able to ensure a stable position on a planar horizontal surface when it is properly supporting a human body.

This definition is too vague, because it can be clearly matched by a number of objects that are not chairs. Further restrictions on structure and shapes can be introduced to distinguish a chair from, for instance, a cube (that is to distinguish the sittingness from the chairness), with a proper definition of the prototype thought of the chair. So structural and geometric features still play a basic role in building descriptions, but only those that are really distinguishing must be included, so that unspecified details do not cause mismatching when an observed object is compared with a "prototypical" description.

Restrictions can be functional as well. In figure 5 chair b can be put into a separate class of chairs if a prototypical frame exists where the "component: leg" has, for example, a functional restriction: vertical_axis_rotation; this restriction must be interpreted by a suitable theory of rotation which infers the relational properties of any given geometric structure.

Given a set of more or less sophisticated functional descriptions of objects, we need a strategy to match the observed patterns of connected volumes to these prototypes. The first problem is where to look for a specific object in the image of the scene. For any searching procedure to be cost effective, must take advantage of some a-priori knowledge of the current environment, which can raise expectations about the most likely positions of objects. So, a search for a chair in a room should look first for objects on the floor, with proper constraints on the overall dimensions, and secondly for objects on tables or other pieces of furniture where there could be reasons to put chairs. Knowledge about the most likely relations between objects is embedded in the object frames and is used to build indirect searching procedures. For example, if the goal is to find a lamp, the first step of the search consists in looking for a table or for some other object on which the lamp frame says that a lamp may be. Each object in this chained search defines a local context for the next step.

Object ordering depends on "identifiability", which could be simply related to their overall dimensions or to distinguishing geometric features such as large horizontal or vertical surfaces. The criterion of first looking for expected "big" objects also works when the goal is not to locate a specific object but to build a general description of the current scene. In this case the selection of the prototypes to match with each observed object is guided by the content of the frame structure describing the prototype of the assumed environment; when an object is recognized, this can raise new expectations because of the relations that link it to other objects in the current contexts and so on. Only those objects that are placed far from their typical positions, such as a chair hanging from the ceiling, or which have no typical positions, like a fly, will escape this phase of analysis and require a further extensive matching.

The second relevant problem is the matching criterion. A set of apparently connected volumes could really be a unique object. In such cases a prototypical description could be

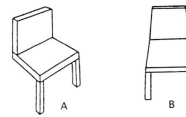

Figure 14. Example of chairs in upright position, with two visible legs: A) formulation of the hypothesis of a rigid connection with the floor; B) formulation of the hypthesis of the existence of a more complete set of legs.

matched by a subset of the observed volumes. This situation results in a hypothesis of instantiation which can be rejected later if it is not possible to find a consistent interpretation for the unmatched volumes. Inconsistency can be caused not only by failure to find a proper prototype, but also by the impossibility of justifying the actual object position using the knowledge the system has of the equilibrium rules or other physical rules.

On the other hand, some relevant details may be completely missed because of occlusions or perspective. To handle this situation the matching requirements must be suitably relaxed. Let us consider, for instance, the image in figure 14b. Since we only see two legs, the functional description of the leg component of the chair frame is not matched. However the object does stand on the floor, and the system knowledge of the equilibrium rules could not justify this fact without assuming the existence of a more complete set of legs: so, the existence of two occluded back legs is assumed, for reasons of symmetry , and this hypothesis can be accepted since it is consistent with the current point of view. For the same reason this assumption were rejected in the case of figure 14a, and the system would formulate a different hypothesis, as that of a rigid connection with the floor, to justify the object's upright position.

Even if many ideas discussed in this paper must be considered working hypotheses, and even if the surface of most problems addressed have only scratched, I claim that spatial and common sense reasoning about facts that form part of our every day experience is a necessary tool to improve object description, hypothesis generation, consistency checking, and object identification.

It is tempting, at this point, to list some major areas that need fuller investigation, such as environment setting, shape constraints from functionality, relations between top-down and bottom-up expectancy generation and so on. However, the whole area of cognitive modelling is still largely open, and a great effort is necessary in this field, since vision is basically a cognitive process sharing its knowledge domain with other basic human activities like language interpretation and action planning. I feel that a significant improvement in the performance of artificial vision systems will only be achieved through a thorough investigation of this cognitive background.

REFERENCES

1. I.E. Sigel, The development of pictorial comprehension, in *Visual Learning, Thinking, and Communication*, B.S. Randhawa and W.E. Coffman eds., Academic Press, New York, NY (1978).
2. A.L. Yarbus, *Eyes movements and vision*, Plenum Press, New York, NY (1967).
3. R.M. Cooper, The control of the eye fixation by the meaning of spoken language: a new methodology for the real-time investigation of speech perception, memory, and language processing, *Cognitive Psychology*, Vol.6, pp.84-107 (1974).

4. N.H. Mackworth and A.J. Morandi, The gaze selects informative details within pictures, *Perception and Psychophysics,* Vol.2, pp.547-551 (1967).

5. J.R. Antes, The time course of picture viewing, *Journal of Experimental Psychology,* Vol.103, pp.62-70 (1974).

6. D.E. Berlyne, The influence of complexity and novelty in visual figures on orienting responses, *Journal of Experimental Psychology,* Vol.55, pp.289-296 (1958).

7. G.R. Loftus and N.H. Mackworth, Cognitive determinants of fixation location during picture viewing, *Journal of Experimental Psychology: Human Perception and Performance,* Vol.4, pp.565-572 (1978).

8. I. Biederman, Perceiving real-world scenes, *Science,* Vol.177, pp.77-80 (1972).

9. I. Biederman, On the semantics of a glance at a scene, in *Perceptual Organization,* M. Kubovy and J.R. Pomerantz eds., Lawrence Erlbaum Associates, Hillsdale, NJ, p.215 (1981).

10. I. Biederman, R.C. Teitelbaum, and R.J. Mezzanotte, Scene perception: a failure to find a benefit from prior expectancy or familiarity, *Journal of Experimental Psychology: Learning, Memory, and Cognition,* Vol.9, No.2, pp.411-429 (1983).

11. J.M. Mandler and R.E. Parker, Memory for descriptive and spatial information in complex pictures, *Journal of Experimental Psychology: Human Learning and Memory,* Vol.2, pp.38-48 (1976).

12. J.M. Mandler and N.S. Johnson, Some of the thousand words a picture is worth, *Journal of Experimental Psychology: Human Learning and Memory,* Vol.2, pp.529-540 (1976).

13. J.D. Bransford and J.J. Franks, The abstraction of linguistic ideas, *Cognitive Psychology,* Vol.2, pp.331-350 (1971).

14. K. Pezdek, Recognition memory for related pictures, *Memory and Cognition,* Vol.6, pp.64-69 (1978).

15. J.R. Anderson, Arguments concerning representations for mental imagery, *Psychological Review,* Vol.85, pp.249-277 (1978).

16. A. Pavio, *Imagery and verbal processes,* Holt, Rinehart and Winston, New York, NY (1971).

17. G. Atwood, An experimental study of visual imagination and memory, *Cognitive Psychology,* Vol.2, pp.290-299 (1971).

18. L.R. Brooks, Spatial and verbal components of the act of recall, *Canadian Journal of Psychology,* Vol.22, pp.349-368 (1968).

19. R.N. Shepard, The mental image, *American Psychologist,* Vol.33, pp.125-137 (1978).

20. G. Adorni, A. Boccalatte, and M. DiManzo, Object representation and spatial knowledge: an insight into the problem of men-robot communication, *Proc. 7th. Conference of the Canadian Man-Computer Communication Society,* Waterloo, CDN (1981).

21. M. Minsky, A framework for representing knowledge, in *The psychology of computer vision,* P.H. Winston ed., McGraw-Hill, New York, NY (1975).

22. R.C. Schank and R.P. Abelson, *Scripts, Plans, Goals, and Understanding,* Lawrence Erlbaum, Hillsdale, NJ (1977).

23. G. Adorni and M. DiManzo, Top-down approach to scene interpretation, *Proc. Convencion de Informatica Latina,* Barcelona, E, pp.591-605 (1983).

24. D. Marr and H.K. Nishihara, Representation and recognition of the spatial organization of 3-D shapes, *Proc. Royal Soc. Lond.B.,* pp.269-294 (1978).

25. R.C. Schank ed., *Conceptual information processing,* North Holland, Amsterdam, NL (1975).

26. G. Adorni, Some notes on a cognitive model for scene description, *Technical Report,* Istituto di Elettrotecnica, Università di Genova (1982).

27. G. Adorni, A. Boccalatte, and M. DiManzo, Cognitive models for computer vision, *Proc. 9th. COLING,* Prague, pp.7-12 (1982).

28. B. Kuipers, Modeling spatial knowledge, *Cognitive Science,* Vol.2, pp.129-153 (1978).

29. H. Clark, Space, time, semantics, and the child, in *Cognitive Development and the Acquisition of Language,* T.E. Moore ed., Academic Press, New York, NY (1973).

30. N.K. Sondheimer, Spatial reference and natural language machine control, *Int. Journal of Man-Machine Studies,* Vol.8, pp.329-336 (1976).

31. L.C. Boggess, Computational interpretation of English spatial prepositions, Coordinated Science Lab., *Tech.Rep. T-75,* Urbana, IL (1979).

32. M. Bierwisch, Some semantic universal of german adjectivals, *Foundations of Language,* Vol.3, pp.1-16 (1967).

33. G.S. Cooper, A semantic analysis of English locative expressions, *BBN Tech.Rep.,* No.1587, Cambridge, MA (1968).

34. N. Goguen, A procedural description of spatial prepositions, *M.S. Thesis,* Univ. of Pennsylvania, Moore School of Electrical Engineering, Philadelphia, PN (1973).

35. D.Waltz and L.C. Boggess, Visual analog representations for natural language understanding, *Proc. 6th. IJCAI,* Tokyo, J, pp.926-934 (1979).

36. D.Waltz, Understanding scene descriptions as event simulations, *Proc. 18th. Annual Meeting of ACL,* Philadelphia, PN, pp.7-12 (1980).

37. D. Waltz, Toward a detailed model of processing for natural language describing the physical world, *Proc. 7th. IJCAI*, Vancouver, CDN, pp.1-6 (1981).

38. G. Adorni, M. DiManzo, and F. Giunchiglia, Some basic mechanisms for common sense reasoning about stories environments, *Proc. 8th. IJCAI*, Karlsruhe, D, pp.72-74 (1983).

39. G. Adorni, M. DiManzo, and F. Giunchiglia, From descriptions to images: what reasoning in between?, *Proc. 6th. ECAI*, Pisa, I, pp.359-368 (1984).

40. M. DiManzo, G. Adorni, and F. Giunchiglia, Reasoning about scene descriptions, *Proceedings of the IEEE*, Vol.74, No.7, pp.1013-1025 (1986).

41. A. Herskovits, *Language and Spatial Cognition: an Interdisciplinary Study of the Prepositions in English*, Cambridge University Press, Cambridge, UK (1986).

42. P.J. Hayes, Naive physics I: ontology for liquids, *Working Paper,* No.35, Univ. of Geneve, ISSCO, Geneve, CH (1978).

43. R.A. Brooks, Symbolic reasoning among 3-D models and 2-D images, *Artificial Intelligence*, Vol.17, pp.285-348 (1981).

44. R.B. Fisher, Using surfaces and object models to recognize partially observed objects, *Proc. 8th. IJCAI*, Karlsruhe, D, pp.231-234 (1983).

45. M. DiManzo, E. Trucco, F. Giunchiglia, and F. Ricci, FUR: understanding functional reasoning, *International Journal of Intelligent Systems,* Vol.4, pp.431-457 (1989).

46. M. DiManzo, G. Adorni, F. Ricci, A. Batistoni, and C. Ferrari, Qualitative theories for functional description of objects, *Esprit Project P419, Report TK4-WP2-DI1* (1986).

47. G. Adorni, Causal analysis: a case study in a vectorial domain, *Proc. 1st. Conf. of AI*IA*, Trento, I, pp.158-164 (1989).

48. J. De Kleer and J.S. Brown, A qualitative physics based on confluences, *Artificial Intelligence,* Vol.24, pp.7-83 (1984).

PANEL SUMMARY
IMAGE INTERPRETATION AND AMBIGUITIES

Piero Mussio[1] *(chairman)*, Nicola Bruno[2], and Floriana Esposito[3]

[1] Dipartimento di Scienze dell'Informazione, Università degli Studi "La Sapienza"
Via Salaria 113, I - 00198 Roma, Italy
[2] Dipartimento di Psicologia, Università di Trieste
Via dell'Università 7, I - 34123 Trieste, Italy
[3] Dipartimento di Informatica, Università di Bari
Via Orabona 4, I - 70126, Bari, Italy

POSITION

The goal of the panel is to discuss if ambiguity can be a means for performing cognitive tasks.

This position contrasts with the traditional view which considers ambiguous situations as pathological ones to be avoided or recovered as soon as they arise. Evidence exists that the human visual system attempts to process its inputs according to two different representations, the distal and the proximal one. N. Bruno as a cognitive scientist introduces the argument and speculates on the cognitive function served by the proximal interpretation. On the other side, P. Mussio explores - from the point of view of an image interpreter - some situations in which scientists and technicians exploit ambiguous image interpretations to better understand the situation under study. Programs - developed to help them in these interpretation activities- explicitly create and exploit ambiguous descriptions of a same image. However, in many cases ambiguity is still a problem. F. Esposito as an AI scientist explores how to face the intrinsic ambiguity in learning models of visual objects.

A Preliminary Agreement and Hot Points

Preliminary to the discussion the panellists agreed:
- to recall that "ambiguous" stems from the Latin "ambo"→two together + "agere"→lead: leading in two directions
- to adopt the Oxford Dictionary definitions for *ambiguity:*
1) state of being ambiguous; 2) expression, etc. that can have more than one meaning; and *ambiguous* - of doubtful meaning; uncertain.
- to focus the discussion on the following hotpoints:
  - In what sense is visual experience fundamentally ambiguous?
  - What is the evidence for parallel image representations in visual processing?

- Can the visual system be said to use multiple representations when interpreting images?
- Some automatic systems are said to exploit ambiguous descriptions to interpret images, but has here "ambiguous" the same meaning?

TWO INTERPRETATIONS FOR ANY IMAGE (N. Bruno)

The Fundamental Image Ambiguity

In a special sense, any image is ambiguous. Given an image, a system can interpret it as a representation of objects laid out in the environment (the *distal* interpretation), or as a representation of flat regions onto a projection plane (the *proximal* interpretation). This fundamental image ambiguity underlies an interesting divergence between models of human and machine vision in their present form. Present computational models focus on recovering distal properties from the image, a task which is most often modelled as a process for solving an underconstrained inverse problem. Although the human visual system also recovers distal properties in some way (not necessarily the same as the computational models), it seems that human observers can also engage in a different type of perceptual activity, one which attempts to represent patterns and intensities as they are projected on the retina. The idea that there are a distal and a proximal interpretation for the visual input has been discussed by Gibson[1], Rock[2,3], Natsoulas[4,5] and several others. Therefore, the notion of two interpretations is not new. Surprisingly, however, psychologists have paid little attention to it, and computer scientists have paid none. For instance, proponents of the distinction satisfied in describing the phenomenology of the interpretations[3,4], but no serious attempt has been made to look at the hard evidence for it. Further, some investigators have claimed that proximal viewing is either irrelevant for natural vision, or that it is experienced only in special laboratory conditions[1]. The time may be ripe for a change of theoretical attitude. I will discuss data suggesting that proximal vision is common and more important than previously recognized. It does not seem to require training or adopting a special attitude, and it may be a component of natural vision, not just of laboratory performance. In fact, there are reasons to believe that the human visual system avoids committing itself to either the proximal or the distal interpretation, but always attempts to process both although one may then become phenomenally more salient. This feature of human visual function, which is not captured by present computational work, raises a wealth of interesting questions for future research in human and machine vision.

Terminology. Here we use "image" as a general descriptor term for the input to a visual system, biological or mechanical. This use is customary in the machine vision literature, but can be confusing for those familiar with the cognitive sciences. For instance, psychologists talk about mental images (experienced when visualizing), retinal images (patterns of activation on retinas), as well as about images in pictures. A lot of attention has been devoted to pictorial ambiguity, defined as the simultaneous double experience of perceiving a flat picture and a representation of a three-dimensional layout. However, pictorial ambiguity has nothing to do with the fundamental ambiguity as discussed here. The terms "distal" and "proximal" derive from the time-honoured tradition of distinguishing between distal stimuli, three dimensional objects which reflect light, and proximal stimuli, the pattern of activations on an observer's retina after light impinges on it.

Evidence for Proximal Vision: Two Case Studies

Although controversial, there is evidence suggesting that the visual system engages in two parallel types of visual activity. In one, distal vision, processing of the image yields a set of object-centered descriptors which are invariant or almost invariant under various kinds of transformations caused by changes in of context. Distal vision yields a stable representation of the environment which we can use to control action and to pursue cognitive goals. For instance, we can negotiate obstacles because we can appreciate that the size of objects does not change as we approach them. Similarly, we can recognize objects because their surface colours remain more or less constant despite changes in the illumination. In the other kind of visual activity, proximal vision, processing of the image yields a set of viewer-centered descriptors which vary with the conditions of observation. When engaging in proximal vision, observers in some sense are "looking at their retinal image". As an example, consider the paradox of the converging railroad tracks. One sees parallel tracks receding in depth, but at the same time one also sees that they converge, reflecting the pattern that is projected on the retina.

Mach Card Demonstration. A simple way of demonstrating the switch from distal to proximal vision is the following. Fold a white card in half to form a dihedral and place it on a table with its edge pointing up, as in a tent (see figure 1). Have the source of illumination on top but somewhat to the side. Look at the two faces of the card. Depending on how one looks at them, they will look as the same white under two different illuminations, or maybe as white and light grey under different illuminations. Now look at the upward-pointing edge with a steady gaze and with one eye only. After a few seconds, the "tent" flattens and appears as two coplanar trapezoids. At the same time (and most interestingly), the two faces change colour in a dramatic fashion. Now the lighted face appears almost luminous, whereas the shadowed face becomes very dark. Both colours become more "solid", more opaque. Your perception now correlates with the amount of light reaching your retina, a psychophysical quantity called luminance.

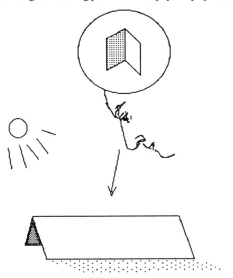

Figure 1. Setup for the "Mach card" demonstration.

First Case Study: Perceiving Luminance. There is an abundant computational literature on surface colour perception. One way to illustrate the gist of the approach is the so-called
fundamental photometric equation. If we constrain our discussion to achromatic colour (the
grey scale from black to white, neglecting chroma), a much simplified version of the
fundamental equation (neglecting considerations of viewing geometry, wavelength, and the
spectral sensitivities of the cones) can be written as follows:

$$L = R \times I \tag{1}$$

where L is luminance, the amount of light provoking a receptoral response, R is surface
reflectance, the dimensionless ratio of reflected to incident light, and I is the illumination at the
surface. Computational models focus on the recovery of the invariant quantity, which in the
equation is R. So in some loose sense we can say that surface colour can be understood as
apparent reflectance and its invariance as a process of separating the contributions of R and I
to L. Equally abundant is the psychophysical literature on surface achromatic colour, which
psychologists call lightness. Surface lightness is defined as the surface property that stays
invariant under various changes in the image. When I look at a white sheet of paper, it remains
white under bright sunlight, under a shadow (change in the intensity of the illumination),
indoors under a tungsten lamp (change in the intensity and in the spectral composition of the
illumination), outdoors in a foggy day (change in the intensity of the illumination and reduction
of border contrast), and so on. However, investigators of lightness have always been aware that
the invariance of lightness, although very good, is not perfect. Under appropriate conditions,
it is not difficult to observe incomplete lightness invariance and systematic effects of
luminance. These luminance effects could be due to imperfections in the neural machinery to
compute lightness, or to experimental limitations. However, an alternative possibility is that the
system can work in two different modes. In the distal mode the system attempts to represent
surface lightness. In the alternative, proximal mode, it attempts to represent luminance. If the
observations failed to control which mode was adopted, it is possible that both were performed,
or that observers alternated between the two. As a consequence, interference from the
luminance representation hindered perfect invariance.

Recent work by Arend[6,7] has provided support for the notion of a representation of
luminance. Typical experiments use computer-generated Mondrian patterns (see figure 2),
which are taken as useful compromises between natural images, which are too complex to
analyze and control, and simple geometric displays which may contain too little information
to support unambiguous colour perception. Observers are shown two similar Mondrians side
by side, but under different illuminations. One of the patches on the left Mondrian is selected
as the standard, and observers are then asked to adjust the corresponding patch on the right until
it appears equal in either apparent reflectance (lightness) or apparent luminance (brightness).

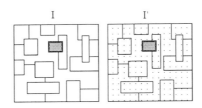

Figure 2. Example of "Mondrian" patterns for color constancy experiments. Illumination I is different than
illumination I'. The observer adjusts one of the patches under I' until it matches the corresponding patch under I.

The important fact here is that instructions are given explicitly as to which property to judge. The results suggest that, depending on the instructions, observers can make judgements that are correlated with surface reflectance, and therefore stay constant under changes of relative illumination, or they can make judgements that are correlated with luminance. Furthermore, judgements of achromatic surface colour are influenced by apparent spatial arrangement only when observers are asked to judge lightness. Taken together, these two results (see figure 3) provide convincing evidence for a perceptual representation of luminance.

Second Case Study: Perceiving visual angle. Computational models have always stressed the importance of using invariant descriptors for object recognition. This approach parallels work on the perception of size, which focuses on the invariance of perceived size under changes in viewing distance. Similarly to the case of luminance, this work conceptualizes the problem with a simple equation, the so-called law of visual angle

$$\alpha = \operatorname{atan}\left(\frac{s}{d}\right) \tag{2}$$

which states that the angle subtended by an object at a viewpoint, α, is a direct function of object actual (linear) size, s, but an inverse function of its distance from the viewpoint, d. The analogy with the case of luminance is straightforward. As in the case of luminance, two distal quantities, size and distance, are confounded in the measure of visual angle. Thus, work on the perception of size has focused on separating the contributions of these two variables to yield a descriptor of object size that will remain invariant under changes in distance (and therefore of visual angle). However, size invariance seems to be only part of what the visual system does. Although the perceptual size of objects tends to remain constant when these are placed at varying distances, the fact that the displacement also changes the visual angle is by no means without representation in consciousness. This has been demonstrated in experiments giving explicit instructions to match on the basis of visual angle rather than size[8-10].

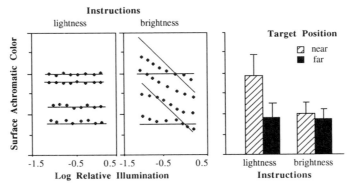

Figure 3. When judging surface color, appropriately instructed observers can produce judgements that correlate with distal reflectance (horizontal lines on leftmost graph) or judgements that correlate with proximal luminance (diagonals on middle graph). Spatial information has different effects on the two kinds of judgement (rightmost graph).

The Nature of Proximal Vision: A Third Case Study

The experiments described in the preceding sections suggest that, at least under appropriate conditions, observers achieve the proximal interpretation of an image. Proximal vision does not seem to require attentional effort or special training, but the ability to adopt a certain kind of "attitude" which is facilitated by reducing the amount of depth information in the image, but can be achieved also in natural viewing conditions. This point was demonstrated well by an elegant experiment of Rock and McDermott[11], who presented luminous patterns in a dark room and asked observers to match them either on the basis of objective size or on the basis of visual angle. The patterns were placed at a 90 deg angle from the viewpoint so that observers could not see them simultaneously. Under these conditions, observers were quite accurate in judging visual angles. After measuring perceived visual angles, however, Rock and McDermott also showed observers the sizes that had been judged to subtend equal visual angles, but this time they asked them to indicate the apparent distances at which the two patterns appeared to be. Practically none of the observers reported that the triangles appeared to be at the same distance. This result therefore supports the idea that the proximal representation is not simply obtained by eliminating depth information by means of focused attention. As seems to be the case when one looks around in the environment, there seems to be an awareness of both the objective size of an object at a given distance, and of the visual angle subtended by that object. However, the perceptual salience of these two representations, the degree of importance they have in the conscious, phenomenal experience of the world can vary. An additional problem is whether these fluctuations in the perceptual salience of either interpretation reflects a hierarchy of processing stages or rather a redirecting of attention to one of the interpretations, both being constantly available. Evidence concerning this problem comes from work on recognition of partly occluded surfaces.

Third Case Study: The "Mosaic" Interpretation of Occluded Surfaces. Occlusions present major challenges to computational models that attempt to recover the shape and the layout of objects from images[12,13]. Computational models have addressed these challenges in several ways, including the analysis of t-junctions, optic flow, and the use of neural networks[14]. Although some of these models have had a degree success in completing occluded surfaces in ways that mimic surface completions experienced by human observers, none of them has addressed the role of the alternative interpretation in which the surfaces are not taken as occlusion between superposed, complete surfaces, but as a mosaic of juxtaposed, coplanar surfaces. However, evidence from recognition and speeded matching tasks suggests that this type of interpretation can be achieved as easily as the occlusion alternative[15-17]. Typically, this work uses response tachistoscopic reaction times as indicators of the time taken by mechanisms necessary to complete recognition tasks. For instance, one can present a figure followed by a composite pattern which contains two figures, side by side. The task of the observer is to indicate, as quickly as possible, whether the figures in the composite pattern were the same or different, by pressing appropriate keys. A computer detects the keypresses and records the times. The first figure, called a *prime,* serves the function of activating a representation of the image before the matching is performed. For instance, the prime can be a pattern such as the one on top in the left part of figure 4 (a circle partly occluded by a square), whereas the composite patterns could be either two circles, or two "notched" circles. Of course, the notched circles are identical to the visible part of the occluded circle in the prime. Now imagine that you have been instructed to answer "same" whenever the two figures were circles, notched or not. You would be faster in producing this response when the circles are complete, despite the fact that it is the notched circles, not the complete ones, that are geometrically identical to the occluded figures in the prime.

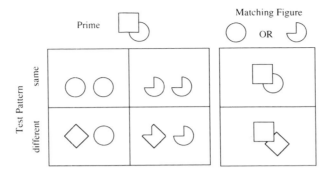

Figure 4. Typical stimuli used in speeded matching tasks involving occluded figures. In the primed matching paradigm (left), observers match two simultaneous figures presented after a prime. In the speeded recognition task (right), observers match the test pattern to a previously presented or simultaneous matching figure.

There are several interpretations for this result. One possibility is that the occluded shape is completed immediately and without any intermediate or parallel representation. According to this interpretation, matches to the full circle are faster than matches to the notched circle because the occluded shape is immediately represented as complete, so that responding to the notched one requires an additional operation, presumably of categorizing. An alternative possibility, however, is that there is a two-stage process, which first represents the visible shape "literally", as a notched circle, but then proceeds to complete it and looses trace of the first representation. If this happened fast enough, then the system could end up with the completed representation while the figure is still being presented. This the interpretation favoured by Sekuler and Palmer[17], who varied the duration of the matching pattern and found that at very short durations the advantage of matching the full circle disappears. It is possible that this happens because, at very short stimulus durations, there is no time for the system to proceed to the second stage of the process. However, data from a different version of this task suggest a different story[15,16]. The main difference is that the first figure can be either a full circle or a notched circle (see figure 4) rather than an occluded circle. In this way, the first figure has a rather unambiguous interpretation, and the speed of the match becomes a direct measure of the time required to reach that interpretation in the occlusion pattern. Under these conditions, it takes essentially the same time to match the full circle than it takes to match the notched circle. This is surprising, because in ordinary vision the distal interpretation of a completed surface behind the occlusion seems more natural. However, these data suggest that the proximal interpretation, the mosaic of juxtaposed surfaces, can be produced as fast as the distal one. This is consistent with a notion of two mechanisms working in parallel, rather than a two-stage model. If the proximal interpretation were simply the initial stage, then matching the notched circle should have been faster than matching the full circle, for the former would have required one operation, the latter two.

Conclusion

The evidence for a special feature of the human visual system, its treatment of the proximal interpretation, not mirrored by features of computational models. Divergencies between human and machine vision raise questions that challenge our understanding of visual function. What is proximal vision? What mechanisms are responsible for it? At least from the present evidence, it seems that proximal vision is a type of perceptual activity which observers can perform naturally given the appropriate conditions. Possibly, it is even a perceptual process which runs in parallel to distal vision, although it may not be always as salient in phenomenal experience. Why do we have proximal vision? What could be the cognitive function, if any, of processing two interpretations for an image instead of just one? On this last point, opinions have differed. Gibson[1,18] proposed that the proximal interpretation stems from adopting a rather special, "pictorial" type of attitude which is learned after exposure to pictures. However, this point of view presents a potential circularity, for depiction in its turn requires that one learns to have perceptual experience of the proximal interpretation. Rock[3] claimed that proximal vision is essentially the first stage in a more complex process that leads to the perceptual constancies. Against this idea, some evidence suggests that the two interpretations may be attained in parallel, not in succession. Natsoulas[4,5] proposed that the activity of proximal vision, which he calls "reflective" seeing, may play a crucial role in our inner awareness of having perceptual experiences. In other words, he hinted that proximal vision may be one of the bases of consciousness. Although all these ideas have some merit, a perhaps simpler function for proximal vision could be that of providing a form of viewer-centered representation to be used in exploratory behaviour. Consider occlusions. In a single view, there simply is not enough information to specify the direction of the occlusion, or how the occluded surface may continue behind the occluder. Typically, however, enough information becomes available if more views are processed over time. After a short visual exploration of the relevant surfaces, motion constraints disambiguate the direction of the occlusion, while parts of the occluded surface may become visible posing additional constraints on its complete shape. However, in order to plan an exploratory action which aims at revealing parts of an occluded object, a system needs to start with a representation of which parts are visible at this point in time and from this point of view. The function of representing how the world appears from a given viewpoint and at a certain time could be included in a three-dimensional model of the environment. However, to compute such a model the system would need time and resources. Perhaps the same function could be served, much more economically, by the proximal interpretation. Maybe the cognitive goal achieved by processing two alternative interpretations for any image could be that of preserving a form of "visual memory" of what is seen from a certain viewpoint at a certain point in time, to be used as a template against which to check how the image has evolved, which surface parts have come into view, how luminances have changed, and so on, for the purpose of planning and controlling explorations.

ON POSITIVE USE OF AMBIGUITY IN MACHINE
VISION AND IMAGE UNDERSTANDING (P. Mussio)

I offer to discussion some findings from different scientific and technical experiments in which experts exploit ambiguous image interpretations to reach a more reliable or a deeper understanding of the observed scene. Programs have been implemented to help experts in their interpretation activities in the different fields. Some observations on the behaviours of both humans and programs suggest that situations exist in which the word "ambiguity" changes its meaning and is no longer associated to uncertainty becoming itself ambiguous.

The Assumed Point of View

I start from the second definition of ambiguity proposed by the Oxford dictionary: a figure is ambiguous when several different interpretations can be derived from it, i.e. a same primitive structure is incorporated in different interpretations "- with different labels - and perhaps having undergone different transformations"[19].

A label is a too synthetic description for our purposes. Here, a description of an image (or structure) is enriched to be an attributed symbol, a triple d = (name; descriptors; sub-description_names) where "name" is the label identifying the structure, "descriptors" denotes a set of couples <name of an attribute - value of the attribute> and "sub-description_names" a set of names of sub-structures which can be identified as components of the structure itself.

Figure 5 shows a remote sensed image of the Italian Argentario promontory. A geographer describes it by the triple (Argentario_promontory; area= area_value, perimeter=perimeter_value; Orbetello_peninsula, Orbetello_lagoon) (see figure 6a). "Orbetello_peninsula" and "Orbetello_lagoon" turn are names of descriptions, which in their can be defined in terms of their components and hence described by attributed symbols.

Obviously this recursive style of description implies the definition of some atomic structures that cannot be further subdivided into components. In the descriptions of atomic components the third component of the description - the set of "sub-description_names" - is empty. Programs describing digital images often assume pixels as atomic structures.

Ambiguity in Image Interpretation

In image interpretation, ambiguity arises when an image is interpreted in different ways by different observers or by a same observer acting in different contexts and pursuing different goals - i.e. acting in different situations.

For example, a geologist describes the same Argentario promontory of figure 5 as two tombolos connecting a rocky island to a continental body (see figure 6c) contrasting the previously introduced geographer description. (figure 6a).

A same entity (the Argentario promontory) is ambiguously described because several of its substructures - sets of pixels in the digital image - are incorporated into different geographical or geological components as outlined in figure 6b and denoted by different labels.

Even a program can obtain these two different descriptions, as shown in figure 7. This is

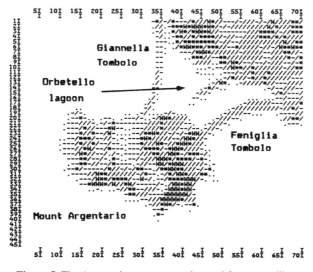

Figure 5. The Argentario promontory observed from a satellite.

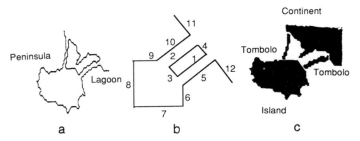

Figure 6. A geographic (a) and a geological (c) description of the Argentario promontory and a sketch (b) outlining the substructures ambiguously labelled by the two different descriptions.

possible if the strategies to derive the two descriptions have been formalised and translated in an executable form.

In the example both the geographer and the geologist strategies were translated into rules, interpretable by an existing interpreter Image Interpreter (II), described in [20]. II interpreted the two sets of rules separately to obtain the results shown in figure 8. With Arbib and Hanson one can say that II-like a human aware of the two interpretation strategies- "produces multiple consistent interpretations and can be thought of being uncertain about what it 'sees'..." [21]. The program can be thought to be in a state of uncertainty because when required to describe the Argentario promontory it has no reason to choose the first or second description [20]. To produce

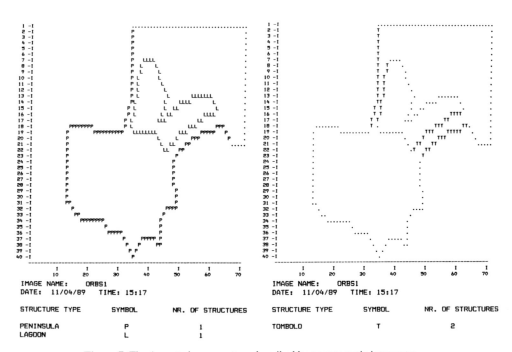

Figure 7. The Argentario promontory described by an automatic interpreter.

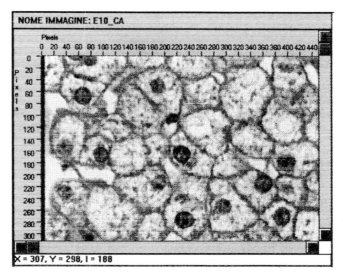

Figure 8. A liver biopsy.

these results it flips from one set of rules to the other and stores two contrasting descriptions for each one of the structures sketched in figure 6b. The two meanings of "ambiguity" are both necessary to describe its behaviours and results.

Ambiguity as a Tool for a Better Image Understanding

There are situations in which observers exploit ambiguous interpretation of an image to reach a better understanding of a state of affairs. For example humans exploit both the geographical and the geological description to establish some navigation strategies around the Argentario promontory. In this activity they exploit three different sources of knowledge: the geographical, the geological and a navigational one. The third allows them to reason on the two descriptions obtained by the other two and to deduce the correct route.

This is not an isolated case.

In our experience, scientists and technicians often exploit ambiguous descriptions in their communication and decision processes to achieve more reliable results or a deeper understanding of a state of affairs. In these cases, they first consciously obtain several - possibly ambiguous - interpretations of a same image according to apparently contrasting sub-goals. Then they reason on the different interpretations to derive a sequence of actions (a strategy) which allows reaching a more abstract or more general goal[22].

This possibility of obtaining and managing different interpretations stems from the capability of 1) interpreting the scene pursuing different sub-goals exploiting different knowledge sources; 2) reasoning on the current status of the interpretation strategy to decide the next sub-goals to be reached and the knowledge bases to be exploited; 3) merging the different descriptions into more abstract ones.

Activities 2) and 3) require to exploit knowledge about the meaning of the different contrasting descriptions and the ability to reason on the findings and on the rules and strategies exploited in the first activity. Hence they are developed at a meta-level[23] with respect to the first activity.

An automatic image interpreter can reach similar results and simulate these types of (meta-)strategies if it is enriched with the ability of: a) metareasoning on the results obtained so far, the successes or failures of past activities, the strategy which was followed, b) deriving from the results of this metareasoning the strategy to be followed next and c) executing the individual actions in this strategy.

For example ABI[22] (Assistant for Biopsy Interpretation) exploits this style of operation to interpret digitized liver biopsies as the one shown in figure 8.

ABI first examines the biopsy from a) a topographic point of view, the coloured image is seen as a 3-D landscape; b) from a geometrical point of view, and tries to enhance three different kinds of anatomical structures (membrane, the nuclei, sinusoids).

In this way ABI obtains five (possibly partial or uncertain) descriptions of the original image and then (meta-)reasons on them to recognize and count high level biological structures such as cells of different kinds, nuclei etc. (see figure 9a).

To reach these results, ABI has been endowed with metareasoning abilities and not only with reasoning capabilities as the original II was.

These capabilities allow ABI to evaluate contextually the different facts pertaining to the present state of the image interpretation and to deduce step by step the activity to be performed, i.e. to construct its strategy of interpretation.

The adoption of a parallel style of execution of the (meta-)strategies makes feasible the synthetic description of this evaluation activities and allows the practical software implementation of these programs[22].

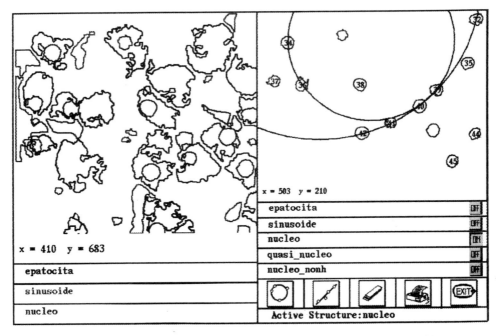

Figure 9. ABI at work: a) some of the biological structures identified in figure 8; b) two out of five emergent structures found in the same image.

Ambiguity in the Exploration of Unknown Phenomena

Facing unknown phenomena human experts go further on in exploiting ambiguities in image interpretation.

For example new biopsy techniques allow physician to study for the first time at microscopic level phenomena (a liver disease) whose effects they observe at a macroscopic level but whose dynamic is unknown. Physicians may suspect that some pathological situations organize the active hepatocites along some nearly geometrical disposition.

Hence they apply the approach that Afzelius[24] calls 'active reading': they read the biopsies looking for some particular disposition of active liver cells which explain some of the observed macro effects.

Active reading implies the conscious exploitation of ambiguous interpretation: every track of an entity is interpreted as belonging to every possible emergent arrangement. It even implies the knowledge necessary for the detection of virtual structures which emerge from the disposition of already classified substructures.

Bianchi and Rubbia[25] summarise this situation in the following way: "The analysis of (cytological) images does not only consists of the recognition of tracks of known biological entities but also requires the discovery of unforeseen arrangements of structures, which can be clues for the detection of unknown phenomena".

An automatic interpreter can help humans in their active reading activity provided that the knowledge and the ability to detect virtual emergent structures are added to its (meta-)reasoning abilities.

For example, ABI can help in checking the hepatologist hypothesis looking for sets of active hepatocites which lie within an annulus of given characteristic. This implies that ABI first recognises the structures present in the biopsy. Figure 9a shows the graphical counterpart of ABI interpretation of the biopsy field of figure 8. Then ABI looks for the emergent structures in the whole biopsy. Figure 9b displays two out of the five emergent structures found in the biopsy. ABI exploits ambiguity to reach the classification of figure 9a while the result in figure 9b is ambiguous in itself because some nuclei belongs to more than one emerging structure.

Is it Still Ambiguity?

ABI is capable of managing emergent structures because it is able to switch from one interpretation (nuclei 40 and 39 belong to circle 3) to a different one (nuclei 40 and 39 belong to circle 4) following a well defined strategy prescribed to it as a set of metarules. In other words ABI is able to choose the semantics - i.e. the association of one structure with one description - appropriate to each situation. Interpreting its metarules ABI understands which set of rules defining a semantic association are better suited to a given situation. In a linguistic perspective[26], the old interpreter II follows a semantic approach in determining good descriptions- its descriptions remain invariant across situation. ABI on the contrary follows a pragmatic approach because it focus on the situation to choose the semantics to be applied. Adopting a pragmatic approach, designers prescribe to ABI the strategy by which to take into account those "marginal condition" (in Tondl language)[23] which determine the correct meaning of a substructure. In its activity ABI creates descriptions which are ambiguous following the first definition of the Oxford dictionary but not following the second one, because ABI is never uncertain about what it 'sees'.

The word "ambiguous" is now used associated to a different set of meanings, becoming itself ambiguous. Or not?

AMBIGUITY IN LEARNING MODELS OF VISUAL
OBJECTS: THE AI VIEW (F. Esposito)

In learning symbolic representations of visual objects by means of conceptual methods, ambiguity is the possibility of generating rules which produce multiple classification. This can be due principally to noise and incompleteness in the observations or to the fact that the concepts themselves change in the time. The properties of completeness and consistency generally required in rule generalization do not guarantee that during the classification phase no ambiguity arises. A first technique to eliminate ambiguity in classification is that of realizing a flexible matching. A second method consists in adjusting incrementally by specialization the rules which have been produced inductively.

Introduction

Inference lies at the heart of perception: since the state of the world is not logically entailed by sensory data, perceptual inferences are inductive. The sensory data play the role of premises and the assertions about the state of the world are the conclusions. Inductive learning aims at finding the evidential relations between the perceptual premises and the conclusions.

The importance of learning from sensory information has been widely discussed in Psychology literature and in Pattern Recognition. The traditional pattern recognition approach typically is based on computing shape descriptions that are supplied in form of feature vectors. Learning mostly consists of defining and modifying the weights with which each feature enters in the discrimination and classification. Frequently the observations have some natural structure: objects have components and relations among those components. Syntactic pattern recognition embraces the important idea of structural descriptions but has mostly been concerned with the definition of formal grammars used to describe and recognize the shape of objects.

The AI methods allow for learning symbolic representations of visual objects that can be extracted from real visual data. Development in segmentation techniques and edge detection have provided solid methods to generate shape descriptions in form of assertions, so playing the role of pre-processing techniques preluding the real learning phase. It is possible to learn concepts using visual shape representations computed by an image understanding system[27]: for example, it is possible in a supervised approach to generalize from pre-classified examples the symbolic description of a concept, namely the intension of a class, and to predict the class which a new observation belongs to by symbolic pattern matching routines. The process of inductive generalization starts from well selected significant examples or instances, generally expressed into complex representation languages. The representation of instances by attribute values and the representation of concepts by sufficient attributes dominated quite along time, but the need for a richer representation language has been put forward by many authors. A powerful representation language like a first order predicate calculus is necessary in order to learn structural descriptions, although some difficulties in generalization arise when handling quantification and its semantics.

The AI inductive learning methods are defined "empirical"[28] and can use background knowledge as a bias in selecting the inductive hypotheses: it is useful to define the importance of some descriptors, the non applicability of others in some situations, or to provide the inference rules for deducing not observed properties or for creating new descriptors. As to specific problem of generating and generalizing models of visual objects the background knowledge can be used in defining a hierarchical ordering on conjunctive sets of attributes, or components, useful to discover functional similarities between objects that do not share common structure. In cognitive science the importance of form/function distinction in classification has been recognized: the functional knowledge guides the search for aggregations while structural information guides the search for characterizations[29]. Generally speaking, this knowledge may be useful to move from the perceptual phase to the high level abstract thinking: supplying a defined coherent theory of the dominion, this knowledge is useful to avoid the ambiguity, related to the possibility of multiple interpretations of an image. As to the concept of ambiguity related to vagueness or uncertainty or noise some considerations concerning learning in noisy environment must be done.

Learning in Noisy Environments

In principle, in AI approaches solving the ambiguity is firstly a problem of the preprocessing step, when the transformation from real data into symbolic description is performed. In fact, most of the implemented learning systems presuppose the existence of ideal conditions which limit their applicability. In particular, input examples are often supposed to be selected by a human teacher rather than by the environment through external sensors. In this last case, noise always affects the object descriptions thus it should be taken into account both in the learning phase and in the subsequent classification process[30].

Quinlan[31] reported results of several experiments of learning decision trees in noisy environments, from which he drew out the following empirical consideration: it is not worthwhile expending effort to eliminate noisy descriptions from the training set if there is going to be a significant amount of noise in the testing set. Indeed, at high levels of noise, rules learned from controlled data perform poorly on noisy testing sets. Nevertheless it is important to eliminate noise or ambiguity regarding class membership of the training examples, if possible, since classification error increases linearly with the percentage of noise on the class information. Although it seems that the problem of noise can be partially solved in the learning phase, it is still open for the classification phase as there is no guarantee that new observations will be noiseless.

During the classification step, production rules or decision trees are used representing the intentional definitions of the concepts which have been generalized by the inductive learning algorithm. The learned concept C_1 is, in the most general case, a disjunctive concept: it can be represented as a DNF (Disjunctive Normal Form), being this a disjunction of several conjunctions of conditions on descriptors. Many learning systems learn a list of concepts C_1, C_2,..., C_n which compose a partition of the representation space (i.e. they are supposed mutually exclusive, and any example belongs to only one of them). However it is frequent that the output knowledge (set of rules, clauses, decision trees etc.) of the learning algorithm does not describe exactly a partition: for some descriptions no rule is fired while for some others several rules concluding on different concepts are fired. Generally properties of completeness and consistency are required in the inductive learning process. Completeness means that each hypothesis for a concept C_i should subsume all the training instances of the concept C_i (positive examples), while consistency means that the same hypothesis should not subsume any training instance of the concept C_j, $i \neq j$ (negative examples). This is enough to avoid, in principle, problems of ambiguity during the classification step. During the testing phase, when a new example E that belongs to a concept C_i occurs, different situations can be observed:

a) the new example does not trigger any rule of the current learned concept (no classification);

b) the new example triggers one or several rules related to a specific concept C_i, and no rule concluding on another concept (single classification, no ambiguity);

c) the new example triggers several rules concluding on several concepts (multiple classification, ambiguity); the ambiguity can be caused by very different reasons, such as the weakness of the representation language, noise or lack of precision in the description of some examples (for instance, the excessive use of "don't care" values).

Case b) does not need the triggering of any mechanism, since a unique classification already exists. In case a) and c) different solutions may be adopted depending on the possibility of modifying during the testing phase the classification rules or not. In fact, when new training examples become available, the properties of completeness and consistency may be no longer satisfied. In this case, these properties can be recovered by activating either a new learning process from scratch or a knowledge repair mechanism such as incremental learning.

Nevertheless, if no possibility exists of modifying the rules it is still possible to recover case a) and c) avoiding to adopt a strict (canonical) matching during classification.

Solving the Ambiguity by Flexible Matching

Indeed, since the main activity of a symbolic classifier is pattern matching, a first possible solution could affect the way by which comparing two descriptions. Again the problem is connected to the representation language: being this a propositional calculus (attribute-value pairs) or a First Order Predicate Calculus the problem of pattern matching is that of unifying two formulas. In particular, matching two structural descriptions, that is FOPL well formed formulas (wff's), t and s, is complex: t matches against s if there is a substitution s for the variables in s such that $t \Rightarrow \sigma(s)$. The problem is NP complete: thus, a method must be found to optimize the matching process. Without changing the learned rules it is possible to cope with noise, which generates ambiguity in classification, relaxing the conditions of matching the example description against the concept description. In fact, in computer vision and, generally, in noisy environments, a symbolic classifier producing a boolean answer (match/no match) is inappropriate to take into account possible distortions. A way of realizing such a "flexible matching" could be a numeric measure of similarity between observations whose value indicates the goodness of the matching itself. This is also equivalent to define the complementar measure of distance between two wff's in a given representation language.

A variety of distance measures have been proposed both from the field of Pattern Recognition [32, 33] and from the area of Machine Learning [34, 35, 36]. Basically they can be grouped into two categories according to the adopted representation language:
• distance measures defined on feature vectors or languages from propositional calculus, which deal only local deformations (the differences between the two matching descriptions concern global/local attributes);
• distance measures defined on graphs or languages from predicate calculus, which are concerned also with structural deformations (the differences between the descriptions are at the level of relations).

A distance measure between two formulas can be defined based on the probabilistic interpretation of the matching predicate.

Let S denote the space of wff's in the selected representation language (FOPL or its extensions), and let Match represent the canonical matching predicate defined on S

$$\text{Match} : S \times S \rightarrow \{\text{false,true}\} \tag{3}$$

Then, it is known as Flexible Matching any function:

$$\text{Flex\_Match} : S \times S \rightarrow [0,1] \tag{4}$$

such that for each $s,t \in S$:

$$\text{Flex\_Match}(s,t) = 1 \Leftrightarrow \text{Match}(s,t) = \text{true}$$
$$\text{Flex\_Match}(s,t) \in [0,1] \text{ otherwise.} \tag{5}$$

The function Flex_Match(s,t) represents a degree of similarity between two descriptions s and t. To each pair of wff's in S the probability of precisely matching the two formulas can be assigned, provided that a change is made in the description t. Formally:

$$\text{Flex\_Match}(s,t) = P(\text{Match}(s,t)) \tag{6}$$

It's possible to define a probabilistic distance measure d(s,t) as:

$$\delta(s,t) = 1 - P(\text{Match}(s,t)) = 1 - \text{Flex\_Match}(s,t) \tag{7}$$

A complete, detailed description of this approach useful to define a distance measure between two structural descriptions expressed into VL21, an extension of predicate calculus, is in [37]. It is also discussed the utility of such a measure in handling noisy and incomplete structural information. Two methods are presented in [38] in order to cope with computational complexity.

Being a distance measure available, a multilayered framework may be adopted in order to classify a new example: different strategies are triggered when matching the new example against a set of rules. When no classification occurs (case a), the conditions of matching may be relaxed activating the computation of the distance measure between each generalized rule in the set and the new example. Classification is performed assigning the example to the class that gives the minimum value of the distance. Case c) means that a classification ambiguity occurs. Such an ambiguity is solved by choosing the class that produces the minimum value of the distance between the example and the concept description C_i (note that the order of the two descriptions is reversed). In this case the computation of the distance measure quantifies the weight of the non unifying subgraph $E-C_i$.

The method has been adopted in the classification of digitized documents basing on layout characteristics.

Solving the Ambiguity by Incremental Learning

Another way of solving the ambiguity in classification may be related to the incremental correction of inductively generated knowledge base of the symbolic classifier.

Incremental learning is necessary when either incomplete information is available at the time of initial rule generation or the nature of concepts evolves dynamically.

The basic operations needed for incremental learning are generalization and specialization. A **generalization problem** can be summarized as follows:

GIVEN

a hypothesized concept C, which is complete and consistent with respect to a set of training examples, and a new positive example E such that C is incomplete with respect to a E

DETERMINE

a new hypothesis C' (a new class intension) that is complete and consistent with respect to both the previous set of training examples and the new positive example E

This problem has been widely investigated by Plotkin[39], Vere[40], Helft[41], and Haussler[42]. A **specialization problem** can be cast as:

GIVEN

a hypothesized concept C, that is complete and consistent with respect to a set of training examples, and a new negative example E' such that C is inconsistent with respect to E'

DETERMINE

a new concept C' that is complete and consistent with respect to both the previous set of training examples and the new negative example E'   .

Having the possibility of dynamically modifying the rules some considerations must be done concerning the operators to activate: case a) means that the decision rule for C_i is too specific and needs to be generalized; case c) means that the decision rule for C_j is too general and needs to be specialized. This typology does not depend on the kind of data representation neither on the particular learning algorithm.

Different specialization operators may be proposed, generally dependent on the representation languages in which descriptions are expressed. A general operator for specializing the hypotheses inductively generated by any system that learns structural descriptions from positive and negative examples can be based on negation[43]. Given a hypothesis C that erroneously covers a testing example E, specialization of C can be accomplished by adding the negation of a non-matching literal of E to the body of C. Each possible substitution according to which C matches against E defines a base of consistent specializations and the intersection of all such bases provides a set of literals that can recover the consistency of C when added to its body. However, completeness has always to be checked. A computational analysis and an initial experimentation in the domain of geometric figures have shown that the computational costs of learning decision rules generated through an incremental process are lower than those required by a single step learning process. A more complex experimentation in the field of digitized document classification is in progress.

A completely different problem arises when the initial classes are not mutually exclusive. In this case ambiguity is related to the problem of learning dependent concepts or learning contextual rules. In order to learn properties that characterize each concept as well as dependencies between concepts (constraints) it is necessary to consider the possibility of changing the description language while learning. When concept dependencies are intrinsically acyclic it is possible to use dependency hierarchies to represent them. The order by which concepts should be learned is completely defined by the dependency hierarchy. In particular, the concepts at the bottom of the hierarchy, the minimally dependent concepts, must be learned first by traditional learning systems. Then it is possible to learn those concepts that depend directly on the minimally dependent concepts, after having properly augmented the language of observations with the learned predicates. This strategy allows for learning contextual rules and the classification may be seen as a constraint satisfaction problem (or a labelling problem). The approach has been applied to the problem of document understanding[44]. It is possible to observe that contextual rules solve many ambiguities and have a higher predictive accuracy than non-contextual rules. Moreover, it has been proved that it is possible to discover the order of learning dependent concepts by means of statistical routines.

Conclusions

Although the symbolic approaches to learning use background knowledge, induction is intrinsically ambiguous. To require properties of completeness and consistency in the generalized concepts does not guarantee that during the classification phase no ambiguity arises. This can be due principally to noise and incompleteness in the observations or to the fact that the concepts themselves change in the time. In order to solve the first problem it is possible to eliminate ambiguity in classification relaxing the conditions of matching a new observation against the concept descriptions, that is to adopt a flexible matching. In order to solve the second question the learning system must be incremental: so it is possible to adjust the rules which have been produced inductively, without activating a new learning process from scratch. Incremental learning may be seen as a knowledge repair mechanism able to restore the consistency property without losing the equally important property of completeness.

REFERENCES

1. J.J. Gibson, *The perception of the visual world*, Hughton Mifflin, Boston, MA (1950).
2. I. Rock, In defense of unconscious inference, in *Stability and constancy in visual perception*, W. Epstein ed., Wiley (1977).
3. I. Rock, *The logic of perception*, Ma: MIT press, Boston, MA (1983).
4. T. Natsoulas, Reflective seeing: an exploration in the company of Edmund Husserl and James J. Gibson, *Journal of Phenomenological Psychology*, Vol.21, pp.1-31 (1990).
5. T. Natsoulas, The tunnel effect, Gibson's perception theory, and reflective seeing, *Psychological Research*, Vol.54, pp.160-174 (1990).
6. L.E. Arend and R. Goldstein, Simultaneous constancy, lightness, and brightness, *Journal of the optical society of America A*, Vol.4, pp.2281-2285 (1987).
7. J. Schirillo, A. Reeves, and L. Arend, Perceived lightness, but not brightness, of achromatic surfaces depends on perceived depth information, *Perception and Psychophysics*, Vol.48, pp.82-90 (1990).
8. W. Epstein, Attitudes of judgment and the size-distance invariance hypothesis, *Journal of Experimental Psychology*, Vol.66, pp.78-83 (1963).
9. A.S. Gilinsky, The effect of attitude upon the perception of size, *American Journal of Psychology*, Vol.68, pp.173-192 (1955).
10. J. Shallo and I. Rock, Size constancy in children: A new interpretation, *Perception*, Vol.17, pp.803-814 (1988).
11. I. Rock and W. McDermott, The perception of visual angle, *Acta Psychologica*, Vol.22, pp.119-134 (1964).
12. N. Bruno and N. Gerbino, L'occlusione dinamica: implicazioni per i modelli di visione artificiale, in *I sensi dell'automa*, G. Adorni, W. Gerbino, and V. Roberto, Edizioni Lint, Trieste, I (1992).
13. D. Marr, *Vision*, Freeman, San Francisco, CA (1982).
14. S. Grossberg and E. Mingolla, Neural dynamics of perceptual grouping: Textures, boundaries, and emergent segmentations, *Perception and Psychophysics*, Vol.38, pp.141-171 (1985).
15. N. Bruno and N. Gerbino, Amodal completion and illusory figures: An information-processing analysis, in *The perception of illusory contours*, S. Petry and G. Meyer, Springer, New York, NY (1987).
16. W. Gerbino and D. Salmaso, Un'analisi processuale del completamento amodale, *Giornale Italiano di Psicologia*, Vol.12, pp.97-121 (1985).
17. A.B. Sekuler and S.E. Palmer, Perception of partly occluded objects: a microgenetic analysis, *Journal of Experimental Psychology: General*, Vol.121, pp.95-111 (1992).
18. J.J. Gibson, *The ecological approach to visual perception*, Hughton Mifflin, Boston, MA (1979).
19. A.P. Witkin and J.M. Tenenbaum, On perceptual organisation, in *From pixels to predicates*, Pentland ed., Ablex Publishing, pp.149-169 (1986).
20. P. Mussio and M. Protti, Attributed parallel rewriting in vision, in *Active Perception and Robot Vision*, NATO ASI Series, Springer-Verlag, Berlin, D (1993).
21. M.A. Arbib and A.R. Hanson, Vision in perspective, in *Vision, Brain and Cooperative Computation*, Arbib and Hanson eds., MIT Press, pp.1-83 (1987).
22. P. Mussio, M. Pietrogrande, P. Bottoni, M. Dell'Oca, E. Arosio, E. Sartirana, M.R. Finanzon, and N. Dioguardi, Automatic cell count in digital images of liver tissue sections, *Proc. 4th IEEE Symposium on Computer-Based Medical Systems*, IEEE Computer Society Press, pp.153-160 (1991).
23. L. Tondl, *Problems of Semantics*, Reidel Publishing Company, Dordrecht, D (1981).
24. B.A. Afzelius, Interpretation of Electron Micrographs, *Scanning Microsc.* 1, pp.1157-1165 (1987).
25. N. Bianchi, G. Rubbia, R. Di Lernia, and P. Mussio, Automatic discovery of emergent patterns in citological images, to appear in *Journal of Hystochemistry* (1993).
26. B. Katzemberg and P. Piola, Work Language Analysis and the naming problem, *Comm. Acm*, Vol.36, No. 4, pp.86-92 (1993).
27. J.H. Connell and M. Brady, Generating and Generalizing Models of Visual Objects, *Artificial Intelligence*, No. 31, pp.159-183 (1987).
28. R.S. Michalski, A theory and methodology of inductive learning, in *Machine Learning, an Artificial Intelligence Approach*, R.S. Michalski, J.G. Carbonell, and T. Mitchell eds., Tioga, Palo Alto, CA, pp.83-134 (1983).
29. S.J. Hanson, Conceptual Clustering and categorization, in *Machine Learning: an Artificial Intelligence Approach*, Y. Kodratoff and R.S. Michalski eds., Vol. III. Morgan Kaufmann, San Mateo, CA, pp.235-268 (1990).
30. F. Bergadano, A. Giordana, and L. Saitta, Automated Concept Acquisition in noisy environments, *IEEE Transactions on Pattern Analysis and Machine Intelligence*, vol.10, pp.555-578 (1988).
31. J.R. Quinlan, Induction of Decision Trees, *Machine Learning*, No. 1, pp.81-106 (1986).
32. A. Sanfeliu and K.S. Fu, A distance measure between attributed relational graphs for Pattern Recognition, *IEEE Transactions on Systems, Man and Cybernetics*, Vol.13, pp.353-362 (1983).

33. A.K.C. Wong and M. You, Entropy and Distance of Random Graphs with Application to Structural Pattern Recognition, *IEEE Transactions on Pattern Analysis and Machine intelligence*, Vol.8, pp.599-609 (1985).

34. R.S. Michalski, I. Mozetic, J. Hong, and N. Lavrac, The AQ15 inductive Learning system: an overview and experiments, *Int. Rep. Dept. of Computer Science*, University of Illinois, Urbana, IL (1986).

35. Y. Kodratoff and G. Tecuci, Learning based on conceptual distance, *IEEE Transactions on pattern Analysis and Machine Intelligence*, Vol.10, pp.897-909 (1988).

36. J. Nicolas, J. Lebbe, and R. Vignes, From Knowledge to similarity, in *Symbolic-Numeric Data Analysis and Learning*, E. Diday and Y. Lechevallier eds., Nova Science Pub., New York, NY, pp.585-597 (1991).

37. F. Esposito, D. Malerba, and G. Semeraro, Classification in noisy environments using a Distance Measure between structural symbolic descriptions, *IEEE Transactions on Pattern Analysis and Machine Intelligence*, Vol.14, No.3, pp.390-402 (1992).

38. F. Esposito, D. Malerba, and G. Semeraro, Flexible Matching for noisy structural descriptions, in *Proc. of 12th Int. Conf.. on Artificial Intelligence*, Morgan Kaufmann, Sidney, AU, pp.658-664 (1991).

39. G.D. Plotkin, A Note on Inductive Generalization, in *Machine Intelligence 5*, B. Meltzer and D. Michie eds., Edinburgh University Press, pp.153-163 (1970).

40. S.A. Vere, Induction of Concepts in Predicate Calculus, *Proc. IJCAI*, Tblisi, USSR, pp.281-287 (1975).

41. N. Helft, Inductive Generalization: a logical framework, in *Progress in Machine Learning, Proc. EWSL 87*, Bled, YU (1987).

42. D. Haussler, Learning Conjunctive Concepts in Structural Domains, *Machine Learning*, Vol.4, pp.7-40 (1989).

43. F. Esposito, D. Malerba, and G. Semeraro, Negation as a Specializing Operator, in *Advances in Artificial Intelligence*, P. Torasso ed., Lectures Notes in AI, No. 728, Springer-Verlag, pp.166-177 (1993).

44. F. Esposito, D. Malerba, and G. Semeraro, Automated Acquisition of Rules for Document Understanding, *Proc. of ICDAR93*, Tsukuba, J, pp.650-654 (1993).

FROM IMAGERY TO IMAGINATION

Ornella Dentici Andreani

Istituto di Psicologia, Università di Pavia
P.zza Botta 6, I - 27100 Pavia, Italy

1. INTRODUCTION

Imagery is a very common experience in everyday life: we hear a sound, and we imagine what might be its source, we try to remember where we have left our keys, and we see, with the mind's eye, all the actions we have done and the rooms where we have entered before going out; we are moving into a new house and we imagine the disposition of the furniture, the colours of the walls and so on; we have an examination, and we imagine the scene, the questions, our reactions. We use imagery to remember, to anticipate, to answer a question or to invent something new; but for many years psychology has neglected this problem because of the difficulty to find objective methods which were not based on introspection. Only recently the cognitivist approach has experienced techniques to study mental representations, proposing models that permit experimentation and computation: twenty years ago we had already good models of propositional representations, but none of other formats like image.

Usually we start our lectures with a definition: but let me be a little trasgressive, and permit me to delay the definition to the end. Any way, since we must communicate, I am giving you a preliminary and provisional definition of the words, based on the common use.

We call *Mental Image* a figural *representation* which preserves information (in this sense, MI should be equivalent to memory trace but not all the Authors agree on this point).

We call *Imagery* the *activity* which generates images and manipulates them by inspection, comparison, transformation.

We call *Imagination* the *use of imagery for constructing something new*, like a product, an invention, a phantasy, a dream.

But if we pass from the linguistic common use to the meaning in psychology we find that the definition is a puzzle, since at the moment the term MI still means different things to different theorists: for most cognitive psychologists Mental Images are *internal representations with a particular format*, and we can assess its properties and the processes which transform them, while for others they are not distinct from the process of transformation, but are *part of general thinking process*, which is symbolic in its nature.

Definitions and methods of studying imagery varied through time between the two poles of perception and high level thinking (see figure 1): in 1964 Holt[1] defined imagery as a faint

Perception	High Level Thinking
Representation with specific properties	Symbolic function
Figural code	Propositional code
Analogic model - Quasi-perceptual copies of reality - Measurement: reaction time during generation, inspection, transformation	Propositional model - Symbolic constructions - Measurement: qualitative analysis of productions

Figure 1. Representational continuum.

representation of perception without sensorial input; in 1967 Neisser[2] suggested that it is a *"quasi perceptual experience"*, which anticipates perception, derives from precedent sensorial input and can facilitate or interfere with the perception of a new stimulus; between 1970-1980 Kosslyn, Shepard and others have been strenuously defending the thesis that image is a *specific mode of representation,* that mental images really exist and are distinct from the process which operate or transform them. According to Kosslyn[3-5] they are not photographs or picture copy of reality, but they are similar to patterns on a TV screen; they occur in a visual buffer, that is a matrix of locations for storing information, which is functioning as if it had spatial properties. The distance among the regions of an object are preserved in imaged representation, and similarly the shape and appearance of the object, that can be inspected, scanned, analysed in details, compared with other object or transformed by shift or blink transformation: the only difference with perception is that it can be generated, stored and retrieved.

On the other side since 1966 Piaget and Inhelder[6] conceived a well organized model which does not relate imagery to perception, but to *symbolic function,* taking its origin from imitation and internalized action.

For these authors Mental Imagery is a product of accommodation of sensory-motor schemes to objects and events: the schema of grasping or looking, when it is applied to an object, imitates the object figural properties, but during growth the infant becomes capable of imitative accommodation, of differed imitation, and later of intellectual operations which are based on mental anticipation, transformation, construction. So the *Image is a symbolic function*, and with language it contributes to the essential nature of intelligent thinking.

The divergence of positions reflects itself in different methods of studying Imagery: the cognitivists measure the *time* required to generate and inspect the images (by scanning and rotation tasks), while Piaget and Inhelder study *qualitative aspects* of productions like drawings, gestural or verbal descriptions and claim that reaction time measures miss the extreme variety of images, static and dynamic, reproductive or anticipatory.

The development of researches, which rapidly expanded in the '80, presents a passage from the *structuralist models,* which study the analogies between MI and visual perception, arguing a correspondence or identity of subjacent mechanism, to *the functional models,* which focus on the correspondence or equivalence of functions[7,8]. The structuralist researchers use tasks in which the subject is instructed to form an image which then is scanned, analysed for details, compared with other objects, rotated, and the times of these operations are measured, while the functionalists study the way in which MI operates when subjects perform a task which requires the generation of imagery: in this case the imagery is not given in the instructions, and its presence and functions are deduced by the results in the task and by the verbal reports of the subject.

In a third phase the focus is on the *active interaction of MI with other functions,* stressing the role of the subject interpretation of instructions, individual differences, and studying the role of imagery in the productive processes of creative thinking and problem solving (see figure 2).

2. IMAGERY AND PERCEPTION: THE ANALOG-PROPOSITIONAL DEBATE

The debate has been opened by Pylyshyn[9] in a paper in which he sharply criticizes the photographic metaphor, observing that if each MI should correspond to an object or event, the memory storage should contain an infinite number of images isomorphic to real objects, and these should be static and rigid. As an alternative he suggests a *propositional model* according to which all informations are coded in an abstract format, defined by the nature of the elements and by the relationships between them: these relationships are governed by propositional rules, and the abstract format enables us to symbolize both figural and verbal informations. The propositional mode is more economic, since it assumes only one format for the storage of information, with the possibility to activate different codes. "This representation comprehends words and images, it is a sort of inter-language"[10].

On the other end, the supporters of the *analogue model,* like Kosslyn, Shepard and coworkers adfirm that MI is very similar to perceptual experience: the "weak version" implies that the Image of an object maintains the same spatial informations given by the real object (distance between parts, localization of the details, and so on); the "strong version" adfirms that spatial information is processed and stored in a spatial medium, so that the proximity of two objects is physically represented in the mind, and the representations in Imagery is very similar to the representation in perception*.

Many experiments seem to support the analogic model: for example Shepard and coworkers (1975 → 1982) have shown that when we compare a tridimensional geometrical figure with another rotated in the space, the time required to compare the images is proportional to the angle of rotation of the figure; Kosslyn, Ball and Reiser[3] have shown that during the mental exploration of the imaged map of an island, the time required to scan distant details in mental imagery is proportional to the distance of the objects in the real map (drawing). Among

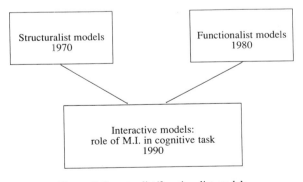

Figure 2. Structuralist/functionalist models.

---

* The term "analogue" derives from computers in opposition to digital. It designs the representation which preserves certain physical features of the object, like in the Electrocardiogram.

the many experiments we could quote, I'll chose an italian one made by Simion, Bagnara and Umiltà[11] (see figure 3a).

The large figure ⊟ is presented to the subjects, than they must imagine the letter and compare it with other symbols that result from the subtraction of one or more segments from the original figure. In the condition "perception" the figure and the new stimula are presented in pairs, in the imagery condition only the new stimulus is presented, and the subject compares it with the image of ⊟ figure: the time required by the answer is registered. The time for "same" response is shorter than for "different"; in the latter case the time decreases with the number of missing segments.

Figure 3b shows that the reaction times have the same profile for the condition of imagery and perception: they are faster when the two stimuli are equal, they became longer when they are different. This suggests that the processes for comparing two objects are similar in imagery and perception, and that images possess a detailed structure, which permits to analyse elementary components: but the generation of an image requires more time than the perception of an object, and also the time for the analytical inspection of details is longer.

The results support the hypothesis by Finke[8] of *functional equivalence:* images and perceptions would be equivalent for their properties and for operations that can be done.

2.1. Neuro-Psychological Evidence

Moreover, some researchers in the seventies[14,15] tried to demonstrate that subjects with selective brain damage which impair perception show corresponding impairments in visual imagery. De Renzi and Spinnler[16] in a large group of unilaterally brain damaged patients found an association between impairment on colour *vision* tasks and colour *imagery* tasks (such as verbally reporting the colour of common objects from memory). Bisiach and Luzzatti[15] found that patients with hemispatial neglect for perception of visual stimuli presented the same deficit in mental images: they asked two patients to *imagine* the Dome square in Milan sitting *before* the Dome, or *in front of* the dome, and they found that details in the left side of the scene were omitted in imagery, although all the patients know very well the place. The result seems a strong

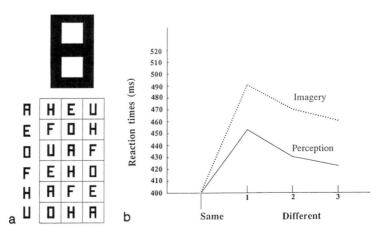

Figure 3. Experiment of Simion, Bagnara, and Umiltà: a) material, b) results (reproduced from Simion et al.[11], Taylor[12], and Bagnara et al.[13], with kind permission from Psychonomic Society Inc.).

evidence for localization of mental imagery, but recent research has introduced a variety of possible localizations, and has shown that in other cases there is loss of imagery following focal brain damage, but perception is still intact. Many neuropsychological researches have tried to show that the right hemisphere is specialized in image generation, but others have produced variable results which question the right hemisphere hypothesis[17].

Farah[18,19] after an exhaustive review of the literature concludes that probably there are different types of image generation, with different hemispheric loci.

2.2. Experimental Evidence

Logie and Reisberg[20] defend the hypothesis of a dissociation between visual and spatial imagery: the first system acts as a form of *inner eye* closely associated with visual-perceptual system, which would store static images, while the spatial imagery system acts as an *inner scribe* closely associated with motor planning, and would be able to redraw images and allow manipulation and transformation. Visual and spatial imagery act together in the visual working memory: the inner eye holds a pattern, and is closely linked with visual perceptual system, while the inner scribe may be linked with retention and control of movement sequences. Both cooperate in feeding reciprocal information: visual information guides the moving hand towards a target, but the motor system informs the visual system when planning and retaining sequences of movement.

Massironi[21] in many ingenious experiments which do not measure imagery using reaction times, but which infer imagery from the productions of subject in parallel tasks with stimulus present or absent, critically examines analogies between perception and imagery, and finds some differences: for instance in imagery the ambiguous figures do not show reversibility like in perception; the evaluation of vividness of a critical letter in a matrix is different in the conditions of imagination and perception: pre-attentive processes in the two conditions are not identical, because in imagery *pop-out* effect and *priming* are absent; moreover, the results very strongly reflect the importance of instructions even in same or similar tasks: that is, while perception depends on physical properties and organization of the visual field, imagery depends on cognitive strategies and on precedent knowledge of the world (*tacit knowledge* by Pylyshyn).

We can see, then, that recent experimental research on the relationship imagery/perception and imagery/memory on one end confirms the specificity of imagery, its existence as a specific modality of representation but on the other end it is incorporating the model of *tacit knowledge* suggested by Pylyshyn, showing that imagery processes are near always penetrated by precedent knowledge: the subject when is scanning an image or performing a mental rotation task, in effect is transferring his knowledge of physical objects and events: he knows by experience that a 180° degree rotation requires more time than a 60° degrees, that a larger distance requires more time in exploration, and in the imagery tasks he reproduces these relationship. Moreover mental scanning and mental rotation can be influenced by varying the instructions given to the subject, who adjusts his belief and goals to the task. So the function of imagery is *cognitively penetrated* by other cognitive processes more than guided by specific properties of an analogue mechanism[10].

2.3. Developmental Contributions

This position is very similar to the Piagetian model, proposed in 1966 and recently confirmed with abundance of empirical evidence by Dean[22] who studies the development in children.

According to Piaget-Inhelder[6] MIs are not copies of external reality, but symbols of objects as known by the subject. The child imitates objects more and more complex, passing

from static to moving patterns and internalizing the actions; anticipatory images are constructions which facilitate the development of thinking and help the passage from concrete to
abstract operations, since they are based on symbolic function which enables to evoke an absent
object or event and to plan new actions. Figural aspects and operational aspects are developing
together, but the figural aspects are present in perception and static images, while imitation and
language are linked to symbolic function, which supports their mobility, flexibility, the
capacity to transform representations from reproduction of reality to new constructions, new
possibilities, new models. In some way these ideas anticipate the most recent theories of mental
models by Johnson-Laird[23] and the studies on analogical thinking.

Even the Piagetian observation that the source of imagery is internalized action looks a
very close forerunner of the experiments of Logie on the relationship of visuo-spatial system
with the motor system, and indicates very interesting new areas of research as interference
studies of working memory, visuo-spatial coding of movement, effect of motor imagery on
motor activity in sports and in neurological impairment, response programming and executive
mechanism in neuro-motor rehabilitation (for a large discussion see the papers by Annett[24],
Engelkamp[25], and Logie[26]).

The provisional conclusion of the last 25 years studies is that it is near impossible to
exclude from the experimental research the role of tacit knowledge and to trace exact limits
between Imagery and perception, Imagery and memory, Imagery and thinking because in
complex and realistic tasks they continuously interact. Mental Imagery seems to have a large
variety of forms, from quasi-pictorial, realistic images, rich in vividness, details, colours and
near isomorphic to perceptions and abstract images, like geometrical shapes, graphs, diagrams,
maps, models. They have a general common aspect because they are a representation of an
absent object or event, but they might use different codes according to the perceptual modality
that originated them, stored them in Long Term Memory (LTM) and promoted retrieval (there are
still few researches on acoustic imagery and we are not sure in which code they might be preserved).

For instance, Mozart claimed that he could hear a whole symphony in his imagination all
at once, understanding its structure and beauty, but we do not know if the word "hear" denoted
an auditory experience, with a succession of parts, or an abstract representation: in fact, if his
imagination was, as he said, "all at once" it could not have reproduced the succession of parts
and their duration in a few minutes. So we can think that he referred to an abstract representation
of the structure, with some auditory, concrete imagery mixed to the musical symbols.

The different type of images continuously and dynamically interact with other cognitive
processes, and very often the subjects utilizes them *as a strategy* that is good for certain tasks
but not for other or for certain parts of the task, even though the subjects usually have a prevalent
style and can be differentiated in high and low imagers. So the actual research is moving
towards the study of individual differences and of the relationship between MI and other
cognitive functions.

3. IMAGERY AND MEMORY

The study of this relationship starts in the same period of the research on Imagery and
Perception (Seventies), and shows a similar trend, passing from discussion about the structure
to discussion about the function and arriving finally to an interactive model. The first actor in
the scene is Paivio[14], who was probably the first to revaluate the role of Mental Imagery in
cognitive process, and to demonstrate the possibility of studying its functions by objective methods.

He introduced the concept of *Imagery value* of words, and showed that pictures are
recalled better than words and that High Imagery words are recalled better than pictures, even

if we partial out the effect of concreteness and high associative value. His theory of *dual code* says that Imagery and verbal structures are two modalities of representing and storing information that cooperate in storing LTM but they are different ways of processing because visual coding works in parallel, while verbal code works in sequential way; visual code is continuous, analogue, synchronous while verbal code is made by discrete units which can be analysed. Stimuli which evoke imagery ale better stored and retrieved, because they use a dual code.

A basic premise of Paivio model[14] is that mental representations retain some of the concrete qualities of the external experience from which they are derived: but in the most recent reformulation of the model[27] he emphasizes that the two systems develop multiple connections *within and between systems* (*associative* networks which are dynamic and variable): so they are influenced by instructions, context, pragmatic goals of the subject who can activate one of the two codes or both.

This positions bring us again to the possibility of a storage of information in amodal format, which reminds us the propositional model; and, like in the study of perception, we find some results that weaken the more realistic interpretation of the model as a stimulus driven representation: the *Imagery value* of words is not fixed, but changes according the personal experience of the subjects and even, in the same subject, in different level of vigilance (in reverie condition the imagery value is higher than in wake)[28]; the *concreteness effect is* evident in remembering list of words and associated pairs, but it disappears in free recall of prose where the effect of imagery is joined to relational process for the comprehension of the text[29-31]. It is interesting to remark that Marschark started his researches with Paivio, and now he adfirms that "there is a common, *conceptually based* mechanism involved in the comprehension and memory for high imagery and low imagery material[32]. In short, it seems that also in memory, like in perception, the function of Imagery is "cognitively penetrated" and, while it shows some specific character, it is very difficult to be discriminated from verbal code, although it may be prevalent in the first phase of remembering and in flash-bulb memories.

4. IMAGERY AND THINKING

4.1. Imagery in Deductive Reasoning and in Problem Solving

The relationship of MI to thinking has been studied in the last years from different points of view, which depend on various models of imagery at one end, on models of thinking at the other.

A lucid review of this field is provided by Denis[33-35] who summarizes the most important contributions in the last decade, starting from the definition of the two terms.

Imagery can be conceived as a *set of representational process which have a structural or functional isomorphism* (of 1st or 2nd order) *with objects* or as a *set of transformations and construction of possible reality.* Thinking in some cases has been described as a manipulation of abstract representations and logical rules opposed to the concrete representation of Imagery, but in other cases it has been identified with imagery processes. For the *conceptualist theories* the deep nature of thinking is conceptual and abstract, although it can use figural symbols for the surface structure, as an expressive medium: so Image is not the core of thinking processes but rather a potential medium; on the other end for the *symbolists theories* thinking is working only through mental symbols, which may have different formats. These formats usually are reduced to verbal and visual representation, but each one includes various forms of notations: words, numbers, musical notes, pictures, maps, graphs, etc.

The most important topic in the last decade has been the study of strategies in deductive reasoning, in problem solving, and finally the construction of mental models.

Many of the experimental studies show that strategies vary between subjects and that spatial strategies are more useful in the early phases of the task, while with increasing practice the subjects prefer linguistic strategies based on propositional coding[5, 36, 37], Johnson-Laird[38] claims that representation based on imagery in transitive inference facilitates the construction of mental models, but these very frequently transform in propositional representations.

Recently there has been an increase in research on the role of imagery on various aspects of thinking problems: researches have moved from the study of syllogistic reasoning to problems of comparative judgments (visual and semantic processing of stimuli in information processing, in memory and in responses), to predictive reasoning[39, 40], to processing of verbal description[35], to mathematical or physical problem solving[41-44]; but the most interesting area of research seems to be analogical reasoning, which shows large potential in order to connect the field of human reasoning with the problems of distributed computer system[45].

Here I give some examples of the topics which can be grouped around the theme of analogical reasoning (they are taken from the book by Vosniadou, and Ortony[46]):
- *similarity, typicality and categorization* (Rips);
- *similarity and decision making* (Smith, Osherson);
- *the development of similarity* (from perceptual similarity of attributes to relational similarities to the production of analogies and the construction of a relational knowledge system) (Vosniadou);
- *the role of explanation in analogy* (Dejong);
- *the mechanisms of analogical learning* (Gentner);
- *computational models of analogical problem solving* (Holyoak, Zagard);
- *analogy, mental models and creativity* (Johnson-Laird);
- *similarity and analogy in development, learning and instruction* (Brown, Vosniadou, Bransford and others).

I shall give a short flash of some results from this area, but first I would like to sum up some properties of images which are directly relevant to thinking processes[34]:
1) *Structural similarities* of images to perceptual representations permit to play a simulation of parts and relationships (map, machine schemes) and of operations (instruction for use).
2) *Economy:* imagery presents in parallel a large number of units in integrate whole. Children learn from images more than from a verbal description, and show more use of imagery in memory. Also illiterates comprehend pictures, strips, TV better than verbal explanations: but the effect of picture may be negative if it contains too many irrelevant details. The image representation must be *simplified* in symbolic code; otherwise, if we conceive the images as quasi pictorial or photographic reproduction of the reality, their number should be infinite.
3) *Flexibility and transformation:* images enable to code the data of the problem and information in efficient way; in a second phase they can be transformed, changing some parts, or representing the initial object by symbols. Transformations too should be symbolized. The use of mathematical formulae, geometrical figures and graphs are examples of transformations of real objects from realistic to abstract representations of things that do not exist in reality (point, graphs, diagram).
4) Imagery can provide an *internal medium for invention and discovery.*
Often the realization of a new idea or invention is largely not anticipated, not planned in advance, but if follows, or it happens, from the exploration of the structure of an imagined form, from the exploration of combinations and possibility (this morning Caianiello provided us with the autobiographic description of this process). MIs are *pre-inventive forms* which result from the combinational play of visualization; but the choice must be done developing the shapes and structures that seem potentially useful[8].
The pre-inventive forms lead the subject to discover the kinds of solution that might bear on a particular class of problems, they can mediate between different problems and solutions

which have an underlying structure in common. The process starts from active exploration and manipulation of physical objects, produces mental representation, passes through combinations, transformations of spatial relationships, discovers inexpected things and concepts, explores possible interpretations restricting them first to a general class of objects, then to a particular object. Invention in this way is a product of combinational play that often shifts into creative thinking.

5) These observations by Finke go in the same direction of the studies on *metaphoric thinking*. Imagery nurtures *metaphoric thinking,* that is very close to pre-inventive form: indeed it could be defined a pre-inventive verbal form.

4.2. Analogical Learning and Reasoning

In recent years studies on learning, reasoning and instruction have directed attention to the role of analogy in the development of higher thinking processes.

Gentner[47] has proposed a model of analogical mapping which decomposes analogical learning in subprocesses, which can be simulated on computer.

The central idea is that an analogy is a mapping of knowledge from one domain (the base) into another (the target), which conveys a system of relations that holds among the base objects and the target objects. Thus an analogy is a way of focusing on *relational commonalities* independently of the objects in which these relations are embedded. In interpreting an analogy, people seek to put the objects of the base in one to one correspondence with the objects of the target, in order to obtain the maximum structural match[47]. To illustrate the structure mapping rules, Gentner offers the example of the analogy between heat flow and water flow (see figure 4).

Temperature in the heat situation plays the role of pressure in the water situation. There is an object correspondence (pressure-temperature) (greater pressure causes water-flow, greater temperature causes heat-flow) (see figure 5). The analogy conveys that a common relational systems holds for the two domains; it is a *mapping of relational predicates* (see figure 6). But this analogy based on relational structure lies on a continuum that goes from literal similarity ("coffee is like water"), mere appearance match based on physical appearance ("the glass gleamed like water"), analogy *(water flow and heat flow)* and abstraction *(heat is a through-variable)** (see figure 7).

Literal match and mere appearance match are based on perceptual similarity, while analogy is based on relational similarity. The figure facilitates the passage to the relational level, and permits to shift the focus from the single objects to the common relationship.

In development we see the passage from early representation, rich and perceptually based, to analogies based on relational structures; the storage through images, retrieved in spontaneous recall, will allow accurate mapping without surface support and facilitate the solution of analog problem (see figure 8). Even more, it will suggest to extend the relational system to other contents, permitting new discoveries. Transferring a structure from an object to another, or from a domain to another, permits to understand analogies that explain different phenomena, although it may bring to errors.

Similar errors by wrong or over-generalization are made by adults too, when they face a new domain; but the process of analogical reasoning, with its shifting from perceptual to abstract domains, is fundamental for the transfer of learning, for restructuring the knowledge base and constructing new models. Sometimes the perceived similarity may be due to a fortuity case, like in the observation of the apple by Newton, of the pendulum by Galileo and of the mould by Fleming who discovered penicillina: but the observations must be developed by deductive and inductive reasoning[48].

---

* Through variable is an abstract principle, "heat" is a concrete variable.

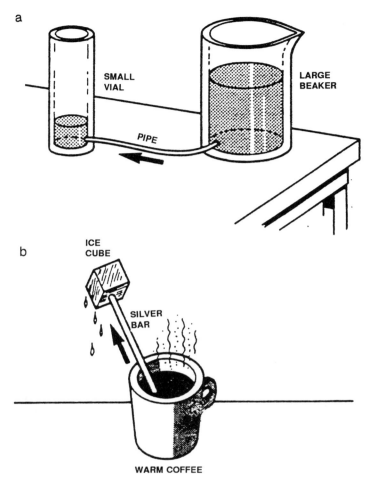

Figure 4. Analogy in physical phenomena. Examples of physical situations involving a) water flow and b) heat flow (reproduced from Gentner[47], with kind permission from Cambridge University Press).

Other times the observation that two systems have the same relational structure derives from the perceived similarity of the equation or the curves: for instance de Broglie noticed that Bohr's equations describing the orbits of electrons in an atom were the same equations used to describe the waves of a violin string; then he proposed a wave theory of matter that brought a revolution in atomic physics and founded quantum mechanics.

So the perceived analogy between two phenomena can lead to restructuring the knowledge base, following the implications of the analogy until the previous theory is changed[49].

4.3. Metaphoric Thinking

A particular interesting form of analogical thinking appears in language and it is metaphor.

Metaphoric thinking has been studied recently by Kogan[50] as an aspect of cognitive style and as a type of divergent thinking which requires the production of new ideas. It refers to

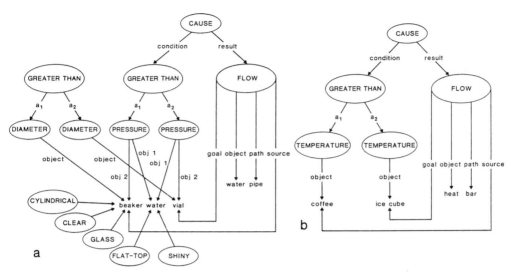

Figure 5. a) A representation of the water situation. Predicates are written in upper case and circled; objects are written in lower case and uncircled. b) A representation of the heat situation that results from the heat/water analogy (reproduced from Gentner[47], with kind permission from Cambridge University Press).

similarity between not related objects or categories, which share a criterial attribute but differ among other dimensions. It is typically expressed by *words which contains some imagery* - visual but also auditory or tactile; the figurative language expresses in new way a similarity, through a conversion from verbal to visual code and vice-versa. The relative facility which

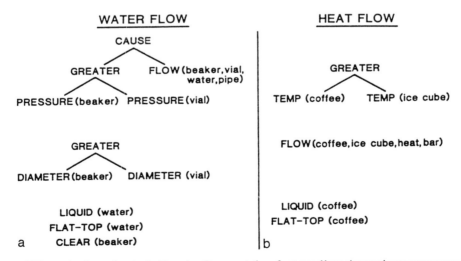

Figure 6. The mechanisms of analogical learning. Representation of water and heat given to the structure-mapping engine (reproduced from Gentner[47], with kind permission from Cambridge University Press).

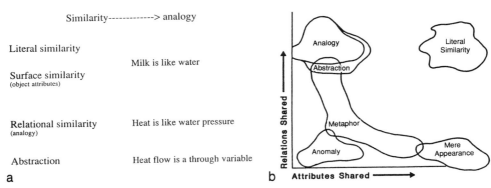

Figure 7. a) Classes of similarities. b) Similarity space: classes of similarity based on the kinds of predicates shared (reproduced from Gentner[47], with kind permission from Cambridge University Press).

people, even if children, have in comprehending and producing metaphors, is an ulterior evidence of the reversibility of verbal and figural code, or, if you prefer, of the cognitive penetrability of imagery, joined with the verbal imagery value of the words. But this mixture

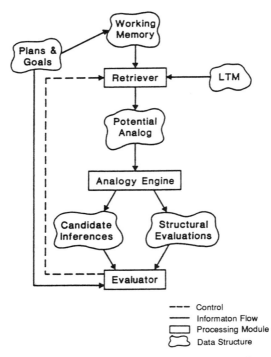

Figure 8. An architecture for analogical processing (reproduced from Gentner[47], with kind permission from Cambridge University Press).

has also a high generation power: Boyd[51] speaks about *theory constitutive metaphors* and Gruber[52] of *images of wide scope* because often they are used as figures of thought, as embryonal models, first step in generating hypothesis and thesis; in addition, they can be used to explain difficult concept and theories in more concrete way, even if they have the same danger of pictures in emphasizing collateral irrelevant details. On the other end, they immediately transmit new ideas, and often direct creative thinking: Osowski[53] nicely analyses the concept of consciousness in William James as a system evolving from the metaphor of STREAM which include change, continuity, movement, flow, alternation among flight and resting.

So we have seen that imagery has some common aspect with perception, to which it adds a more constructive and dynamic character; during the storage in LTM only few of these images preserve their figural and concrete properties, while the largest part is probably scored in an abstract, amodal code, a sort of dynamic scheme which can activate images and words separately or together in concrete or in abstract form; in the process of activation and new constructions, images interact with other cognitive processes and permit to explore new possibilities which are realized in creative products (analogical reasoning, metaphoric and visual thinking) when they are integrated with logical processing, reality testing, and guided by goal and motivation of the subject. Thus we can have the transformation of pre-inventive form in technical design, or in artistic products; we can have the development of metaphoric thinking in the complex network of a model, and we can have the development of analogical reasoning towards new discoveries.

Figure 9. Sky and water I (reproduced from Locher[54], Copyright (1994), M.C. Escher / Cordon Art, Baarn, Holland. All rights reserved).

5. CONCLUSIONS

At this point the research on mental imagery has taken us in the fascinating but quick sand field of creative thinking, in which images can be only a starting point, or perhaps the end point, but they always interact with logical processes. In the history of research we have seen that the previously opposite positions have been softening and converging in many points.

So we have arrived to our final definition: there is a large variety of images which lie on a continuum from perception to imagination and creative thinking: they are a form of representation which may be generated in every phase of the cognitive processes, in initial phase when they anticipate the identification of the stimulus, in working memory when they code the material by visuo-spatial code, in LTM as dynamic schemata which can activate the visual or verbal code, in analogical reasoning, when the perceived similarity is the source of the relationship between objects and systems, in metaphoric thinking which often generates models, in imagination which constructs new worlds. They can be concrete, vivid representations based on sensorial inputs, or they can be abstract as in geometrical or other symbolic form.

In synthesis, they can be fishes swimming under the water, or in the surface, and they can transform in birds flying in the light, as in the picture by Escher of figure 9.

REFERENCES

1. R.R. Holt, The return of the ostracized, *American Psychologist,* Vol.19, pp.254-264 (1964).
2. U. Neisser, *Cognitive Psycllology,* Prentice-Hall Inc., NJ (1967) (It.tr. Martello-Giunti, 1976).
3. S.M. Kosslyn, T. Ball, and B.J. Reiser, Visual images preserve metric spatial information: evidence from studies of image scanning, *Journal of Experimental Psychology: Human Perception and Performance,* Vol.4, pp.47-60 (1978).
4. S.M. Kosslyn, *Image and Mind,* Harvard University Press, Cambridge, MA (1980).
5. S.M. Kosslyn, *Ghosts in the Mind's Machine: Creating and Using Images in the Brain,* Norton and Co., New York, NY (1983) (It.tr. Giunti, 1989).
6. J. Piaget and B. Inhelder, *L'Image Mentale Chez l'Enfant,* Presses Universitaires de France, Paris, F (1966) (It.tr. La Nuova Italia, Firenze, I 1974).
7. R.N. Shepard and L.A. Cooper, *Mental Images and their Transfonnations,* MIT Press, Cambridge, MA (1982).
8. R. Finke, *Creative Imagery,* L. Erlbaum, New York, NY (1990).
9. Z.W. Pylyshyn, What the mind's eye tells the mind's brain: a critique of mental imagery, *Psychological Bulletin,* Vol.80, pp.1-24 (1973).
10. Z.W. Pylyshyn, The imagery debate: analogue media versus tacit knowledge, *Psychological Review,* Vol.88, pp.16-45 (1981).
11. F. Simion, S. Bagnara, and C. Umiltà, Immaginare e vedere: le caratteristiche dei processi di generazione e di confronto delle immagini visive, *Giornale Italiano di Psicologia,* Vol.XV, pp.491-513 (1988).
12. D.A. Taylor, Holistic and analytic processes in the comparison of letters, *Perception and Psychophysics,* Vol.20, pp.187-190 (1976).
13. S. Bagnara, F. Simion, M.E. Tagliabue, and C. Umiltà, Comparison processes on visual mental images, *Memory and Cognition,* Psychonomic Society Publications, Vol.16, pp.138-146 (1988).
14. A. Paivio, *Imagery and Verbal Processes,* Holt, Rinehart, and Winston, New York, NY (1971).
15. E. Bisiach and C. Luzzatti, Unilateral neglect of representational space, *Cortex,* Vol.14, pp.129-133 (1978).
16. E. De Renzi and H. Spinnler, Impaired performance on colour tasks in patients with hemispheric lesions, *Cortex,* Vol.3, pp.194-217 (1967).
17. M. Liotti and D. Grossi, Le immagini mentali. Dati e modelli psicologici e neuropsicologici, *Giornale Italiano di Psicologia,* Vol.15, pp.381-411 (1988).
18. M.J. Farah, Is visual imagery really visual? Overlooked evidence from neuropsychology, *Psychological Review,* Vol.95, pp.307-317 (1988).
19. M.J. Farah, Neuropsychological studies of mental imagery, in *Fewic, Fourth European Workshop on Imagery and Cognition,* Puerto de la Cruz, Tenerife (1992).
20. R.H. Logie and D. Reisberg, Inner eyes and inner scribes: a partnership in visual working memory, in *Fewic, Fourth European Workshop on Imagery and Cognition,* Puerto de la Cruz, Tenerife (1992).

21. M. Massironi, Quante gocce di percezione nel cocktail di un'immagine mentale?, in *Le Immagini Mentali*, L. Vecchio ed., La Nuova Italia, Firenze, I (1992).

22. A.L. Dean, The development of mental imagery: a comparison of piagetian and cognitive psychological perspectives, *Annals of Child Development*, Vol.7, pp.105-144 (1990).

23. P.N. Johnson-Laird, *Mental Models: Towards a Cognitive Science of Language, Inference and Consciousness*, Cambridge University Press, Cambridge, MA (1983).

24. J. Annet, Imaginary Actions, *Fourth European Workshops on Imagery and Cognition*, Puerto de la Cruz, Tenerife (1992).

25. J. Engelkamp and G. Mohr, Memory for ordinal size information and selective interference, *Fourth European Workshops on Imagery and Cognition*, Puerto de la Cruz, Tenerife (1992).

26. R.H. Logie and D. Reisberg, Inner eyes and inner scribes: a partnership in visual working memory, *Fourth European Workshops on Imagery and Cognition*, Puerto de la Cruz, Tenerife (1992).

27. A. Paivio, *Mental Representations. A Dual Coding Approach*, Oxford University Press, Oxford (1986).

28. O. Andreani, B. Bidone, and G. Sangiorgi, Processi mnestici e immaginativi in stato di veglia e in stato ipnagogico, *Archivio di Psicologia, Neurologia e Psichiatria*, Vol.XLVIII, No.1, pp.32-61 (1987).

29. M. Marschark, *The functional role of imagery in cognition?* in *Cognitive and Neuropsychological Approaches to Mental Imagery*, M. Denis, J. Engelkamp, and J.T.E. Richardson eds., Martinus Nijhoff Publishers, Dordrecht (1988).

30. M. Marschark and C. Cornoldi, Imagery and verbal memory, in *Imagery and Cognition*, C. Cornoldi and M.A. Mc Daniel eds., Springer-Verlag, New York, NY (1990).

31. M. Marschark and C.J. Huffman, Memory for prose: the case of the disappearing concreteness effect, in *Fourth European Workshop on Imagery and Cognition*, Puerto de la Cruz, Tenerife (1992).

32. *Fourth European Workshop on Imagery and Cognition. Abstracts & extended summaries*, Puerto de la Cruz, Tenerife (1992).

33. M. Denis, *Les Images Mentales*, P.U.F., Paris, F (1979).

34. M. Denis, Imagery and thinking, in *Imagery and Cognition*, C. Cornoldi and M.A. Mc Daniel eds., Springer-Verlag, New York, NY, pp.103-131 (1990).

35. M. Denis, Imagery and the description of spatial configurations, in *Fourth European Workshop on Imagery and Cognition*, Puerto de la Cruz, Tenerife (1992).

36. R.J. Sternberg, Representation and process in linear syllogistic reasoning, *Journal of Experimental Psychology: General*, Vol.109, pp.119-159 (1980).

37. R.J. Sternberg, Reasoning with determinate and indeterminate linear syllogisms, *British Journal of Psychology*, Vol.72, pp.407-470 (1981).

38. P.N. Johnson-Laird, Images, models, and propositional representations, in *Fourth European Workshop on Imagery and Cognition*, Puerto de la Cruz, Tenerife (1992).

39. M. De Vega and M.J. Rodrigo, Updating mental scenarios in predictive reasoning, paper presented at the *2nd Workshop on Imagery and Cognition*, Padova, I (1988).

40. J. Mayor, J. Gonzalez-Marqués, F.J. Sainz, and A.G. Suengas, Comparative judgments in transitive inferences, paper presented at the *2nd Workshop on Imagery and Cognition*, Padova, I (1988).

41. A. Antonietti and C. Angelini, La ristrutturazione nel problem-solving attraverso disegni schematici, *Contributi del Dipartimento di Psicologia dell'Università Cattolica di Milano*, IV, pp.5-15 (1990).

42. A. Antonietti, Why does mental visualization facilitate problem-solving? in *Mental Images in Human Cognition*, R.H. Logie and M. Denis eds., North Holland, Amsterdam, NL, pp.211-227 (1991).

43. A. Antonietti, Immagine mentale nella soluzione di problemi, in *Le Immagini Mentali*, L. Vecchio ed. (1992).

44. O. Andreani, L. Vecchio, and M. Bardone, Individual differences in imagery and their functional role in divergent and convergent problems, in *Fourth European Workshop on Imagery and Cognition*, Puerto de la Cruz, Tenerife (1992).

45. C. Santamaría, The use of analogical models in reasoning, in *Fourth European Workshop on Imagery and Cognition*, Puerto de la Cruz, Tenerife (1992).

46. S. Vosniadou and A. Ortony, *Similarity and Analogical Reasoning*, Cambridge University Press, Cambridge (1989).

47. D. Gentner, The mechanisms of analogical learning, in *Similarity and Analogical Reasoning*, S. Vosniadou and A. Ortony eds., Cambridge University Press, New York, NY, pp.197-241 (1989).

48. S. Vosniadou, Analogical reasoning as a mechanism in knowledge acquisition: a developmental perspective, in *Similarity and Analogical Reasoning*, Cambridge University Press, Cambridge (1989).

49. S. Vosniadou and A. Ortony, *Similarity and Analogical Reasoning*, Cambridge University Press, Cambridge (1989).

50. N. Kogan, A cognitive style approach to metaphoric thinking, in *Aptitude, Learning and Instruction*, Vol.I, R. Snow, P.A. Federico, W. Montague, and L. Erlbaum eds., Hillsdale, NJ, pp.247-282 (1980).

51. R. Boyd, Metaphor and theory change: What is metaphor?, in *Metaphor and Thought*, A. Ortony ed., Cambridge University Press, Cambridge (1979).

52. H.E. Gruber, The evolving systems approach to creative work, in *Creative People at Work,* D.B. Wallace and H.E. Gruber eds., Oxford University Press, New York, NY, pp.3-24 (1989).

53. J.V. Osowski, Ensembles of metaphor in the psychology of W. James, in *Creative People at Work,* D.B. Wallace and H.E. Gruber eds., Oxford University Press, Oxford, UK, pp.127-146 (1989).

54. J.L. Locher (ed.), *The world of M.C. Escher,* Meulenhoff International, Cordon Art, Amsterdam, NL (1971).

Interested readers could have additional details in the following papers:

55. O. Andreani, Logical and creative thinking in adolescents, in *Proceeding of the 3rd European Conference on High Ability,* München, in press.

56. C. Cornoldi and M.A. McDaniel, *Imagery and Cognition,* Springer-Verlag, New York, NY (1990).

57. R. De Beni and F. Giusberti, Conoscenze esplicite e tacite sulle immagini mentali. Uno studio metacognitivo, *Ricerche di Psicologia,* 2, pp.58-75 (1990).

58. J. González Margués and J. Sainz, The modality of task in comparative judgements, in *Fewic, Fourth European Workshop on Imagery and Cognition,* Puerto de la Cruz, Tenerife (1992).

59. P.N. Johnson-Laird, Analogy and the exercise of creativity, in *Similarity and Analogical Reasoning,* Cambridge University Press, Cambridge, pp.313-331 (1989).

60. S.M. Kosslyn, J. Brunn, K.R. Cave, and R.W. Wallach, Individual differences in mental imagery ability: a computational analysis, *Cognition,* Vol.18, pp.195-243 (1984).

61. R.H. Logie and M. Denis eds., *Mental Images in Human Cognition,* Elsevier, Amsterdam, NL (1991).

62. E. Mellet, N. Tzourio, U. Pietrzyk, L. Raynaud, M. Denis, and B. Mazoyer, Visual perception and mental imagery: a PET activation study, in *Fewic, Fourth European Workshops on Imagery and Cognition,* Puerto de la Cruz, Tenerife (1992).

63. P.E. Morris and P.J. Hampson, *Imagery and consciousness,* Academic Press, London, UK (1983).

64. J.T.E. Richardson, Imagery and brain, in *Imagery and Cognition,* C. Cornoldi and M.A. McDaniel eds., Springer-Verlag, New York, NY (1990).

65. J. Sergeant, The neuropsychology of visual image generation: data, method and theory, *Brain and Cognition,* 55. R.J. Stemberg, *The Nature of Creativity,* Cambridge University Press, Cambridge (1988).

66. L. Vecchio (ed.), *Le Immagini Mentali,* La Nuova Italia, Firenze, I (1992).

67. D.B. Wallace and H.E. Gruber, *Creative People at Work. Twelve Cognitive Case Studies,* Oxford University Press, Oxford, UK (1989).

NEURAL NETWORKS, FUZZINESS
AND IMAGE PROCESSING

Eduardo R. Caianiello and Alfredo Petrosino

Facoltà di Scienze, Università di Salerno
I - 84081 Baronissi (SA), Italy

ABSTRACT

A fuzzy neural network system suitable for image analysis is proposed. Each neuron is connected to a windowed area of neurons in the previous layer. The operations involved follow a method for representing and manipulating fuzzy sets, called *Composite Calculus*. The local features extracted by the consecutive layers are combined in the output layer in order to separate the output neurons in groups in a self-organizing manner by minimizing the fuzziness of the output layer. In this paper we focalize our attention on the application of the proposed model to the edge detection based segmentation, reporting results on real images and investigating the robustness of the system with noisy data.

1. INTRODUCTION

... a thinking machine built for some special purpose may well need organs, or subsystems, organized quite differently from those of the animal brain, although the same elementary laws will be valid.

Without a complete mathematical control of the situation, a machine may perhaps think, but one would hardly know how or why. (E. R. Caianiello, 1961)[1].

Since from early sixties *thinking machines* or, equivalently, *artificial neural network -* based systems, have represented along years the matter of study of many researchers belonging to different fields of science. Pioneering insights and contributions of scientists as Hebb, Rosemblatt, Caianiello have notably contributed to the increasing of neural network research along years. Their basic aim was to emulate the human neural information processing system, making the system artificially intelligent, i.e. able to *learn* sophisticated linear or non-linear relationships and providing processing capabilities beyond the limits presented by existing systems which are usually used to deal with pattern recognition problems. Indeed, artificial neural networks are often reputed to enjoy four major advantages over many classical pattern recognition techniques: *adaptivity*, i.e. the ability to adjust when given new information; *speed* via massive parallelism; *fault tolerance* to missing, confusing and/or noisy data; *optimality* as regards error rates in classification systems.

Several neural network models have been proposed in the literature. Some of the important models are that of Caianiello[2, 3], Hopfield[4], Kohonen[5], Carpenter and Grossberg[6], Rumelhart and McClelland[7], Fukushima[8]. All these models share common features, but they differ in net topology, processing element functionality, learning rules. Peculiarities that allow one model to be more appropriate for solving some specific tasks than others. Specifically, their noise immunity and ability to produce output in real time suggested several attempts for using neural networks in natural language processing, speech recognition and image process-ing (see for example Pao[9] and Lippman[10]). Last studies have also considered the possibility of integrating the advantages of neural networks with those not less known of fuzzy sets, giving rise to systems named *fuzzy neural networks*. Several interesting works have been made in this field[11-13]. Robustness (neural networks) and decision from imprecise and/or incomplete knowledge (fuzzy set theory) compounded together can be of aid in many tasks where one request cannot be satisfied without considering the other one. This may turn out into realizing systems which outperform many existing techniques in quality and computational load.

The present work is in this direction. It reports one of the last research activities Caianiello was involved into and is fully dedicated to his inspiring and fascinating ideas. Specifically, in the paper we analyze the advantages of using the combination of *fuzzy set theory* and *neural networks* in realizing an edge-based image segmentation technique.

It is widely accepted that one of the first and important stages in machine vision is the image segmentation. Image segmentation can be performed in several ways. Among others gray level (histogram) thresholding and pixel classification using global (region growing) and/or local (edge detection) information of an image are certainly the most important. There are many challenges around the segmentation problem like (*i*) the development of a unified approach appliable to different kinds of images, i.e. gray scale, range, MRI, thermal images and so on, and (*ii*) the selection of the right threshold value, crucial both for obtaining a binary edge map after the application of edge detectors, and for thresholding-based segmentation techniques.

Edges, i.e. points in the image where there is a coherent local change in image intensity, play a key role in segmentation, but also in recognition. They simplify analysis of images by reducing the amount of data to be processed, while preserving useful structure information about the scene content. Several edge detection techniques exist[14-17]. They (*i*) use spatial gradients, Laplacian or differences of averages (enhancement operators); (*ii*) match or fit to a preassigned pattern (edge-fitting operators); (*iii*) operate on the image filtered by the first derivative of the Gaussian (Canny) or filtered by the Laplacian of the Gaussian (LOG) and (*iv*) adopt the cost minimization techniques, such as simulated annealing and relaxation. Most of these edge operators generally produce imperfect results. For instance, the edge detectors of kind (*i*) and (*ii*) present a strong degradation in presence of noise, while those related to classes (*iii*) and (*iv*) take advantage of a low-pass filtering, such as the application of a Gaussian operator, which suppresses noise while distorts edges, changing their magnitude and localization.

Neural network techniques have also been studied to realize edge detectors able to take away noise, without preprocessing the image. The massive connectionist (distributed) architecture usually makes the system robust, while the involved parallel processing enables the system to produce real time output. Some examples of neural networks for finding edges can be found in Shah[18], Lu and Szeto[19], Paik, Brailean and Katsaggelos[20]. Shah[18] formulated the problem as energy minimization, capturing only strong edges and wide regions. Lu and Szeto[19] proposed a three-layered neural network for enhancing the edge image obtained by applying gradient based operators. Their results for noisy images are very promising, while the number of neurons required even for an image of moderate size is very large. Paik et al.[20]

use three-state adaptive linear neurons (ADALINES) for realizing an edge detection algorithm. Although the algorithm shows low computational complexity and good location of edges also in presence of noise, it needs to find a good threshold value. It may be obtained "by computing the percentage of edge points detected in the original noiseless image", but "if the original image is not available, the threshold is estimated from an ensemble of similar images".

Thereby, our aim was to design a neural network system where few parameters are to be tuned and the computational complexity is as low as possible, without discarding the inner properties of noise immunity and real time output typical of the neural paradigm. This turns out to be very advantageous, since in automatic machine vision the automation degree increases as the number of parameters of the vision system decreases.

Specifically, we point out on the use of a fuzzy neural network, since the fuzzy set theory has been widely applied to image processing tasks getting good performance in most cases[21-23]. According to the possible multivalued levels of brightness, each pixel of the input image may be considered as a fuzzy singleton. Instead, if the gray levels are scaled to lie in the range [0, 1], we can regard the gray level of a pixel as its degree of membership in the set of high-valued "bright" pixels. Features, primitives, properties and relations among pixels that are not crisply defined can similarly be regarded as fuzzy subsets of images. The rules of extracting these informations follow those of the *Composite Calculus*, ideated by Caianiello in 1973[24] and studied during the last two decades as a mathematical tool for representing and manipulating fuzzy sets (see for example Apostolico et al.[25] and Caianiello et al.[26]). In particular, in Apostolico et al.[25] some algorithms for solving image processing problems by using local information are reported; however, they suffer from the need to exactly know the analysis window. This results in a decrease in performance, as well as in an increase of the computational load for finding the exact window size. Furthermore, robustness of the algorithms with noisy data has not been investigated.

The neural system we designed is able to retain as much of the information content of the input image as possible for making decision at the classification level, i.e. at the level where local features and relations are combined to classify a pixel as belonging or not to the edge map of the image. Local information is extracted and, by a feedback mechanism, propagated for utilizing global information. Once an image is provided to the neural network, it *self organizes* to automatically find out the edge image, by minimizing the *fuzziness* of the output layer. If some convergence criteria are not satisfied, the edge image is fed back to the input layer as new image to be processed. The neural network is completely fuzzy. Indeed, it may be seen as divided into two meaningful stages: the feature selection stage for finding non-uniform regions in the image which uses the *C*-calculus rules; the classifying stage which uses the fuzziness of the output image to decide if a pixel belongs or not to the edge map of the input image.

The paper is organised as follows. Section 2 introduces the fuzzy set theory and the *Composite Calculus* principles; section 3 reports the fuzzy hierarchical neural network-based system designed for edge detection. Experiments and evaluations of the noise tolerance capability on real images are shown in section 4. Conclusions as well as ongoing and future work are reported in the last section.

2. NOTATION AND BACKGROUND

In this section we set up the notation and state definitions that we will use in the following.

2.1 Fuzzy Sets and Operators

Fuzzy sets were introduced by Zadeh in 1965[27] as a mean of representing and manipulating data that was not precise. Conventional (crisp) sets contain objects that satisfy precise properties required for membership: an object belongs or not to a set. Zadeh's theory provided a mechanism for measuring the degree to which an object belongs to a set by introducing the "membership degree" as a characteristic function $\mu_A(x)$ which associates with each point x a real number in the range [0, 1]. The nearer the value of $\mu_A(x)$ to unity, the larger the membership degree of x in the set A. Assume S be the universe of discourse and denoting with $A \in S$ and $B \in S$ two fuzzy sets, i.e. $A = \{(x_i, \mu_A(x_i)), i = 1,..., n\}$ and $B = \{(x_i, \mu_B(x_i)), i = 1,..., n\}$. The operations on fuzzy sets are extensions of those used for conventional sets (intersection, union, comparison, etc.). Some of these operations are defined as follows:

Definition 1. *The membership degree of the* **intersection** $A \cap B$ *is*

$$\mu_{A \cap B}(x) = \min\{\mu_A(x), \mu_B(x)\} \quad x \in S \tag{1}$$

Definition 2. *The membership degree of the* **union** $A \cup B$ *is*

$$\mu_{A \cup B}(x) = \max\{\mu_A(x), \mu_B(x)\} \quad x \in S \tag{2}$$

Definition 3. *The membership degree of the* **set - difference** $A - B$ *is*

$$\mu_{A-B}(x) = |\mu_A(x) - \mu_B(x)| \quad x \in S \tag{3}$$

The previous constitutes only a restricted set of operations appliable among fuzzy sets, but they are the most significant for our aim.

2.2 Measures of Fuzziness

Measures of fuzziness $F(A)$ of a fuzzy set $A = \{(x_i, \mu_A(x_i)), i = 1,..., n\}$ give the average amount of ambiguity in making a decision whether an element belongs to the set or not. $F(A)$ has to have the following properties[9]:

1. $F(A)$ is minimum when $\forall i$ $\mu_A(x_i) = 0$ or 1
2. $F(A)$ is maximum when $\forall i$ $\mu_A(x_i) = 0.5$
3. $F(A) \geq F(B)$ iff

$$\mu_B(x_i) \geq \mu_A(x_i) \quad \text{if} \quad \mu_A(x_i) \geq 0.5$$
$$\mu_B(x_i) \leq \mu_A(x_i) \quad \text{if} \quad \mu_A(x_i) \leq 0.5$$

4. $F(A) = F(\overline{A})$, $\overline{A}$ being the complement set of A, i.e. $\overline{A} = \{(x_i, 1 - \mu_A(x_i))\}$.

Let us introduce A' as the crisp version of A, i.e. $\mu_{A'} = 0$ if $\mu_A \leq 0.5$, 1 if $\mu_A \geq 0.5$. Denoting with n the cardinality of A, some measures of fuzziness defined in Kandel[28] and named *Indeces of Fuzziness* are the following:
Linear Index of Fuzziness (LIF)

$$LIF(A) = \frac{2}{n} \sum_{i=1}^{n} |\mu_A(x_i) - \mu_{A'}(x_i)| = \frac{2}{n} \sum_{i=1}^{n} \min\{\mu_A(x_i), (1 - \mu_A(x_i))\} \tag{4}$$

Euclidean Index of Fuzziness (EIF)

$$EIF(A) = \frac{2}{\sqrt{n}} \sqrt{\sum_{i=1}^{n} \{\mu_A(x_i) - \mu_{A'}(x_i)\}^2} \tag{5}$$

We shall use these measures to compute the unstability degree of the network output, as will be described in the following sections.

2.3   A Fuzzy Model: The Composite Calculus

An alternative approach to the study of fuzziness was introduced by Caianiello in 1973[24] in an attempt to understand whether not crisply defined sets might be used, for a quantitative description of properties of sets endowed with hierarchical structures, proponing a general extension to the basic concepts of fuzzy theory. He gave to this approach the name of Composite Calculus and to the involved sets the name of Composite sets (C-sets briefly). Formally, let X be a totally ordered set. Consider a partition of X, $X^0 = (X_1^0, ..., X_p^0)$, i.e. a finite pairwise disjoint family of subsets of X such that their union coincides with X. Let f be a real function which takes value in X_i^0 and introduce the vectors $m^0 = (m_1^0, ..., m_p^0)$ and $M^0 = (M_1^0, ..., M_p^0)$ where $m_i^0 = \min \{f(x) : x \in X_i^0\}$ and $M_i^0 = \max \{f(x) : x \in X_i^0\}$, $i = 1, ..., p$. The triple $C^0 = (X^0, m^0, M^0)$ is named C-set of X. Based on fuzzy set theory, the fuzzy union and intersection operations are applied to the elements belonging to two different C-sets. In the sequel they will be indicated as the Minkowski addition $\oplus$ and subtraction $\otimes$ respectively.

If we consider another C-set $C^1 = (X^1, m^1, M^1)$ such that $\forall i \; X_i^0 \cap X_i^1 \neq \theta$, the operation defined by

$$C^0 \tilde{\cap} C^1 = (X^2, m^2, M^2) \tag{6}$$

where

$$X^2 = \left(X_1^0 \cap X_1^1, ..., X_p^0 \cap X_p^1 \right)$$

$$m^2 = \left(m_1^0 \oplus m_1^1, ..., m_p^0 \oplus m_p^1 \right)$$

$$M^2 = \left(M_1^0 \otimes M_1^1, ..., M_p^0 \otimes M_p^1 \right)$$

is named C-product.

It has been demonstrated in Apostolico et al.[25] that recursive application of the previous operations provide a refinement of the original sets, so realizing a tool for measurement and signal analysis.

3. THE PROPOSED NEURAL NETWORK MODEL

The system we designed is depicted in figure 1. The Hierarchical Fuzzy Neural Network (HFNN) is demanded to iteratively produce an edge image, where (i) true edges are enhanced; (ii) false, spurious edges are suppressed and (iii) noise is eliminate. The edge image is updated after each iteration and the result is fed back to the HFNN if some convergence criteria are not satisfied. This reflects into using *local information* within the HFNN since the classification of each pixel is based only on its local neighborings, while *global information* is utilized as the information is propagated to neighboring elements in the edge image. When the system stabilizes (the convergence holds), it outputs a binary edge image produced by thresholding the *stabilized* edge image with a *learned* threshold θ. In the following the system functionality will be detailed.

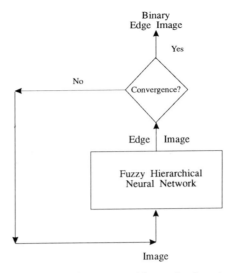

Figure 1. Overview of the proposed fuzzy edge detection system.

3.1 The Hierarchical Fuzzy Neural Network (HFNN)

The HFNN is made up of sets of nodes arranged in layers and only nodes of two different consecutive layers are allowed to be connected by weighted links. The layer where the image is presented is the *input layer*, whereas the output image producing layer is called *output layer*. There is a number of layers between input and output layers (usually named *hidden layers*), each making the *C-product* among elements of the previous layer. All the weights are fixed except those related to links connecting the upper hidden layer and the output layer.

In figure 2 the proposed network architecture is depicted.

The first or input layer is constituted by $M \times N$ nodes, each taking as value a gray level in the input image. Both the second and the third layers are constituted by two cell-planes, each consisting of a number of neurons equal to $(M + w - 1) \times (N + w - 1)$. Let us define a neighborhood of a neuron in the second and third layers. From now on each neighborhood will correspond to a subimage-subset of the *C*-set the image is divided into. Whatever the size of the subimage is, we consider four *C*-sets (they are sufficient to catch information around a pixel). Then, for a $M \times N$ image the $w \times w$ sized neighborhoods of a neuron localized in (i,j) (see figure 3) are defined as:

$$N^1_{i,j} = \left\{ (k,l) \mid i - w + 1 \leq k \geq i, \ j - w + 1 \leq l \leq j \right\}$$
$$N^2_{i,j} = \left\{ (k,l) \mid i - w + 1 \leq k \leq i, \ j \leq l \leq j + w - 1 \right\}$$
$$N^3_{i,j} = \left\{ (k,l) \mid i \leq k \leq i + w - 1, \ j - w + 1 \leq l \leq j \right\} \tag{7}$$
$$N^4_{i,j} = \left\{ (k,l) \mid i \leq k \leq i + w - 1, \ j \leq l \leq j + w - 1 \right\}$$

Besides the *input* and *output* images, each neuron in the *i*-th intermediate layer, $i > 1$, will have $|N^m|$, $|N^m|$ being the cardinality of N^m, links to the $(i - 1)$-th layer. Moreover, the output layer

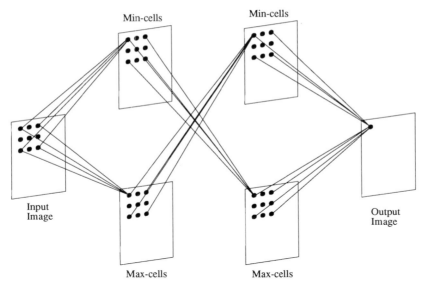

Figure 2. The Hierarchical Fuzzy Neural Network.

is structured such as each neuron belonging to it looks at two ($p \times p$) neighborhoods centered on corresponding neuron in both cell-planes at the upper hidden layer.

3.2 Network Operations

Starting from the input layer, the operations involved in the network are the following. *Hidden Layers:*

According to the definition of neuron neighborhoods (7), in the first hidden layer the first and second cell-planes are determined letting each neuron localized in (i, j) compute respectively the minimum and maximum values over all the activities of neurons belonging to the corresponding neighborhood at the input layer (the input image I). In formulae:

$$o_{i,j}^{1,1} = \min_{\substack{k \in \{i,\dots,i+w-1\} \\ l \in \{j,\dots,j+w-1\}}} I(k,l) \tag{8}$$

$$o_{i,j}^{1,2} = \max_{\substack{k \in \{i,\dots,i+w-1\} \\ l \in \{j,\dots,j+w-1\}}} I(k,l) \tag{9}$$

where $o_{i,j}^{p,q}$ means the output of the neuron localised in position (i, j) at the p-th layer and q-th cell-plane. Seeing each gray level of the input image as a singleton set, the operations involved applying (1) and (2) correspond to the determination of a first *C-product*. Moreover, the operations involved at the second hidden layer are the following:

$$o_{i,j}^{2,1} = \min_{\substack{k \in \{i,\dots,i+w-1\} \\ l \in \{j,\dots,j+w-1\}}} o_{k,l}^{1,2} \tag{10}$$

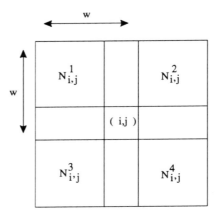

Figure 3. The neighborhood of a pixel (i,j).

$$o_{i,j}^{2,2} = \max_{\substack{k\in\{i,...,i+w-1\} \\ l\in\{j,...,j+w-1\}}} o_{k,l}^{1,1} \tag{11}$$

The previous steps allow to obtain at the upper hidden layer a new and more representative C-set whose $o_{i,j}^{2,1}$ and $o_{i,j}^{2,2}$ values represent relevant structures in the image. Here, by combining these "image properties" in a self-organization manner we can extract the wished information. In particular, minimizing pixel by pixel the fuzziness of the weighted distance between the values $o_{i,j}^{2,2}$ and $o_{i,j}^{2,1}$ of this C-set reflects into extracting great variations in local areas.

Output Layer:

The combination of the minima and maxima values in the cell-planes at the upper hidden layer, is made up at the output layer. For sake of clearness, for units at the output layer we drop out the indication of their localisation, meaning that they are numbered in the row by row order. The output of a node j is then obtained as:

$$o_j = f(I_j) \tag{12}$$
$$I_j = \sum_i w_{j,i}^1 o_i^{2,2} + \sum_i w_{j,i}^2 o_i^{2,1}$$

where $w_{j,i}^q$ indicates the connection weight between the j-th node of the output layer and the i-th node of the previous layer in the q-th cell-plane, $q = 1,2$. Each sum is intended over all nodes i in the neighboorhood of the j-th node at the upper hidden layer. f (the *membership function*) is chosen to be sigmoidal:

$$f(I_j) = \frac{1}{1 + e^{-I_j}} \tag{13}$$

with the accorgement that if o_j takes the value 0.5, a small quantity (usually 0.001) is added; this reflects into dropping out unstability conditions.

To retain the value of each output node o_j in [0,1], we apply the following mapping to each input image pixel g:

$$g' = \frac{g - g_{min}}{g_{max} - g_{min}} \tag{14}$$

where g_{min} and g_{max} are the lowest and highest gray levels in the image.

3.3 Learning

The HFNN has to self-organize by minimizing the fuzziness of the output layer. This is an iterative process with a feedback mechanism which provides the input layer with new pixel values (the neuron activities of the output layer), so continuing the "learning" until convergence (see figure 1). Since the membership function is chosen to be sigmoidal, minimizing the fuzziness is equivalent to minimizing the distances between corresponding pixel values in both cell-planes at the upper hidden layer. To this aim the weights $w_{j,i}^1$ and $w_{j,i}^2$ are settled respectively as excitatory and inhibitory. Since random initialization acts as noise, *all the weights are initially set to unity*. The adjustment of weights is done using the gradient descent search, i.e. the incremental change $\Delta w_{j,i}^l$, $l = 1,2$, is taken as proportional to the sum of the negative gradient $-\eta \dfrac{\partial E}{\partial o_i} f'(I_i) o_j$. The adjustment rule is then the following:

$$w_{j,i}^l = w_{j,i}^l + \Delta w_{j,i}^l \tag{15}$$

Specifically, we compared the following two learning strategies, where E indicates the energy-fuzziness of our system and $n = M \times N$:

1. Linear Index of Fuzziness (LIF) based Learning:

$$E = \frac{2}{n} \sum_{i=1}^{n} \left\{ min\left(o_i, (1 - o_i)\right) \right\} \tag{16}$$

The weight updation rules look as follows:

$$\Delta w_{j,i}^1 = \begin{cases} -\eta_{LIF}\left(1-o_j\right) o_j o_i^{2,2} & \text{if } 0 \le o_j \le 0.5 \\ \eta_{LIF}\left(1-o_j\right) o_j o_i^{2,2} & \text{if } 0.5 < o_j \le 1 \end{cases} \tag{17}$$

$$\Delta w_{j,i}^2 = \begin{cases} -\eta_{LIF}\left(1-o_j\right) o_j o_i^{2,1} & \text{if } 0 \le o_j \le 0.5 \\ \eta_{LIF}\left(1-o_j\right) o_j o_i^{2,1} & \text{if } 0.5 < o_j \le 1 \end{cases} \tag{18}$$

where $\eta_{LIF} = \eta \times 2/n$.

2. Euclidean Index of Fuzziness (EIF) based Learning:

$$E = \frac{2}{\sqrt{n}} \sqrt{\sum_{i=1}^{n} \left\{ \mu_A(x_i) - \mu_{A'}(x_i) \right\}^2} \tag{19}$$

The weight updation rules look as follows:

$$\Delta w_{j,i}^1 = \begin{cases} -\eta_{EIF}\left(1-o_j\right)^2 o_j o_i^{2,2} & \text{if } 0 \le o_j \le 0.5 \\ \eta_{EIF}\left(1-o_j\right)^2 o_j o_i^{2,2} & \text{if } 0.5 < o_j \le 1 \end{cases} \tag{20}$$

$$\Delta w_{j,i}^2 = \begin{cases} -\eta_{EIF}\left(1-o_j\right)^2 o_j o_i^{2,1} & \text{if } 0 \le o_j \le 0.5 \\ \eta_{EIF}\left(1-o_j\right)^2 o_j o_i^{2,1} & \text{if } 0.5 < o_j \le 1 \end{cases} \tag{21}$$

where $\eta_{EIF} = \eta \times 2 / \sqrt{(n)}$. The previous rules hold also for the determination of the exact threshold value, θ, adopted for binarizing the image when convergence is reached.

According to the properties of fuzziness the initial threshold is set to be 0.5; this value allows to determine an hard decision from an unstable condition to a stable one.

As said before, the updating of weights is continued until the network stabilizes. To check convergence the following quantity is calculated

$$O(t) = \sum_{j:o_j>=0.5} o_j \tag{22}$$

The system is said *stable* (the learning stops) when

$$E(t + 1) \le E(t) \tag{23}$$

and

$$\left|O(t + 1) - O(t)\right| < \gamma \tag{24}$$

where $E(t)$ is the system fuzziness computed at the t-th iteration and γ is a prefixed very small positive quantity.

After convergence, the pixels j with $o_j > q$ are considered to constitute the edge map of the image. They are settled to take value 1, in contrast with the remaining which will constitute the background (value 0).

Figure 4. Original aircraft image.

4. EXPERIMENTAL RESULTS

Several experiments were conducted to test the efficacy of the proposed fuzzy neural network-based edge detection system. Some of the experimental results on two real images are reported here. The first image of size 128 x 128 and 214 gray levels, shown in figure 4, is an aircraft image that prevalently contains step edges. The second real image of size 512 x 480 and 174 gray levels (the face of *Lenna*), shown in figure 5, contains several types of edges such as continuous and step edges. During experiments, we focus our attention on three aspects: (*i*) the ability of the system to detect true edges and neglect spurious edges; (*ii*) the robustness of the system when applied to noise corrupted versions of the images; (*iii*) the convergence time of the system with a prefixed value of γ.

In each experiment the system was implemented using a neighborhood size of a neuron in the first and second layers equal to $w = 2$. In this case we analyze the local properties within a 3 x 3 region around each neuron. On the contrary, a neuron in the output layer looks at exactly two neurons, each belonging to a cell-plane at the upper hidden layer. The learning rate η_{LIF} has been taken as 0.2 while the convergence parameter γ has been taken to be 0.001. The reason for these choices resides in a most successful edge detection system, both for detecting edges and for suppressing noise, while requiring the minimum amount of computation or, equivalently, the minimum number of iterations to converge.

Figures 6 and 7 show the edge images obtained applying the system to both images with the fuzziness calculated by using the Linear Index of Fuzziness. The same results are obtained by the Euclidean Index of Fuzziness. As shown in figures, as the number of iterations increases, the edge images are refined and retain only significant edges, both of step type (more evident in figure 6) and continuous type (see figure 7).

The proposed system has been used to detect edges in both images corrupted by noise. The noisy versions were obtained by adding noise from a Gaussian $N(0, \sigma^2)$ distribution with

Figure 5. Original *Lenna* image.

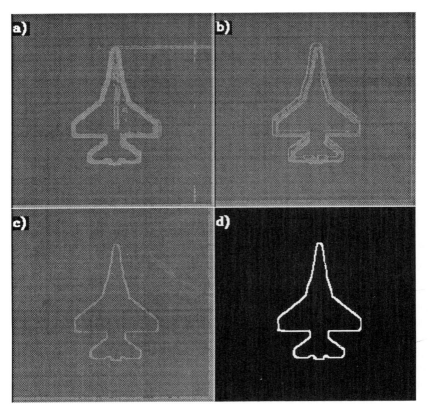

Figure 6. Edge images of the aircraft image obtained by the proposed system using the Linear Index of Fuzziness after a) two iterations; b) four iterations; c) six iterations; d) convergence ($\theta = 0.5$).

different values of the standard deviation σ corresponding to the 3%, 6% and 12% of the dynamic range. The three versions of the aircraft image are shown in figure 8a), 8b) and 8c), while those corresponding to *Lenna* image in figure 9a), 9b) and 9c).

The edge images obtained after convergence by the proposed technique with different fuzziness measures for different noisy versions of the image aircraft can be seen in figure 8. The same experiments made on the image *Lenna* resulted in the images of figure 9. Since the learning rate of the EIF is low compared with the LIF, the number of iterations is the same; only the parameter η_{EIF} for the EIF is selected to be greater than the LIF one (η_{LIF}). In all the experiments made, and not only those reported, the system converged after six iterations, using a parameter value $\eta_{EIF} = 2 \times \eta_{LIF}$. For a fixed η value, the system converges in approximately the same number of iterations for any image we provide as input. Furthermore, the values of the threshold θ were automatically found by using the updating rules (17-18) and (20-21); they are all around the value 0.5 which corresponds to the total unstability condition for output neurons.

Examining the results, as the noise increases, the quality of the edge image, as expected, deteriores, but true edges are mostly saved. The main advantages of the system comes from the fully automatic procedure to choice the exact threshold θ and from operating directly on the luminance images. This allows to detect fine edges, suppressing false edges usually present in

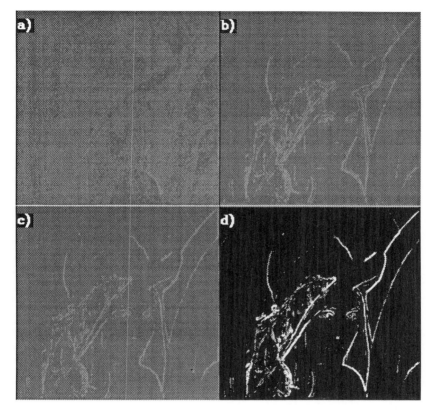

Figure 7. Edge images of *Lenna* obtained by the proposed model using the Linear Index of Fuzziness after a) two iterations; b) four iterations; c) six iterations; d) convergence ($\theta = 0.5$).

real images. Other advantages are the fixed analysis window size at any level of the neural network structure and the computational load quite fixed also in presence of noise.

5. CONCLUSIONS

Any existing edge detection-based image segmentation technique suffers of unsatisfactory results due to noise, gradual variations in gray-level, imprecision in selecting the threshold to be used for binarizing the edge image.

Fuzzy neural networks are a viable solution to cope with these problems, since they resume the peculiarities of the neural networks (tolerance to noise, an intrinsic parallel processing and modifiability when some contour conditions are altered) and the fuzzy set theory for treating imprecision into classifying (edges in our case).

On the basis of the results achieved in both fields of interest (neural networks and fuzzy sets), we derived and analyzed a Hierarchical Fuzzy Neural Network capable to segment a scene a) suppressing false and spurious edges and b) eliminating noise as automatically as possible. The network operations follow the *C*-calculus rules, a fuzzy model ideated by

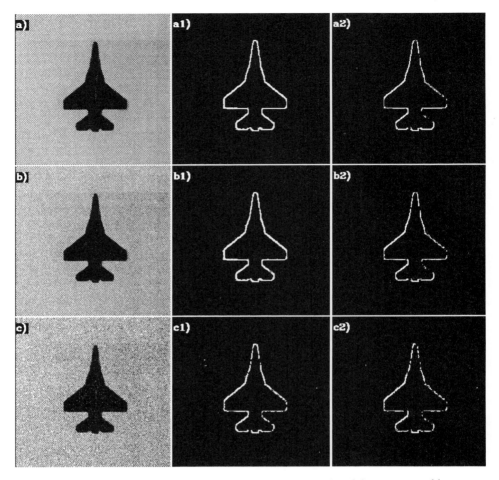

Figure 8. Experimental results by the neural network system on the aircraft image corrupted by zero mean Gaussian noise with standard deviation equal to a) 3%, b) 6% and c) 12% of the dynamic range. The corresponding edge images a1), b1) and c1) have been obtained by using the Linear Index of Fuzziness ($\theta = 0.5, 0.5, 0.5026$), while a2), b2) and c2) by using the Euclidean Index of Fuzziness ($\theta = 0.5, 0.5, 0.501$).

Caianiello to model and analyze hierarchical systems. Since training data for image analysis involves the knowledge of images belonging to the same class of that under investigation, *no training data* have been considered; the neural system operates on a single image at a time. The system is designed for *self-organizing* to automatically find out a peculiar structure in the input data, by providing a feedback mechanism from the output layer to the input layer at each iteration.

Based on our experiments, we concluded that a 3 x 3 analysis window ($w = 2$) is sufficient in succesfully finding out true edges and suppressing false ones. The major advantage of the proposed system is the ability to suppress noise with the same parameter values than for noiseless images. This reflects in a low computational requirement, allowing real time implementation be feasible. The system is being effectively and succesfully applied to other image analysis tasks like image restoration and region-based image segmentation[29].

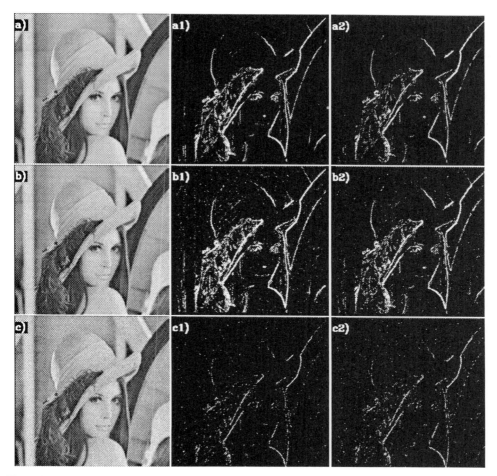

Figure 9. Experimental results by the neural network system on *Lenna* corrupted by zero mean Gaussian noise with standard deviation equal to a) 3%, b) 6% and c) 12% of the dynamic range. The corresponding edge images a1), b1) and c1) have been obtained by using the Linear Index of Fuzziness ($\theta = 0.5, 0.5030, 0.5038$), while a2), b2) and c2) by using the Euclidean Index of Fuzziness ($\theta = 0.5, 0.5, 0.5078$).

All the advantages of the proposed system reinforced the common idea that neural structures and learning are very useful to accomplish tasks in pattern recognition. However, according to us, the study about how combining these human-like mechanisms with strong mathematical frameworks and classical techniques should be more and more investigated to get convenient and satisfactory systems able to solve ill-posed problems too.

ACKNOWLEDGEMENTS

We thank, first and foremost, Prof. Marinaro for her constant and valuable discussions and suggestions. Special thanks are also due to Prof. Cantoni for his kindness and appreciation in our regard.

This work was supported in part by Italian CNR PF 'Sistemi Informatici e Calcolo Parallelo' research grant 93.01.591.PF69, by MURST 40% -INFM Section Univ. of Salerno and by 'Contratto Quinquennale' IIASS-CNR.

REFERENCES

1. E.R. Caianiello, Outline of a theory of thought processes and thinking machines, *Journal of Theoretical Biology*, Vol.1, p.209 (1961).
2. E.R. Caianiello, A. De Benedictis, A. Petrosino, and R. Tagliaferri, Neural associative memories with minimum connectivity, *Neural Networks*, Vol.5, pp.433-439 (1992).
3. E.R. Caianiello, M. Ceccarelli, and M. Marinaro, Can spurious states be useful?, *Complex Systems*,Vol.5, pp.1-12 (1992).
4. J.J. Hopfield, Neural networks and physical systems with emergent collective computational abilities, *Proc. National Academy of Sciences USA*, Vol.79, pp.2445-2558 (1982).
5. T. Kohonen, The self-organizing map, *Proc. IEEE*, Vol.78, No.9, pp.1464-1481 (1990).
6. G.A. Carpenter and S. Grossberg, A massively parallel architecture for a self-organizing neural pattern recognition machine, *Computer Vision, Graphics and Image Processing*, Vol.37, pp.54-115 (1987).
7. D.E. Rumelhart, G.E. Hinton, and R.J. Williams, Learning internal representations by error propagations, *Parallel Distributed Processing: Explorations in the Microstructure of Cognition*, D.E. Rumelhart and J.L. McClelland eds., Vol.1, MIT Press, Cambridge, MA (1986).
8. K. Fukushima, A hierarchical neural network capable of visual pattern recognition, *Neural Networks*, Vol.1, pp.119-130 (1988).
9. Y.H. Pao, *Adaptive Pattern Recognition and Neural Networks*, Addison-Wesley, Reading, MA (1990).
10. R.P. Lippman, Review of neural networks for speech recognition, *Neural Computation*, Vol.2 (1989).
11. S. Lee and E. Lee, Fuzzy neural networks, *Math. Biosci.*, Vol.23, pp.151-177 (1975).
12. G. Carpenter, S. Grossberg, and D. Rosen, Fuzzy ART: fast stable learning and categorization of analog patterns by an adaptive resonance system, *Neural Networks*, Vol.4, pp.759-771 (1991).
13. P.K. Simpson, Fuzzy min-max neural networks - Part I: classification, *IEEE Trans. Neural Networks*, Vol.3, pp.776-786 (1992).
14. A. Rosenfeld and A.C. Kak, *Digital Picture Processing*, Academic Press, New York, NY (1992).
15. R.C. Gonzalez and P. Wintz, *Digital Image Processing*, Addison-Wesley, Reading, MA (1977).
16. L.S. Davis, A survey of edge detection techniques, *Computer Vision, Graphics and Image Processing*, Vol.4, pp.248-270 (1975).
17. N.R. Pal and S.K. Pal, A review on image segmentation techniques, *Pattern Recognition*, Vol.26, No.9, pp.1277-1294 (1993).
18. J. Shah, Parameter estimation, multiscale representation and algorithm for energy-minimizing segmentation, *Proc. Int. Conf. Pattern Recognition*, pp.815-819 (1990).
19. S. Lu and A. Szeto, Hierarchical artficial neural networks for edge enhancement, *Pattern Recognition*, Vol.26, pp.1149-1163 (1993).
20. J.K. Paik, J.C. Brailean, and A.K. Katsaggelos, An edge detection algorithm using multi-state ADALINES, *Pattern Recognition*, Vol.25, No.12, pp.1495-1504 (1992).
21. S.K. Pal and D.D. Majumder, *Fuzzy Mathematical Approach to Pattern Recognition*, John Wiley, Halsted Press, New York, NY (1986).
22. J. Serra, *Image Analysis and Mathematical Morphology*, Academic Press, New York, NY (1982).
23. R.M. Haralick, S.R. Sternberg, and X. Zhuang, Image analysis using mathematical morphology, *IEEE Trans. Pattern Anal. Mach. Intell.*, Vol.9, pp.32-550 (1987).
24. E.R. Caianiello, A calculus of hierarchical systems, *Proc. Int. Conf. Pattern Recognition*, Washington DC (1973).
25. A. Apostolico, E.R. Caianiello, E. Fischetti, and S. Vitulano, C-calculus: an elementary approach to some problems in pattern recognition, *Pattern Recognition*, Vol.10, pp.375-387 (1978).
26. E.R. Caianiello and A. Ventre, A model for C-calculus, *Intern. Journ. General Systems*, Vol.11, pp.153-161 (1985).
27. L.A. Zadeh, Fuzzy sets, *Information and Control*, Vol.8, pp.338-353 (1965).
28. A. Kandel, *Fuzzy Mathematical Techniques with Applications*, Addison-Wesley, New-York, NY (1986).
29. A. Petrosino and F. Pan, A fuzzy hierarchical neural network for image analysis, *Proc. IEEE Conf. System, Man and Cybernetics*, Vol.2, pp.657-662 (1993).

PANEL SUMMARY
PERCEPTUAL LEARNING AND DISCOVERING

Salvatore Gaglio[1] *(chairman)*, Floriana Esposito[2], and Stefano Nolfi[3]

[1] Dipartimento di Ingegneria Elettrica, Università di Palermo
Viale delle Scienze, I - 90128 Palermo, Italy
[2] Dipartimento di Informatica, Università di Bari
Via Orabona 4, I - 70126 Bari, Italy
[3] Istituto di Psicologia del CNR
V.le Marx 15, I - 00137 Roma, Italy

ABSTRACT

The problem of learning and discovering in perception is addressed and discussed with particular reference to present machine learning paradigms. These paradigms are briefly introduced by S. Gaglio. The subsymbolic approach is addressed by S. Nolfi, and the role of symbolic learning is analysed by F. Esposito. Many of the open problems, that are evidentiated in the course of the panel, show how this is an important field of research that still needs a lot of investigation. In particular, as a result of the whole discussion, it seems that a suitable integration of different approaches must be accurately investigated. It is observed, in fact, that the weakness of the most part of the existing systems is imputed to the existing gap between the rather ideal conditions under which most of those systems are designed to work and the very characteristics of the real world.

TOPIC INTRODUCTION (S. Gaglio)

Hot Points

The following questions are identified as a basis for discussion:
1. Basic questions (The role of learning in vision)
- a) What is learning in vision?
- b) Are present machine learning paradigms relevant to vision?
- c) How can context information be appropriately used in perceptual learning?
- d) How to treat exceptions in generalisation?
- e) Many learning paradigms could be better described as tuning strategies. Is true learning

more related to adaptation? Which is the difference?

    f) Do novelty detection and generalisation require different approaches in perceptual learning and discovery?

2. Which learning paradigm?

    a) Are current machine learning paradigms more useful in the case of active perception (for instance, in developing plans and strategies)?

    b) Is the dichotomy of supervised and unsupervised learning related to the degree of interaction with the environment (a system either may or may not modify the environment)?

    c) Is concept learning only related to supervised learning?

    d) Is symbolic reasoning adequate for visual learning (it usually refers to an idealised simple world without imprecision)?

    e) Structural learning has to do with formal languages and formal reasoning. How can primitive symbols be chosen in structural learning?

    f) In concept and structural learning visual information is only relegated to primitive symbols. How significant could be a form of "visual reasoning"? How could it be performed?

    g) Does it make sense to ignore function in perceptual learning of form? How knowledge about function could be used?

    h) To what degree neural network models open new perspectives in perceptual learning?

    i) Can studies on chaotic systems and evolutionary processes shed light on perceptual learning processes (e.g., novelty detection) and discovery?

3. Performance questions

    a) How to control the combinatorial explosion arising in inductive learning in the case of perception?

The above questions are presented by S. Gaglio to stimulate a discussion concerning our current understanding of learning in perception and the significance of machine learning models. Many of the participants, besides the panellists, take part to the discussion. As it is summarized in the following sections, after a brief introduction to machine learning made by S. Gaglio, the subsymbolic approach and the symbolic approach are presented and analysed by S. Nolfi and F. Esposito, respectively. The most interesting observations made during the general discussion are reported at the end of the paper.

AN INTRODUCTION TO MACHINE LEARNING PARADIGMS (S. Gaglio)

Decision Theoretic Techniques

Decision theory has played a very important role in machine learning and has been widely used in applications in which the main task is classification. Classification may be performed either in a deterministic way or in a probabilistic way, according to Bayesian inference. Specific techniques are usually based on decision trees or discriminant functions, which are particularly suited to be automatically inferred from set of examples.

A decision tree (as shown in figure 1), allows for the automatic classification of a set S0 of objects into a set C of classes, by considering in an ordered fashion the possible discrete values of a set of attributes attached to every object.

The learning task consists of inferring a decision tree, given selected sets of examples. In supervised learning the desired classification is provided by the instructor, while in unsupervised learning the classes are automatically proposed according to a given fitness criterion[1].

A discriminant function[2] $d = d(\underline{x};\underline{w})$ describes an hypersurface in a feature space, in which $\underline{x}$ stays for a vector in that space, corresponding to the characteristics of a given object, and $\underline{w}$

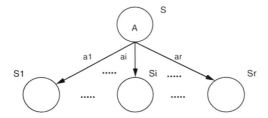

Figure 1. A 1-level decision tree.

for an n-ple of parameters that characterize the function. The hypersurface divides the space into two regions, i. e. the regions in which d > 0 and the region in which d < 0. A set of discrimination functions tessellate the space into many regions, in such a way that each region contains the vectors corresponding to objects belonging to the same class.

In this case, the learning problem may be stated as follows: given a set of examples, find the best set of of n-ples of parameters $\underline{w}$ such that the tessellation of the feature space corresponds to the desired classification. The parameters are iteratively updated according to a gradient descent method which tries to minimize the mean square error. In figure 2 the case of a 2-D feature space, in which two linear discriminant functions define four classes S1, S2, S3, S4, is shown.

Neural Modeling and Genetic Algorithms

Neural networks[3] are extensively used as selforganizing systems to perform association or classification tasks. Their architecture is loosely inspired to the neural apparatus of living beings. They are made of simple processing elements (the neurons), characterised by an analog activation function a = a(t) and a possibly non-linear output function u = u(a(t)). Neurons are highly interconnected through synapses characterised by strength weights w_{ij} (see figure 3). The activation function of the neuron i depends on the weighted sum of the inputs through

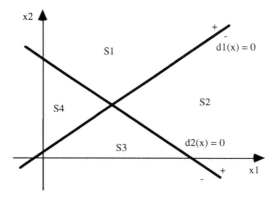

Figure 2. 2-D feature space tessellated by linear discriminant functions.

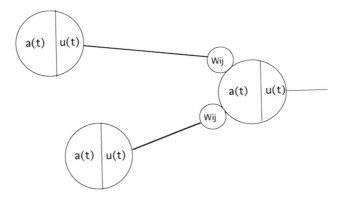

Figure 3. A neural network.

incoming synapses $\sum_j w_{ij} u_j(t)$. The weights can be either positive or negative to allow for mutual activation or mutual inhibition.

A neural network learns to perform a given classification or association task by adjusting its weights according to various versions of the Hebb's rule. This rule establishes that the weights corresponding to pairs of simultaneous active neurons must be increased, while the weights corresponding to pairs of not simultaneous active neurons must be decreased. Learning can be either supervised or unsupervised.

Another important class of selforganizing systems particularly suitable for optimization problems is based on genetic algorithms[4]. As shown in figure 4, a genetic algorithm assemblies a desired solution from a population of partial solutions that are generated from existing ones through the application of cross-over and mutation operators. The generated partial solutions survive if they satisfy a fitness criterion.

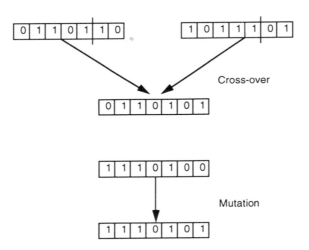

Figure 4. Cross-over and mutation operations in genetic algorithms.

Symbolic Concept-Oriented Learning

In the symbolic learning paradigm a system tries to build internal representations corresponding to general concepts that analytically describe a given set of test cases. The derived descriptions attempt to capture and to abstract the underlying structure of the examples.

Well-known techniques in this field are grammatical inference for syntactic pattern recognition[5], the inference of structural descriptions proposed by P. H. Winston[6], and inductive learning[7].

Inductive learning exploits inductive reasoning mechanisms to find general hypotheses that explain observed data. According to Michalsky[7], it can be expressed as follows:

GIVEN

the observational statements (facts), F, a tentative inductive assertion (which may be null), the background knowledge

FIND

an inductive assertion (hypothesis), H, that tautologically implies the observational statements:
H |> F   (H specializes to F)   or   F |< H   (F generalizes to H)

In the formalization introduced by Michalsky the language is an extension of predicate calculus, i.e. Annotated Predicate Calculus.

The problem is formulated as an automatic search problem in which *States* are symbolic descriptions (the initial state is the set of observational statements); *Operators* are generalization, specialization and reformulation rules; the *Goal State* is an inductive assertion that implies the observational statements.

Examples of rules are:
CTX & S ::> K |<   CTX ::> K (Dropping condition rule)
F[a], F[b], ..., F[i] |< ∀ v, F[v] (Turning constraints into variables rule)
etc.

Knowledge Intensive Learning Systems

Knowledge intensive learning systems extend the symbolic learning paradigm with the esplicit representation and use of domain knowledge, usually contained in a large knowledge base.

An example is Explanation Based Learning (EBL)[8]. An EBL system tries to explain why a particular example is an instance of a concept. In the case of game playing, for instance, after falling in a trap, the system analyses why the trap succeeded and it identifies a sequence of rules. By computing the weakest preconditions of that sequence of rules, the system can identify the general constraints that enable the trap to succeed. The final result is a Macro-Rule that can be directly applied when a situation of the same kind is encountered again.

Another example is automatic discovery[9]. Typical applications (e.g. BACON, FAHREN-HEIT, ABACUS, FERMI) concern the discovery of physical or chemical laws "explaining" sets of numerical data in terms of general formulas of the following kind:
dependent variable = f(independent variables)

SELF-ORGANISED METHODS (S. Nolfi)

Self-organised methods are particularly suited for building artificial systems that interact with an external environment.

Traditional approaches to the development of artificial systems have made only modest progress. We think that this is largely due to the traditional implicit assumption of functional decomposition - the assumption that perception, planning and action can be analysed independently from each other. Recent work at MIT used a different approach based on the idea of decomposing behaviour instead of functions[10,11]. This approach tries to identify different independent types of behaviour so that each form of behaviour can be "wired in" all the way from sensor input to motor output. Simple forms of behaviour are wired first, and then more complex forms of behaviour are added as separate layers, affecting earlier layers only by means of suppression or inhibition mechanisms. On the other hand, even in this approach it is the designer of the system that decides how to structure the system itself. We will try to show that systems which interact with an external environment are too difficult to design and how this problem can be overcome by using self-organising methods.

Natural systems are self-organising systems. They are free to develop their way to survive and reproduce and actually nature found a great number of ways to do that (i.e. a large number of different species evolved). On the contrary, the way in which artificial systems should perform their task is usually predetermined by the designer of the system itself. The engineer or the researcher that develops an artificial system which should perform some task usually predetermines the answer that the system should give to each particular stimulus it can encounter. After having decided that, he designs the system itself in order to have it produce the desired answer for each input stimulus. This means that the way in which the system should accomplish the task is not free but is determined in advance.

In principle, at least from an engineering point of view, pre-determining the way in which the system should accomplish the task (i.e. adding constraints to the system) cannot be considered negative. On the contrary one can argue that having pre-determined good constraints can help or facilitate the design of the system itself. On the other hand, one should consider that if the constraints are not well specified this can not only be disadvantageous but can make the design of the system impossible. In fact, while it can be possible to design a system that performs a certain task, it could be impossible to design a system that performs the same task in the way specified by some badly determined constraints. We think that while in the case of passive systems (i.e. systems that passively receive input stimuli and do not act in an environment) determining the correct answer for each stimulus can be feasible; in the case of systems that interact with an external environment (i.e. for example in the case of system that perceive actively modifying its position with respect to the external environment) determining good constraints is impossible.

On the basis of their behaviour, systems that behave in an environment partially determine the stimuli they will receive in input[12]. Suppose that we want to built an artificial system that collects garbage on the ground and suppose that such a system is perceiving a particular object on the left of its visual field; the kind of stimulus it will perceive next will depend on the motor behaviour of the system. It will perceive different stimuli if it moves in forward, turns plain left or remains still. This implies that each motor action that the system performs has two different effects: a) it determines how well the system performs with respect to the given task (the system should approach objects in order to recognise garbage and pick it up); b) it determines the next input stimuli the system will perceive and as a consequence if it will be able to recognise and eventually pick up the object (the system may be able to pick up the object only by approaching it from a particular direction).

Determining the correct motor action that the system should perform in order to experience good input stimuli is extremely difficult because each motor action can have long term consequences, because the effect of each motor action is a function also of the preceding and the successive actions that also have long term consequences and because the environment and the effect of the interaction between the system and the environment may not be perfectly known. Even knowing what are good input stimuli for a given task may be impossible because they partially depend on the characteristics of the system itself and therefore cannot be known in advance. It should be also noted that it can happen that the two needs, determining an action that allows the system to solve its task, and determining an action that allows the system to experience good stimuli, can be in conflict and, in that case, one should try to find a solution that make the most of the two needs. Using self-organised methods[13,14] one can develop systems that are free to find their way of solving the task. Using this approach the designer of the system should specify a criteria for evaluating the performance of the system but not the way in which the system itself should satisfy such a criteria. The architecture and the behaviour of the system is not determined by the experimenter but is progressively modified during an adaptive process.

As Nolfi and Parisi show[15] the fact that each motor action of the system determines the kind of stimuli that the system itself receives as input is not an obstacle but an advantage for self-organised systems. The authors developed systems that "live" in a two dimensional world and that should approach objects randomly distributed in it. These systems receive sensory information from such objects and must decide how to move in order to approach them. Systems are implemented with neural networks in which architecture and connection weights are progressively modified in order to efficiently solve the task. Each network is evaluated on the basis of the number of food elements it is able to approach in a fixed amount of time, but is free to develop its way to react to each input stimulus. Analysing the behaviour of the systems that are progressively selected the authors show that trained systems behave in a way that increases the number of input stimuli to which they are able to react more efficiently and decreases the remaining stimuli (i.e. systems that are able to select, through their motor actions, favourable stimuli). Moreover it appears that such systems progressively increase their performances in two different ways: a) by acquiring a capacity to react better to input stimuli; b) by acquiring a capacity to behave in a way that increases the frequency of stimuli to which they are able to react efficiently. In other words, the fact that each motor action modifies the next experiences is used by such self-organising systems in order to self-select favourable input stimuli and to increase performance.

In other research Nolfi and Parisi[16,17] show how self-organising systems that learn appear to self-select input stimuli in order to improve their learning ability. Simulative results show that system which should learn the same approaching task described above develop a behaviour that allow them to have good learning experiences (i.e. to experience input stimuli that are able to increase their learning ability).

THE ROLE OF SYMBOLIC LEARNING IN VISION (F. Esposito)

Visual pattern recognition can be thought of as the process of transforming a pattern function to a more structured representation of information, that is the representation of properties of the input data using a representation that is different from the input data. Learning can be thought of as the process of generating a representation of information that is suitable for accomplishing some goals; it concerns the creation of a new representation of the data using the input and some knowledge.

The integration of pattern recognition and learning techniques aims at generating and transforming the representation of information. In computer vision, the main goal is that of designing systems that are able to receive input sensorial data, to apply successive transformations in order to generate structural symbolic descriptions, to apply symbolic learning techniques in order to generalize concepts, to classify, to make abstractions, to discover regularities and causal relations, finally, to understand and to interpret a scene. The relevance of subsymbolic techniques is, in my opinion, restricted to the perceptual task, while symbolic learning techniques should have a more important role in the higher phases of vision involving the abstraction and the recognition of concepts.

Machine learning algorithms and systems have attained in the last years a remarkable degree of sophistication; besides the well known inductive algorithms for learning from examples and conceptual clustering[18], and the related set of Similarity-Based methodologies, other techniques have been developed. Learning can also be a deductive process aimed at making explicit the useful knowledge embedded in a body of a priori knowledge about the given domain. The explanation-based generalization[19] has a reduced need for training examples: under ideal conditions, one example is sufficient and its role is only that of focusing the attention of the deductive process. These ideal conditions require that the available domain theory be perfect, i.e. complete and consistent. In order to cope with an imperfect theory, integrated approaches have been realized, combining inductive and deductive techniques: in "structural matching"[20] the inductive learning process is augmented with deductive inference. Now, between the two extremes of pure induction, with a poor a priori domain knowledge, and deduction, with almost no data, there is a continuum of integration[21-23]. As to the application to vision, the pure inductive approach is generally preferred: a system proposed by Winston[24], ANALOGY, has been used slightly modified[25], to acquire disjunctive characteristic descriptions of images, in the form of semantic networks, starting from positive examples only. Segen[26,27] proposes supervised and unsupervised techniques to learn structural models of shape from noisy image data.

However, the number of reported practical applications did not experience an increment correspondent to the number and the variety of the symbolic learning methodologies. One of the reasons can safely be imputed to the existing gap between the rather ideal conditions under which most of those systems are designed to work and the very characteristics of the real world. The information contained in the data is sometimes blurred by noise and excessive variability or hidden behind too many details. Attention must be focused on the pre-processing phase and different aspects must be investigated.

A Good Pre-processing for Efficiently Learning

A first problem is the selection of "informative" features, for example by using well known pattern recognition techniques as the quantitative evaluation of the feature information content. Also a symbolic method can be appropriate to select relevant characteristics, for example, a conceptual clustering approach, which seems particularly useful when the descriptions exhibit a complex internal structure[28]. Moreover, the availability of a background knowledge can help and clarify the feature selection process. Some researches, for instance in model-based diagnosis[29], have pointed out the relevance of deep knowledge for reducing system brittleness. In order to deal with deep models of knowledge in Machine Learning, including causal models, general models of structure and goal dependency, explicit representation of time and processes, these must be represented in a suitable way. Suitable means that a representation can be effectively learned (by empirical-inductive techniques) or that it can be used for learning operational knowledge (by analytical-deductive techniques).

The second problem deals with the choice of a suitable level of details in describing the data with respect to the current task: this aspect spans a wide range of problems, from abstracting the continuous-valued numerical attributes to defining mechanism to describe and handle composite internal structure of complex data. So far, most learning algorithms in AI are based on a symbolic representation, so that numerical values of attributes must be discretized into intervals prior to start the learning process. This "classification" influences the learning results considerably; objective criteria to perform the discretization process may be found, making it explicit[30].

Another question concerns the extraction of interesting information from a set of data affected by noise[31,32]. The existence of noise, that may be also intended as uncertainty and incompleteness in the training and/or the testing cases, generally causes ambiguity in the classification phase. This is related to the problem of "outliers", i.e. examples that are singular points in the distribution of the example properties. It is important to state if an outlier is a normal case which has not been recognized or it is an exceptional case. The former possibility evidences the limits of the symbolic classifier, due to different reasons, such as the weakness of the representation language, noise or lack of precision in the description of some examples (for instance, the excessive use of "don't care" values). The last case is related to the problem of dealing with incremental learning, related to the capability of adjusting the learned classification rules in the light of new evidence supplied during the testing phase or the operational phase too, when many failures take place. Generally, exceptional cases are not considered in the initial training set, representing rare modalities, although significant. This causes a gradual reduction of the system performance, which is mostly measured as recognition rate. While it is possible to make the symbolic classifier robust with respect to local and little deformations by relaxing the matching conditions, it is difficult to handle with a real outlier: in fact, the intrinsically imperfect classification theory which has been generalized by the inductive methods, must be modified in order to treat the exceptions. This goal may be achieved without starting ex novo the learning process using inductive or deductive approaches.

If an exception represents a case that must not be covered by an existing concept, censors can be added to the old rules. Censored production rules have the form:

<premise> $\cap$ NOT<censor> $\rightarrow$ <decision>

which is logically equivalent to

<premise> $\rightarrow$ <decision> $\cup$ <censor>

where the censor is similar to the UNLESS condition[33]: it is a predicate or a disjunction of predicates that, when satisfied, block the rule.

The possibility of dynamically modifying the rules in an inductive learning process which works incrementally is achieved by the use of generalization and specialization operators. Censored rules may be obtained by specialization. In fact, a general operator for specializing the hypotheses inductively generated by any system that learns structural descriptions from positive and negative examples can be based on negation[34]. If the presence of an outlier Ex causes a concept C erroneously covering it, specialization of C can be accomplished by adding the negation of a non-matching literal of Ex to the body of C. Each possible substitution according to which C matches against Ex defines a base of consistent specializations and the intersection of all such bases provides a set of literals that can recover the consistency of C when added to its body. However, completeness has always to be checked.

The problem of revising the production rules which constitute the classification theory may be attacked by a deductive learning technique too, the Explanation Based Generalization that, working as a repair mechanism, allows us to refine knowledge by domain specific criteria. It is possible to improve the recognition function of a symbolic classifier by an explanation of its failure to recognize an exception[35]. In such a way the knowledge base of the classifier can be refined using the domain theory in an abductive process. Learning by failing to explain[36] is widely used when an incomplete or intractable domain theory reduces the applicability of explanation based techniques. It has also been presented as an inductive extension of EBL, named Learning by Failure to Prove, and it has been applied as an approach to rule base refinement[37].

All the preceding issues are even more relevant when structural descriptions are required, since the observations show an internal structure and the target knowledge must be expressed in a First Order Predicate Logic. Most of the realized symbolic learning systems do not show an adequate robustness in face of noise, deformations and exceptional cases neither a sufficiently accurate treatment of numeric continuous-valued attributes.

How to Evaluate the Learning System Performance?

A first consideration about the performance of learning systems is related to efficiency: in inductive learning, since the search space to be considered is the powerset of the set of known objects, the number of candidate concepts grows exponentially with the number of objects. This implies the use of strong constraints (similarity based, feature correlation and structure based constraints) during generalization. Moreover the background knowledge can be used as a bias in reducing the space of candidate hypotheses. The problem is more complex when the observations are represented by first order languages allowing to represent object components and relations among those components. A powerful structured representation language like a first order predicate calculus is desirable, but there are difficulties with handling quantification and its semantics.

The learning system performance cannot simply be related to efficiency. Learning systems which have the task of classification, as the empirical supervised methods, generally use the learning rate and the classical prediction accuracy in order to evaluate performance, here intended as a measure of the learning quality. When new observations are presented in a stream and when responses to these observations are required in a timely manner some new considerations on performance must be done. In fact, assuring incremental learning capabilities is not the only requirement; the design must take into account constraints along dimensions of assimilation cost and concept quality, such as the amount of time which is required to respond the external stimuli and the accuracy of these responses.

There are four dimensions along which to evaluate the performance of supervised incremental learning systems[38]:

• the learning rate or the number of observations that are necessary to reach specified accuracy levels;
• the asymptotic level of prediction accuracy that is achieved;
• the cost of assimilating a single observation;
• the total cost required to obtain a specific level of accuracy.

Unsupervised systems are subject to many of the same complexities as the supervised systems in terms of performance. Conceptual clustering is often used to create classifications that can be used for precisely the same performance task: to predict membership with respect to a priori known classes as these are defined by human experts. In numerical taxonomy this is related to relative validation methods[39], which compare discovered classes with those that were expected to exist in the reality. This "rediscovery", possible when class labels are known, cannot be the unique evaluation method for unsupervised learning.

Unsupervised learning supports prediction not simply of a single concept or class label but of arbitrary information that is missing from observations. How to evaluate the best concepts, i.e. the best partition? In terms of number of classified observations, distribution of population within the classes, quality of the produced concept descriptions, relevance as to the goal of classification?

In sum, evaluation of performance in unsupervised learning systems is relevant but it is often not strongly linked to an overt performance task as are supervised systems. Missing attribute prediction is recognized to be better exploited as an evaluation strategy for purposes of comparing concept formation and clustering systems. Mahoney and Mooney[40] compared the performance of Fisher's COBWEB and an unsupervised connectionist approach known as competitive learning. A prediction based performance task may induce family resemblance sorting, useful even at the pre-processing level when a preliminary division of data is sometimes required.

In general, assumptions about performance are vital in the design of conceptual clustering algorithms, though they are often left implicit.

Conclusion

Current machine learning paradigms, both inductive and deductive, may be useful in computer vision, but attention must be focused on the pre-processing techniques in order to make the symbolic algorithms more robust as to noise, incomplete descriptions and exceptional cases. The background knowledge and the domain theory can be used to better the process of recognizing and understanding images. Moreover, the domain knowledge may be used as a bias in reducing the search space of inductive hypotheses, so augmenting the efficiency of symbolic learning algorithms. The evaluation of the performance of a learning system is a central issue: however, in unsupervised learning algorithms it is often not strongly linked to an overt performance task as are supervised systems.

OBSERVATIONS

B. Zavidovique (Université Paris XI, Paris Orsay):

1) Not physically said: everybody should learn Hockerson's "Formal Theory of Learning" and think how to adapt it to control without oracles.
2) Action is really what should be learned in perception. Probably, if machines don't know how to recognize handwritten characters it is because they never wrote any.

   Now about supervised/unsupervised learning type of things: of course learning actions might be dangerous and damaging, that is exactly why knowledge and imagination are made for (simulation replaces action and it allows to predict). If we don't leave them, we need to define perfectly the control objective and the learning objective. Remaining on a circuit is not the same for a car as remaining on the road and different again from knowing how to drive.

   So learning to control a system is probably what we can have our robots to accomplish. We can have this hierarchy: fix model is automatic control, parameterized model is adaptive control, changing the model (by mere clustering in a "state" space) is learning. So the last one profits from the two previous ones, specially in terms of what performance could be. If the action is well defined then the robot accomplishes the task or not!

S. Ribaric (Faculty of Electrical Engineering, University of Zagreb):

When we talk about evaluation criteria for successful learning and performance evaluation of computer vision system, I would like to tell you a joke. Let us imagine that we are in cinema watching a film together with an intelligent computer vision system. After the projection, the computer vision system has to generate correct answers to some questions, for example: "Who is the director of this film?", "Who wrote the music?", "What is the message of the film?", or "Was Robert Redford as good as in "Three Days of Condor"?". This is Turing-like test for computer vision systems. I am a pessimist here, and I don't believe that in near future computer vision will be capable of generating correct answers to these questions. We are still very far from this kind of systems.

S. Damiani (Centro Ricerche FIAT, Torino)

Brooks at the MIT builds simple machines that immediately react to simple stimula. In the field of artificial vision we are collecting many experiences from different fields (trying to copy from nature, which nature?). Are we in the risk of loosing already identified simple solutions? Which architecture could we devise, given the complexity (and the complex metrics) of those present in nature?

REFERENCES

1. J. H. Gennari, P. Langley, and D. Fisher, Models of incremental concept formation, *Artificial Intelligence*, Vol.40, pp.11-61 (1989).
2. J. T. Tou and R.C. Gonzalez, *Pattern Recognition Principles*, Addison Wesley, Reading, MA (1974).
3. D.E. Rumelhart and J.L. McClelland, *Parallel Distributed Processing,* Vol.1, Foundations, MIT Press, Cambridge, MA (1986).
4. L.B. Booker, D.E. Goldberg, and J.H. Holland, Classifier systems and genetic algorithms, *Artificial Intelligence*, Vol.40, pp.235-282 (1989).
5. K.S. Fu, *Syntactic Methods in Pattern Recognition*, Academic Press, New York, NY (1974).
6. P.H. Winston, *Artificial Intelligence*, Addison Wesley, Reading, MA (1977).
7. R.S. Michalsky, J.G. Carbonell, and T.M. Mitchell, *Machine Learning - An Artificial Intelligence Approach*, Springer-Verlag, Berlin, D (1984).
8. S. Minton, J.G. Carbonell, C.A. Knoblock, D.R. Kuokka, O. Etzioni, and Y. Gil, Explanation-based learning: a problem solving perspective, *Artificial Intelligence*, Vol.40, pp.63-118 (1989).
9. P. Langley and J.M. Zytkow, Data-driven approaches to empirical discovery, *Artificial Intelligence*, Vol.40, pp.283-312 (1989).
10. R.A. Brooks, Achieving artificial intelligence through building robots, *A.I. Memo 899,* MITAI Lab. (1986).
11. R.A. Brooks, Intelligence without representations, *Artificial Intelligence*, Vol.47, pp.139-159 (1991).
12. D. Parisi, F. Cecconi, and S. Nolfi, Econets: Neural networks that learn in an environment, *Network*, Vol.1, pp.149-168 (1990).
13. S. Nolfi and D. Parisi, Growing neural networks, *Technical Report*, Institute of Psychology, Rome, I (1992).
14. I. Harvey, P. Husbands, and D. Cliff, Issue in evolutionary robotics, in *Proceedings of SAB92, The Second International Conference on Simulations of Adaptive Behaviour*, MIT Press Bradford Books, G.A. Meyer, H. Roitblat, and S. Wilson eds., Cambridge, MA (1993).
15. S. Nolfi and D. Parisi, Self-selection of input stimuli for improving performance, in *Neural Networks and Robotics*, G.A. Bekey, Kluwer Academic Publisher, Den Haag, NL (1993).
16. S. Nolfi and D. Parisi, Desired responses do not correspond to good teaching input in ecological neural networks, *Technical Report*, Institute of Psychology, Rome, I (1993).
17. D. Parisi and S. Nolfi, Neural Network Learning in an Ecological and Evolutionary Context, in *Intelligent Perceptual Systems,* V. Roberto ed., Springer-Verlag, Berlin, D, pp.20-40 (1993).
18. R.S. Michalski, A theory and methodology of inductive learning, in *Machine Learning, an Artificial Intelligence Approach*, R.S. Michalski, J.G. Carbonell, and T. Mitchell eds., Tioga, Palo Alto, CA, pp.83-134 (1983).
19. T. Mitchell, R. Keller, and S. Kedar-Cabellu, Explanation-based generalization: a unifying view, *Machine Learning*, Vol.4, pp.47-80 (1986).
20. Y. Kodratoff and G. Tecuci, Learning Based on Conceptual Distance, *IEEE Trans. on Pattern Analysis and Machine Intelligence*, PAMI-10, No.6, pp.897-909 (1988).
21. M.J. Pazzani, Integrated Learning with incorrect and Incomplete Theories, *Proc. of the 5th Int. Conf. on Machine Learning*, Morgan Kaufmann, Ann Arbor, MI, pp.291-297 (1988).
22. F. Bergadano and A. Giordana, A knowledge Intensive Approach to Concept Induction, *Proc. of the 5th Int. Conf. on Machine Learning,* Morgan Kaufmann, Ann Arbor, MI, pp.305-317 (1988).
23. G. Widmer, A Tight Integration of Deductive and Inductive Learning, *Proc. of the 6th Int. Workshop on Machine Learning*, Cornell University, Itaca, NY (1989).
24. P.H. Winston, Learning Structural Descriptions from Examples, in *The Psychology of Computer Vision*, P.H.Winston ed., McGraw Hill, New York, NY (1975).
25. J.H. Connell and M. Brady, Generating and Generalizing Models of Visual Objects, *Artificial Intelligence*, Vol.31, pp.159-183 (1987).

26. J. Segen, Graph Clustering and Model Learning by data compression, Pro*c. of the 7th Int. Conf. on Machine Learning*, Morgan Kaufmann, Austin, TX, pp.93-101 (1990).

27. J. Segen, GEST: a Learning Computer Vision System that Recognizes Hand Gestures, in *Machine Learning IV*, R.S. Michalski and G. Tecuci eds., Morgan Kaufmann, Austin, TX (1993).

28. G. Mineau, J. Gecsei, and R. Godin, Improving Consistency within Knowledge Bases, in K*nowledge, Data and Computer-Assisted Decisions,* M. Schader and W. Gaul eds., Springer-Verlag, Berlin, D (1990).

29. P. Torasso and L. Console, *Diagnostic Problem Solving,* Van Nostrand Reinhold, The Netherlands, NL (1990).

30. M. Moulet, Accuracy as a new information in law discovery, *Proc. of the Conf. Symbolic/numeric Data Analysis and Learning,* E. Diday ed., Nova Science Pub. (1991).

31. F. Bergadano, A. Giordana, and L. Saitta, Automated Concept Acquisition in noisy environments, *IEEE Trans. on Pattern Analysis and Machine Intelligence*, PAMI-10, pp.555-578 (1988).

32. F. Esposito, D. Malerba, and G. Semeraro, Classification in noisy environments using a Distance Measure between structural symbolic descriptions, *IEEE Trans. on Pattern Analysis and Machine Intelligence*, PAMI-14, pp.390-402 (1992).

33. P.H. Winston, Learning by augmenting rules and accumulating censors, in *Machine Learning II*, R.S. Michalski, J.G. Carbonell, and T.M. Mitchell eds., Morgan Kaufmann, Los Altos, CA (1985).

34. F. Esposito, D. Malerba, and G. Semeraro, Negation as a Specializing Operator, in *Advances in Artificial Intelligence*, P. Torasso ed., Lectures Notes in AI, No. 728, Springer-Verlag, Berlin, D, pp.166-177 (1993).

35. F. Esposito, D. Malerba, and G. Semeraro, Machine Learning Techniques for Knowledge Acquisition and Refinement, *Proc. of the 5th Int. Conf. on Software Engineering and Knowledge Engineering*, San Francisco, CA (1993).

36. J.R. Hall, Learning by Failing to Explain: Using Partial Explanations to Learn in Incomplete or Intractable Domains, *Machine Learning*, Vol.3, pp.45-77 (1988).

37. J. Mostow and N. Bhatnagar, Failsafe: a Floor Planner that Uses EBG to Learn from its Failure, *Proc. IJCAI87*, Milano, I, pp.249-255 (1987).

38. J.C. Schlimmer and D. Fisher, A case study for incremental concept induction, *Proc. of the 5th Nat. Conf. on Artificial Intelligence*, Morgan Kaufmann, pp.496-501 (1986).

39. A.K. Jain and R.C. Dubes, *Algorithms for Cluster Analysis*, Prentice Hall, Englewood Cliffs, NJ (1988).

40. J.J. Mahoney and R.J. Mooney, Can Competitive Learning Compete? Comparing a Connectionist Clustering Technique to Symbolic Approach, *Tech. Rep. AI89-115*, University of Texas, Austin, TX (1989).

EVENING PRESENTATION
THE STATE OF THE ART IN VIRTUAL REALITY

Stefania Garassini[1] and Maria Grazia Mattei[2]

[1] Virtual
Via C. Ravizza 53/A, I - 20149 Milano, Italy
[2] Mediatech
Via Rimembranze di Lambrate 31, I - 20134 Milano, Italy

1. INTRODUCTION

Virtual Reality (VR) is the result of the combination of different existing technologies, which focus on the ability of man to interact with the machine in a new and revolutionary way. It is the result of the evolution of computer graphics, of visual simulation, of telepresence technologies. The pioneer in this field was Ivan Sutherland who introduced the concept of "Ultimate display" in 1965. He described it with these words: "A display connected to a digital computer gives us a chance to gain familiarity with concepts not realizable in the physical world... The screen is a window through which one sees a virtual world. The challenge is to make that world look real, act real, sound real, feel real"[1].

But VR is also different from any existing technology because it focuses its attention on the human experience. It is an experience mediated and reconstructed by the computer "Virtual environments are computer generated worlds that users may enter and take action in via telepresence technology. For example a person might directly explore a molecular model, an ancient city, or an imaginary planet and its inhabitants", says Scott Fisher[2], one of the first researchers to concentrate on this field.

Consequently VR is a technology which deeply alters the meaning of experience and also the traditional concept of vision. It is possible to change the perspective from which we see things, to fly through worlds, as in the first experiments of Vpl, the company founded in '85 by Jaron Lanier, which first introduced VR systems on the market, to see persons and things at distance as if they were present, like in the experience of Cluny, it is possible to superimpose different realities: synthetic images and real images, like in the latest researches of the University of North Carolina, which is one of the most important research centers in the field of VR.

2. FIRST PERSON VR

All these experiences are interactive, in real time and as they involve the body of the participant in an immersive way they are considered as "First person Virtual Realities": they usually involve goggles gloves or other devices (like datasuits), which can permit all the body to enter in the virtual world. As an aspect of the first person point of view, users control their individual viewpoints through the movements of their heads and/or eyes[2].

These experiences are also described as "immersive VR" or "first person VR", because the user is completely surrounded by the images of the simulated world and he can interact with them. The feeling of being completely immersed in the simulated world can be obtained with different devices, such as Head Mounted Displays (HMD), or Booms. They both place a pair of display screens right in front of the user's eyes. In the HMD the screens are mounted on a helmet that users wear while moving in the virtual world. The Boom is like a pair of binoculars that users hold in front of their eyes to explore the virtual space. Two kinds of technologies are used for the screens Liquid Crystal Displays (LCD) and Cathode Ray Tubes (CRT). CRT provide a higher resolution and better display quality than LCD but have the disadvantage of being too heavy to be mounted on a user's head (but a new head mounted display from Virtual Research uses CRT for the first time: the Eyegen 3); the Boom uses CRT.

LCD are most used in HMD because they are lighter and flatter and therefore easier to mount on a HMD. They have the disadvantage of Low resolution (usually less than 300x200 pixels in colour) and poor display quality due to problems with contrast and brightness. Both HMD and Boom are equipped with a tracking system to determine the viewer's position in space. The environment responds visually in a manner similar to what we are used to see in the real world.

A variant of HMDs are the see-through head mounted displays. In these devices, each one of the two screens is placed on each side of the user's head and the images projected on half silvered mirrors placed in front of the viewer's eyes. The viewer can see the virtual object and also the real world; the virtual objects seem to merge with the real surroundings as in the researches at the University of North Carolina (UNC), which are also called «X ray vision».

The ways to create a better feeling of immersion are objects of studies in different research centers. Immersion can also be obtained with devices that the user does not need to wear. For example with projection-based systems. In these systems, user's positions and actions are tracked, and the corresponding virtual scene is projected on large screens. One example is the CAVE system developed at the Electronic Visualization Laboratory of the University of Chicago by Carolina Cruz Neira under the supervision of Dan Sandin and Tom DeFanti. In the CAVE the illusion of immersion is created by projecting stereoscopic computer graphics into a 10x10x10 foot cube composed of display screens that completely surround the viewer. The current implementation projects images on three walls and a floor. The viewer explores the virtual world by moving around inside the cube. The CAVE blends real and virtual objects naturally in the same space.

3. THIRD PERSON VR

There is also another kind of experience and of man-machine relationship, which has been studied since the Seventies and which permits a reflected involvement of the body of the participant. An example is the Videoplace system, "a conceptual environment with no physical existence", implemented by Myron Krueger in 1975, and called "Artificial Reality" to distinguish it from Virtual Reality which involves goggles, gloves, or other devices. In this system the user can see his image projected as a silhouette on a screen where he can interact with a series of graphical objects or with the silhouettes of other users who are in different places. The canadian system Mandala is based on the same principle, but in this case the user can see his image as if it was reflected in a mirror: it is his video image, not a silhouette and he sees himself interacting with different objects and environments. The latest achievements in this field is ALIVE a system implemented by Pattie Maes at Mit, which gives the participant the chance to interact on the screen with a synthetic character which has some kind of an autonomous, intelligent behaviour[3].

All these experiences are called third person VR because the user sees himself from the outside, he can see his actions as if he was another person. There is also a third kind of VR which is called "Desktop Vehicle" which is much more similar to traditional computer graphics: "Users view 3D worlds through a portal, steering through the worlds as in a flight simulator"[4,5]. The interaction of the user with the images on the screen is obtained through a mouse or other devices, the most well-known of which is called "Spaceball" and permits to change the point of view in a 3dimensional Virtual Space.

4. POTENTIAL APPLICATIONS

As we mentioned above, VR (especially the immersive or first person one) is a technology which deeply alters the character of human experience, mainly because it gives us the chance to have a direct "experience" of invisible or remote situation, not only by seeing, but also by touching distant or invisible objects and having the real feeling "to be there". VR creates a "synthetic experience"[5], following the definition of Warren Robinett, one of the pioneers of the field, former researcher at the UNC. This kind of "surrogate" experience can be very useful in different kinds of applications.

Currently the main field of application is medicine. Researches in the field of telesurgery, of surgical simulation and of minimally invasive surgery are going on in different universities and laboratories. "The virtual reality surgical simulator will be as important to a surgeon as a flight simulator is to a pilot"[6]. The idea is to obtain a sort of "digital physician" no longer limited by space or time, or by the inadequacies of his physical senses, because he can obtain a sort of "X ray vision" as it is called. In this sense researches conducted at the University of North Carolina are very interesting. They built an HMD in which virtual images are superimposed on the real ones. This result is also called "Ultrasound vision" because you can have the images from an ultrasound examination superimposed on the ones of the real body of the patient, so you can "see inside the body" while looking at him.

Another interesting application developed at UNC, which is not immediately related to medicine is the Nanomanipulator, "an atomic scale teleoperation system that uses a head-mounted display and force-feedback manipulator arm for a user interface and a Scanning Tunneling Microscope as a sensor and effector"[7].

Another interesting trend of application of VR is Architecture. Researches are going on in the field of "Architectural Walk-through" to simulate the structure of a building before its construction and to be able to explore it: examples of this kind of VR applications have been realized for example at the UNC and at the Fraunhofer Institute of Stuttgart in Germany.

Another kind of "Architectural Walk-through" which recreates in a virtual reality a real existing building has been made by Francesco Antinucci and a group of research of CNR in Rome. They created a model of the San Francesco Abbey in Assisi. It is a very detailed and very realistic model, made possible by the very high quality of the machines involved.

A televirtuality experiment took place during Imagina edition 1993. The goal of it was to show the concept of televirtuality that combines immersion in a virtual world and communication through standard digital networks.

In televirtuality it does not exist a transmission of images, but a transmission of feelings and actions of each user. Images are created in real time from the user's computer. This shows how the transmission can simply exploit the ISDN low-band networks, or also the most common telephone system.

In Montecarlo, we have seen the ancient Cluny Abbey, rebuilt by IBM through a computer, visited by two remote persons (one in Montecarlo and the other one in Paris).

The experiment at Imagina has shown how two users interconnected through Numeris network could develop and interact in a defined virtual environment, can live in the same tridimensional database, move, for example, in the abbey and follow the indications of each partner virtually present through the rappresentation of synthetic images[8].

A brief overview should be made also of art and entertainment, where VR has already been used quite often. Entertainment is seen as the first real market for this kind of technology. Many attractions for theme parks are trying to achieve a higher and higher involvement by the participant, trying to have his body move while he plays, like in the Cybertron (a game presented by Straylight corporation) or wrapping up the user in a strong audiovisual experience (like in the Pteranodon ride created by GreyStone in collaboration with Silicon Graphics).

For the artist VR is the chance to change the relationship between the public and the work of art from a passive one into an active, where seeing a work of art will mean acting in it and determine its sense together with the author.

REFERENCES

1. I. Sutherland, The ultimate display, *IFIP Proceedings*, pp.506-508 (1965).
2. S. Fisher, Applications of virtual environments telepresence, Course Notes 23, *ACM Siggraph* (1993).
3. P. Maes, ALIVE: an artificial life interactive video environment, *Computer Graphics Visual Proceedings*, *ACM Siggraph*, p.189 (1993).
4. S. Tice, Tomorrow's Realities, *ACM Siggraph*, p.5 (1991).
5. Synthetic experience a proposed taxonomy, *Presence*, Mit Press, Vol.1, No.2.
6. R.M. Satava, Telepresence and virtual reality, A Framework for surgery in the 21st century, *Virtual Reality* Vienna, Abstracts (1993).
7. R.M. Taylor, W. Robinett, V.L. Chi, F.P. Brooks Jr., W.V. Wright, R.S. Williams, and E.J. Snyders, The Nanomanipulator, *Computer Graphics, Acm Siggraph*, Course Notes (1993).
8. H. Tardif, Liaison de televirtualité par numeris entre Paris et Monaco, *Proc. of Imagina*, p.91 (1993).

INDEX